W9-CCY-226

Fundamental Research Statistics
for the Behavioral Sciences

INTERNATIONAL SERIES IN DECISION PROCESSES

Ingram Olkin, Consulting Editor

Fundamental Research Statistics for the Behavioral Sciences

Second edition

John T. Roscoe

Lecturer, Division of
Continuing Education
University of Colorado at Denver

HOLT, RINEHART AND WINSTON, INC.

New York   Chicago   San Francisco   Atlanta   Dallas
Montreal   Toronto   London   Sydney

Copyright © 1969, 1975 by Holt, Rinehart and Winston, Inc.
All rights reserved

Library of Congress Cataloging in Publication Data

Roscoe, John T
Fundamental research statistics for the behavioral sciences.

(International series in decision processes.)
1. Statistics. I. Title.
HA29.R784 1975   519.5'02'43   74–14795
ISBN 0–03–091934–7

Printed in the United States of America
567890 038 09

PREFACE

This text is intended to provide an introductory survey of research statistics for students in the behavioral sciences. The content is sufficient for a rather substantial course of study over a full academic year. However, the method of organizing the chapters permits the professor to edit the content liberally and should encourage the adoption of the text for use in the one-semester course.

A major goal in writing the text has been to achieve a level of readability somewhat lower than that of many competing textbooks. The approach to statistical theory emphasizes the conceptual rather than the mathematical — no background other than elementary algebra is assumed.

Many topics ordinarily reserved for a textbook in experimental design are treated in this text at an elementary level. It is not my intention that these topics be viewed as a substitute for a course of study in experimental design. For the many students who do not go on to a course in experimental design, the survey approach is believed to provide greater insight into the research methodology than is achieved with the traditional statistics sequence. For those students who do go on to study experimental design, my experience is that the survey approach provides an excellent background for the advanced course. The treatment of regression analysis as an alternative to the analysis of variance in experimental design is in keeping with recent trends in behavioral research.

The chapters have been kept very short, and they are more numerous than in the typical statistics textbook. Three benefits are expected from this type of organization: (1) Short learning units are notoriously more popular with students. (2) Professors may conveniently edit the text and control the content of their courses. (3) The materials are conveniently classified for reference purposes.

One of the unique features of the text is the functional treatment of the various techniques of statistical inference. The emphasis is upon the selection of the most appropriate analysis for a given research situation. The most popular nonparametric procedures are presented in a fashion which encourages their consideration as alternatives to the parametric procedures. The statistical tables in the appendix, especially those for the nonparametric statistics, are superior for a text of this level.

Every statistical procedure is illustrated with a practical example of the type of thing encountered in elementary behavioral research. Many short, simple exercises are included, and an answer guide appears in the appendix. The examples and the exercises utilize small numbers of small numbers — this is done because I believe that little insight into research analysis is gained from many hours of arithmetic labor. The student should note, however, that most practical research does indeed involve large quantities of data and many extensive calculations. In our generation, these calculations will ordinarily be handled by an electronic computer.

The user's attention is directed to the existence of a companion volume to this text, *The Funstat Package in Fortran IV*. *The Funstat Package* contains thirty easy-to-use computer programs capable of handling practically all of the statistical analyses in this text. Each program is well documented with simple instructions for its use. It is my conviction that the statistics student should be introduced to the use of the computer as early as possible, and *The Funstat Package* is intended to provide a vehicle for accomplishing that end.

I am indebted to the Literary Executor of the late Sir Ronald A. Fisher, F.R.S., to Dr. Frank Yates, F.R.S., and to Oliver & Boyd Ltd., Edinburgh, for permission to abridge Tables 7 and 12 from their book *Statistical Tables for Biological, Agricultural and Medical Research*.

Finally, I am much indebted to Drs. Jack N. Sparks and Beatrice Heimerl, who reviewed the manuscript of the first edition, and to Dr. Ingram Olkin, who reviewed the manuscript of the second edition. They offered many helpful suggestions, but should not be held responsible for any inadequacies in the text. A note of thanks is also extended to Marvin H. Halldorson, Timothy H. Hoyt, Gary C. Stock, Eric R. Strohmeyer, and David E. Suddick, who were graduate students at the time of writing the first edition, and who prepared the answer guide in the appendix.

October 1974   John T. Roscoe

CONTENTS

Answers to Selected Exercises

Index

Fundamental Research Statistics for the Behavioral Sciences

SCIENCE AND RESEARCH

1.1 SCIENCE AND THE SCIENTIST

The readers of this book live in a scientific age; science is a dominant force in their lives; they are members of a scientific society. Their individual and collective stature as members of a profession is irrevocably associated with their understanding and use of scientific knowledge, methods, and modes of thought.

The central purpose of science is to provide an objective, a factual, and a useful account of the universe in which we live. Science seeks an empirical and verifiable explanation of natural phenomena in contradistinction to artistic, philosophic, and religious ways of explaining things. The behavioral sciences are the product of the comparatively recent notion that scientific methods are appropriate for the task of pushing back the boundaries of human ignorance about man himself.

There is no universally accepted, inflexible, unchanging definition of science. It is a dynamic sociological enterprise that has been defined in a variety of different ways, each reflecting the philosophic biases of the individual who makes the definition. The author likes to think of science as consisting of the following three domains:

1. *Scientific knowledge.* Scientific knowledge is knowledge that has been verified by scientific methods. Our generation is experiencing an unprecedented accumulation of scientific knowledge — knowledge that has had a prodigious impact on modern modes of thought and action. A surprising proportion of this accumulation is far removed from the

ordinary experience of the nonscientist. But modern science is far more than the mere accumulation of facts.

2. *Scientific research.* The methodology scientists use for the accumulation and verification of scientific knowledge is called research. Scientific research is empirical research dealing with observable evidence — evidence which is the common experience of trained investigators. The preeminence of science in our society may be attributed to the unparalleled accomplishments of scientific research as a means of accumulating knowledge. But modern science is more than method.

3. *Scientific theory.* A critical domain of modern scientific endeavor is the building of theories that systematically organize the available facts in a fashion which yields general explanations about a wide range of phenomena. Ordinarily, scientific theories deal not only with verified facts but also with hypothesized facts. Such theories, when constructed according to scientific principles, are subject to evaluation through research. Thus scientific theories tend to generate new and creative research, research accumulates additional knowledge, new knowledge brings about modification of existing theories and the building of new theories, and the cycle begins again. The feature of self-correction is a fundamental property of modern sciences. Figure 1.1 illustrates the interacting relationships among scientific knowledge, research, and theory.

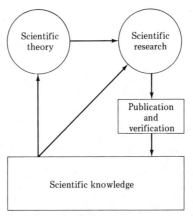

Figure 1.1   Model for science.

1.2 SCIENTIFIC KNOWLEDGE

Scientific knowledge is an enormous body of facts verified by scientific research. Most scientists would reject the notion that scientific knowledge is the only trustworthy knowledge, or that scientific research is the only way to discover and verify new knowledge. The essential distinction between scientific knowledge and other kinds of knowledge is that scientific knowledge is subject to verification by systematic, objective observation of natural phenomena.

Ordinarily, scientific knowledge is public knowledge, and efficient communication among scientists is essential for a number of reasons. First, modern science is an historical enterprise — the work of every scientist is based upon a foundation laid by other scientists. Communication of new findings is essential for scientific progress. Second, scientific findings must be subject to independent examination and verification by other competent investigators. When this procedure leads to consensus among the scientific community, scientists tend to regard the findings as factual. Scientific journals play an important part in this process, for they are the primary means for dissemination and subsequent verification of scientific knowledge.

There are two fundamental assumptions that undergird all scientific endeavor. (1) The behavior of the universe is orderly; it is not capricious, chaotic, or spontaneous. (2) Every natural event has an explanation that may be eventually discovered by intelligent and diligent men. These two assumptions are not subject to empirical verification; nevertheless, they appeal to common sense, and the available evidence appears to support them. Without these assumptions, it is difficult to conceive how one could carry on scientific research, accumulate scientific knowledge, or develop scientific theories. The first is not only essential to science but to human reason, the second is the basis of all human learning.

1.3 SCIENTIFIC THEORY

It was previously suggested that science seeks an empirical and verifiable explanation of natural phenomena. Independent and disorganized bits of scientific knowledge are of little utility to the scientist in his role as explainer until they are systematically integrated into the framework of scientific theory. A theory is an attempt on the part of the scientist to order many facts into a meaningful pattern — to synthesize the existing knowledge in a particular field in such a fashion that he is able to offer logical explanations of the interrelationships among phenomena and to hypothesize additional interrelationships. A knowledge of the interrelationships among specified phenomena can be used to predict events — for the behavioral scientist, the predicted events are incidents of human behavior. For example, if the psychologist knows the relationship between amount of practice and degree of proficiency in a certain learning situation, he can predict the degree of proficiency from a knowledge of the amount of practice. Such relationships are ordinarily imperfect, but this does not necessarily detract from their utility.

The role of scientific theory is not adequately recognized outside the scientific community. Theory not only synthesizes the existing knowledge, but goes beyond the existing knowledge; theory offers *needed* explanations of natural phenomena — explanations not yet subjected to empirical verification. It is important that the explanations put forth in the theory be capable of eventual verification or rejection; if this principle is adhered to, the theory may become the fountainhead of the most valuable research. Research that has its initiative and direction from scientific theory is invariably more systematic and more fertile than other research, for it bridges the gaps in the existing knowledge and broadens the base from which the scientist operates. The building of scientific theories is perhaps the supreme creative work of the scientist.

Scientific theories vary a great deal in their ability to explain or predict natural phenomena. The following are regarded as criteria for a good scientific theory:

1. It should be consistent with the known facts. That a theory appears to be at variance with some established fact is not necessarily cause to modify or discard the theory — the apparent inconsistency may be an opportunity for creative research. Nevertheless the eventual reconciliation of the theory with verified scientific knowledge is necessary if the theory is to receive wide acceptance.
2. The theory should be internally consistent; the various ideas within the theory should not contradict each other. Occasionally, problems of internal consistency can be resolved through research, but more often they are evidence of the inadequacy of the theory.
3. The theory should be sufficiently well-structured that the explanations which it offers can be readily communicated to competent investigators who may derive hypotheses from it and subject these to empirical verification. A theory that does not lend itself to empirical verification is not a scientific theory; it is pseudoscientific at best.
4. Ordinarily, it will be desirable that a theory be as simple as possible yet provide a general explanation of a wide range of phenomena. Of course, it may be necessary to abandon simplicity in favor of consistency, accuracy, or utility. Mathematical statements of relationships among phenomena are preferred over verbal statements.
5. The adequacy of any theory is its ability to sucessfully predict — in the case of behavioral theory, to predict human behavior. Most predictions of human behavior contain substantial amounts of error — better theories yield less error.

1.4 SCIENTIFIC RESEARCH

Research is the method that scientists use to accumulate scientific knowledge. It is usually a more productive endeavor when conducted within the framework of scientific theory, systematically seeking to bridge the gaps in existing knowledge revealed by the theory and/or to test relationships hypothesized by the theory. We will define scientific research as: the systematic and empirical study of relationships among variables. It will be extremely important for the student to learn to conceptualize any research problem in these terms.

The scientific study of human behavior is based upon observations of that behavior. In order for the observations made by the behavioral scientist to have scientific relevance, these observations must be capable of variation from time to time or from person to person. The term, sex, for example, would be meaningless if all persons were of the same sex. Similarly, the expression, "intelligence," would also be meaningless if all persons had the same amount of this trait. Traits such as these, traits which are capable of variation from person to person, are called *variables*.

There are two broad classes of variables: those which vary in quality and those which vary in quantity. Of the examples cited, sex is a *qualitative variable* and intelligence is a *quantitative variable*.

Most research will require the identification of *independent* and *dependent* variables. In the simplest research (and often the most profound), only two variables are studied, one independent and one dependent. The investigator is interested in determining whether or not a relationship exists between the two, and if it does indeed exist, he wants to know about the nature of that relationship. Suppose, for example, that an educational researcher wishes to determine whether the traditional textbook or a new programmed textbook is more effective for teaching elementary algebra to students with low aptitude for studying mathematics. He might arrange to teach one group of students from the traditional textbook and another group from the new programmed textbook. At the conclusion of the learning experience, an examination would be given to determine to what extent the students learned their algebra. Then a determination would be made whether one group learned more than the other, if so, which group, and perhaps how much more. In this simple research project, the two textbooks used constitute the independent variable, and the examination given at the end is the dependent variable.

The terms, independent variable and dependent variable, are extremely difficult to define in a way that fits all situations. However, in a given research setting they are usually quite readily identified, and it will ordinarily be necessary to so classify them in the planning of the research. We will define the two in terms of a *presumed* cause-and-effect relationship: variations in the independent variable are presumed to cause variations in the dependent variable. In the example cited in the previous paragraph, if there is a significant difference in the performance of the two groups of students, it is presumed that this was caused by the fact they studied from two different textbooks. Often, it is not practical to establish a cause-and-effect relationship, and it may not be the intent of the investigator to establish one, but the presumption of a cause-and-effect relationship is a convenient technique for classifying the independent and dependent variables. Another way of looking at the distinction between the two is that the determination of the individual's score (or category in the case of a qualitative variable) on the independent variable will ordinarily precede the determination of his score on the dependent variable.

In the situation in which the investigator is seeking to predict human behavior of some sort, the independent variable is used as the predictor, and the behavior to be predicted is the dependent variable. Suppose, for example, an Air Force psychologist is seeking to predict from an aptitude test whether or not an Air Force cadet will successfully complete pilot training. The aptitude test is the independent variable, and performance in pilot training is the dependent variable.

In an experimental setting, the independent variable is manipulated by the experimenter, and the dependent variable is the criterion by which he determines whether or not anything of consequence happened when (not necessarily because) he manipulated the independent variable. However, a great deal of very important behavioral research is undertaken in which the investigator does not manipulate the independent variable; he merely observes and explains. Such research is called *ex post facto* research in contradistinction to *experimental* research in which the independent variable is manipulated. Suppose that the educator in the example cited earlier had been interested in the relationship between the sex of the student and achievement in the study of algebra. In this case, the sex of the student is the independent variable.

Now let us suppose that the educator mentioned before is primarily interested in comparing the relative effectiveness of the two different textbooks, but he suspects that other variables, such as IQ and sex of the student, may also influence the dependent variable and affect the outcome of the experiment. This suggests that independent variables should be broken into groups: the experimental variables which the investigator is interested in studying to determine their influence upon the dependent variable, and other variables whose influence upon the dependent variable he wishes to control. The problems of organizing, carrying out, and interpreting behavioral research center around these three kinds of variables: (1) independent variables whose effects are to be studied; (2) independent variables whose effects are to be controlled; and (3) dependent variables that are observed in order to determine the consequences. The planning of research, taking into consideration these three kinds of variables, is ordinarily referred to as *experimental design*.

1.5   RESEARCH PROBLEMS, QUESTIONS, AND HYPOTHESES

Research is initiated because the accumulated scientific knowledge is never sufficient to solve all the problems of the day. Practically all scientific research begins with a problem that the investigator wishes to solve. It is highly improbable that he will succeed in his quest for a solution unless the problem is clearly stated in a form which lends itself to research. Ordinarily, one simply cannot solve a problem unless he knows precisely what the problem is. The problem must be delineated in such a fashion that it can be solved, that it can be solved in the manner proposed, that the solution will be recognized when it is found, that the solution can be communicated to other investigators, and that the solution can be defended. Considerable behavioral research suffers due to inadequate delineation of the problem. Typically, the problem will be stated as a question to be answered or a hypothesis to be tested. Whatever form is chosen for the statement of the problem, it is not ready for research until it is expressed in terms of relationships among variables.

The problem may be stated simply as a question about the relationships among two or more variables. For example, the researcher might ask: "Is there a relationship between scores on the Graduate Record Examination and the successful completion of the Ph.D. in sociology?" Or he might ask: "What is the relationship between scores on the Graduate Record Examination and the successful completion of the Ph.D. degree in sociology?" Or: "To what extent can scores on the Graduate Record Examination be used to predict the successful completion of the Ph.D. degree in sociology?" In the case of experimental research involving an experimental and a control group and using academic achievement as the criterion, the investigator might ask: "Is there a significant difference in the mean achievement of the two groups?" It should be noted here that the behavioral scientist will encounter many interesting but nonscientific questions — he should not expect to use scientific methods to find scientific answers to nonscientific questions. For example, the question, "Should the Graduate Record Examination be used to screen candidates for the Ph.D. degree in sociology?," is an important and interesting one, but it is not a scientific question. The distinction between good and poor research more often than any other factor is the distinction between asking a good or a poor research question. The ability to ask good research questions is often the distinguishing characteristic which separates the real scientist from the rest of the crowd.

The problem may be stated in the form of a research hypothesis — a conjectural statement about the relationships that might exist among two or more variables. This approach is often used when the investigator is conducting research within the framework of scientific theory. The hypothesis under these circumstances is a prediction of what should come to pass if the theory is true. For example, if a certain counseling theory were sufficiently well-structured that scientific hypotheses could be derived from it, an investigator should be able to specify the behaviors that would distinguish clients undergoing this type of counseling from those who do not receive this type of counseling. Of course, the statement of the hypothesis must be sufficiently precise that it can be confirmed or contradicted by research.

The investigator will often elect to state the problem in the form of a *null* hypothesis — a statement that no significant relationship exists among the variables. The usual intent in this case is to disprove the hypothesis; it is stated in this form as a convenient approach to statistical analysis. While we will regard the null hypothesis as primarily a statistical tool, the logic is not far removed from scientific thinking with respect to theories and hypotheses. Consider, for example, the following statement: "There is no relationship between the availability of pornography and the incidence of aberrant sexual behavior." In order to prove that this statement is true, the investigator would have to examine *all* of the evidence. He would have to examine every possible relationship between the independent variable (the availability of pornography) and the dependent variable (the incidence of aberrant sexual behavior). Every kind of pornography would have to be identified, every possible use of pornography would have to be explored, and every conceivable incidence of aberrant sexual behavior would have to be recorded. Such a task would prove impossible — most scientists would agree that it is impossible to prove that such a statement is true.

We need not examine all of the evidence, however, to prove that such a statement is false — we need only prove that it is false under a given set of conditions critically examined in a restricted setting. Beginning with the null hypothesis, "there is no relationship between two variables," the research may bring the investigator to one of two conclusions: (1) the null hypothesis is false; there is indeed a relationship between the two variables or (2) the evidence is insufficient to cause rejection of the null hypothesis. The implications of these two conclusions will be considered in greater detail in subsequent chapters.

The identification and quantification of the variables in behavioral research are crucial. Terms such as creativity, intelligence, self-actualization, authoritarianism, anxiety, and success are completely worthless in the research setting unless they are operationally defined. An operational definition is a definition that assigns meaning to an idea by specifying observable phenomena which represent the idea. Operational definitions are simply instructions for assessing human behavior — for translating constructs into researchable variables. Occasionally, an operational definition is an easy thing to come by. For example, it might be completely appropriate in a given research to define intelligence as a score on a certain IQ test, or to define academic success as graduation from college. However, the development of good operational definitions can be one of the most difficult problems encountered by the behavioral researcher. The task is complicated by the fact that many of the terms used by practitioners in the behavioral sciences are so vague that they defy definition in operational terms. The very survival of some of these terms is due to the fact that they cannot be operationally defined and empirically verified. However, there are many important ideas in the behavioral sciences that are still in process of development and that may eventually lend themselves to research verification. These should not be cast aside just because a simple operational definition is not immediately forthcoming. The eventual quantification of these is one of the most important tasks confronting the behavioral researcher.

1.6 SCIENCE AND PROBABILITY

There is a common misconception that science deals with certainties. Nothing could be further from the truth. Science does not deal with certainties, it deals with probabilities. The best the scientist can hope to draw from his research is that, given a specified level of an independent variable and well-defined control over certain other variables, it is extremely probable that certain effects will be observed in the dependent variable. The mathematics of probability enable him to quantify the level of confidence he has in his prediction.

SELECTED REFERENCES

BOHM, DAVID, *Causality and Chance in Modern Physics*. New York: Harper and Brothers, 1957.

BROWN, CLARENCE W., and EDWIN E. GHISELLI, *Scientific Method in Psychology*. New York: McGraw-Hill, 1955.

CONANT, JAMES B., *Modern Science and Modern Man*. New York: Columbia University Press, 1952 (also available in paperback from Doubleday Anchor Books).

HEISENBERG, WERNER, *Physics and Philosophy*. New York: Harper & Row, 1958.

KAPLAN, ABRAHAM, *The Conduct of Inquiry*. Scranton: Chandler, 1964.

KERLINGER, FRED. N., *Foundations of Behavioral Research*, 2nd ed. New York: Holt, Rinehart and Winston, 1973.

MARX, MELVIN H., and WILLIAM A. HILLIX, *Systems and Theories in Psychology*, 2nd ed. New York: McGraw-Hill, 1973.

TRAVERS, ROBERT M. W., *An Introduction to Educational Research*, 3rd ed. New York: Macmillan, 1969.

WHITEHEAD, ALFRED NORTH, *Science and the Modern World*. New York: Macmillan, 1925 (also available in paperback as a Mentor Book).

2

MEASUREMENT
AND STATISTICS

2.1 THE IMPORTANCE OF QUANTITATIVE METHODS

Given a suitable problem to explore, the success of any scientific endeavor is often a function of the extent to which the methods of the endeavor have been subjected to quantification through the use of measurement and mathematics. A two-step process of quantification is implied: (1) measurement, which involves the systematic representation of the data by numbers and (2) mathematics, which involves the systematic representation of the relationships among the measurements with mathematical expressions.

The techniques of measurement can be thought of as means to achieve two important objectives. (1) The history of science and the common experience of contemporary scientists confirm that objectivity is essential to communication of scientific findings and to the achievement of interpersonal agreement as to the validity of those findings. The reader is reminded that scientific findings are regarded as factual only when there is a high degree of interpersonal agreement among the appropriate scientists. One great contribution of the techniques of measurement is the achievement of a high degree of objectivity. (2) Measurement involves the assignment of numbers to data according to some well-established rules. This process of assignment of numbers to data is essential if the scientist is to treat his data mathematically.

The importance of mathematics to any scientific endeavor cannot be overemphasized. A prominent behavioral scientist, J. P. Guilford, in his book, *Psychometric Methods*, has made this observation: "Mathematics is a universal language that any science or technology may use with great power

11

and convenience. Its vocabulary is unlimited and yet defined with rigorous accuracy. Its rules of operation, or its "syntax," are unexcelled for logical precision." Fortunately, for the beginning behavioral scientist the required mathematical procedures will be found in this text.

2.2   QUALITATIVE AND QUANTITATIVE VARIABLES

It was noted in Chapter 1 that there are two broad classes of variables: those that vary in quality and those that vary in quantity. *Qualitative* variables are those which fit research subjects into categories in which the notion that one category is higher than or lower than another category is meaningless. For example, the sex of the subject (whether the subject is male or female) is a qualitative variable. Often, research subjects are divided into two groups: an experimental and a control group — this is also a qualitative variable. Qualitative variables are an important class of variables to the behavioral scientist; it is important for the student to begin to think of phenomena of this sort as variables. *Quantitative* variables require the notion that a subject may have more of or less of a given characteristic than another subject. For example, height, weight, and intelligence test score are quantitative variables. In common usage, the term "measurement" ordinarily applies to quantitative variables.

There is an extremely useful technique for the quantification of qualitative variables that the student should learn to use. It calls for the use of the binary numbers, one and zero. The quantification of a qualitative variable having only two categories is quite simple. For example, consider the subject's sex: If the subject is a male, he is assigned a score of one; if the subject is a female, she is assigned a score of zero. Consider a second example: If the subject is a member of the experimental group, he is assigned a score of one; if the subject is a member of the control group, he is assigned a score of zero. Of course, that a score of one is algebraically larger than a score of zero is quite meaningless under the circumstances, because the determination as to which category receives the higher score is quite arbitrary.

The quantification of a qualitative variable having more than two categories is a bit more involved. Suppose, for example, that a group of subjects is classified according to race as being (1) white, (2) black, or (3) other than white or black. To quantify this three-categoried variable, it will be necessary to generate two binary variables. Let the first variable (call it X_1) represent "whiteness": If the subject is white, he receives a score of one on X_1; if the subject is not white, he receives a score of zero on X_1. Let the second variable (call it X_2) represent "blackness": If the subject is black, he receives a score of one on X_2; if the subject is not black, he receives a score of zero on X_2.

Now, let us summarize: (1) A white subject receives a score of one on X_1 and a score of zero on X_2; (2) a black subject receives a score of zero on X_1 and a score of one on X_2; and (3) a subject classified as neither white nor black receives a score of zero on X_1 and a score of zero on X_2. Thus each subject may be uniquely and quantitatively identified with respect to race by these two scores.

Some researchers may prefer to generate a third variable (call it X_3) such that: A subject who is other than white or black receives a score of one on X_3, and a subject who is either white or black receives a score of zero on X_3. However, this is a redundant variable that adds nothing to the information recorded on the first two variables. The situation is illustrated in Table 2.1, where X_1, X_2, and X_3 represent race, and X_4 represents sex.

Table 2.1 Quantification of qualitative variables

Name (race, sex)	X_1	X_2	X_3	X_4
Joe (white, male)	1	0	0	1
John (black, male)	0	1	0	1
Mary (black, female)	0	1	0	0
Susan (red, female)	0	0	1	0
Phyllis (white, female)	1	0	0	0
Sam (yellow, male)	0	0	1	1

The utility of these binary variables will become increasingly apparent to the student as he becomes aware through experience that numbers so assigned have many of the characteristics of numbers assigned by quantitative variables. In fact, in much research, it will be profitable to manipulate these numbers with the same mathematical procedures ordinarily reserved for quantitative variables. That is, one may add, subtract, multiply, and divide such numbers under certain circumstances.

2.3 DISCRETE AND CONTINUOUS VARIABLES

If a variable can take on only a finite set of values, it is said to be *discrete*. When we say that a variable can take on only a finite set of values, this *usually* means that it is either (1) a qualitative variable or (2) a quantitative variable such that fractional values are not allowed. Variables such as sex, political or religious affiliation, and number of children in a classroom are discrete variables.

If a quantitative variable can take on *any* value (including fractional values) over a range of values, it can take on an infinite set of values. Such a variable is said to be a *continuous* variable. Variables such as height, weight, spelling ability, and intelligence are *believed* to be continuous variables.

Whenever we deal precisely with the notion of continuity, it becomes necessary to draw a distinction between the *variable* and the *measure* of the variable. Let us suppose that we are concerned with how well a third-grade student can spell (spelling ability is the variable), and we administer a 20-word spelling test (the test is the measure). We have every reason to believe that spelling ability is continuous; however, there are just 21 possible scores (zero through 20) on a 20-word spelling test (assuming that we grade each word as either correctly or incorrectly spelled). Notice that the variable is continuous, but the measure is discrete! Even if we permit fractional scores, we will always be restricted to a finite number of possible scores. The conclusion should be obvious: all measures are discrete!

Typically, scores on psychological tests are reported as whole numbers, but this is due to imprecision of measurement rather than any inherent restriction on the continuity of the variable. For example, intelligence test scores are customarily reported as whole numbers, but this does not mean that a person could not have an intelligence of 115.5 or even 115.49087341. It simply means that the measure is an approximation of the variable. More explicitly, an intelligence test score of 115 must be regarded as an approximation of some score between $114\frac{1}{2}$ and $115\frac{1}{2}$.

One of the characteristics of a continuous variable is that no two subjects should have precisely the same score. We would always expect that if the accuracy of recording the variable were to be extended to an infinite number of decimal places that no two people would have precisely the same amount of the trait. Thus, it is customary to regard tie scores as evidence that the assumption of continuity has been violated.

In the process of deriving certain (in fact, most) statistical procedures, theoretical statisticians have often found it necessary to assume that one or more variables are continuously distributed. Most statistical procedures outlined in this text have the underlying mathematical assumption that the variables are continuously distributed. However, when the behavioral scientist makes his statistical calculations, he does not have data from continuous variables, but rather from measures which are only discrete approximations of the variables. In short, all data are discrete!

Recent research by the author and other scientists has demonstrated that the assumption of continuity is not nearly so important in the practical research analysis as it was once believed to be. When it is possible to violate one or more of the mathematical assumptions underlying some statistical procedure (for example, the assumption of continuity) without appreciably affecting the results of the statistical analysis, the statistical procedure is said to be *robust*. This topic will come up frequently in the chapters to follow.

In practice, it appears desirable to think in terms of relatively continuous measures rather than absolutely continuous measures. For example, it has been demonstrated that the 20-word spelling test referred to earlier will behave very much like a continuous variable, provided that the students are fairly well distributed throughout the range of possible scores. If all the students are tied at a score of 13, for example, we do not have a variable at all, but rather a constant. If the students yield scores of 13, 14, and 15, we would probably have to regard the measure as discrete. However, if the scores were fairly well distributed over scores of 13, 14, 15, 16, and 17, for example, there is a possibility that we could treat the measure as a continuously distributed variable. By now the reader may be wondering just what distinction the author draws between a finite and an infinite number of score levels. As a matter of fact, statisticians often define infinity as some large number on the order of 30, but there are times when 10 or 12 will do as well. It is the author's experience that 5 sometimes constitutes a good approximation of infinity. The implications of these conclusions will be shown in later chapters.

Occasionally, it will be desirable to convert a relatively continuous measure into a discrete measure by grouping scores. Referring again to the 20-word spelling test, for example, the classroom teacher might wish to use the test scores to classify the students as (1) those of high ability, (2) those of average ability, and (3) those of low ability. Thus, a 21-point scale is reduced to a 3-point scale. This procedure is ordinarily avoided by researchers, for it is wasteful of valuable information.

2.4   MEASUREMENT SCALES

We usually think of measurement as the assignment of numbers indicative of quantity to some object or trait that we seek to assess. Behavioral scientists ordinarily prefer to think of measurement as involving the systematic representation of data by numbers according to rules which permit the scientist to legitimately use the arithmetic operations of addition, subtraction, multiplication, and division with these numbers. This is not always the case, however, and it will be convenient for the behavioral scientist to think in terms of several different levels of measurement, each with its own rules for the assignment of numbers to data. The various levels of measurement are often called *measurement scales*.

2.5   NOMINAL MEASUREMENT

The lowest level of measurement is *nominal* or *classificatory* measurement, which consists simply of classifying observations into categories (in short, a nominal variable is the same as a qualitative variable). These categories must be mutually exclusive and collectively exhaustive. Each observation must be capable of classification into one and only one category. There must not be any observation that cannot be assigned to one of the categories, and there must not be any observation that can be assigned to more than one category. For example, sex is a nominal variable, and we usually assume that all persons can be accurately classified as either male or female, and that no person can be classified as both. Similarly, all persons might be classified as college graduates or noncollege graduates, married or unmarried, and so on. If the population were limited to secondary school teachers, these might be classified as teachers of English, mathematics, social studies, science, and so on. Similarly, students might be classified on a nominal scale by state of birth or school attended.

Nominal scales are both discrete and qualitative. Occasionally, numbers will be assigned in the process of nominal measurement. For example, students might be assigned to Group I, Group II, or Group III, but arithmetic operations performed with these numbers would be meaningless.

Any nominal measurement may be dichotomized. For example, secondary school teachers can all be classified as either English teachers or teachers of other subjects. Students in college may be classified as residents of the state in which the college is located or as nonresidents of that state. In any event, a nominal scale may be quantified using binary numbers in the fashion described in Section 2.2. Mathematical operations with these binary numbers have proved to be extremely useful to behavioral scientists.

Authors in the field of measurement disagree as to whether nominal measures are measures at all. The utility of treating nominal data as measures will become increasingly apparent as the reader works his way through this text.

As a general rule, statistical procedures designed especially for use with nominal measures may also be used with higher order measures. In applications of this sort, a higher order measure is treated as though it were a nominal measure; this procedure is to be used with caution, for it does not ordinarily make full use of the data. When a higher order scale is available, it will ordinarily be to the investigator's advantage to use a statistical procedure designed to make full use of the additional information contained in the higher order scale.

2.6   ORDINAL MEASUREMENT

An *ordinal* scale is distinguished from a *nominal* scale by the additional property of order among the categories; that is, a category will be thought of as higher than or lower than the adjacent category. Although the idea of relative quantity of a given trait is characteristic of an ordinal scale, nothing is specified with respect to the magnitude of the interval between two measures. For example, a professional musician might rank three pianists in order of their ability to perform a particular composition, but it is improbable that he could speak precisely of how much more talent one has over another. Similarly, a clinical psychologist might speak with confidence regarding a client's improvement, but it is unlikely that he could quantify the amount of improvement. Ordinal measurement may be compared to a foot race without a stop watch; the order in which the runners finish is determined, but the magnitude of the invervals between them is indeterminate. Ordinal measurement is quite common in the behavioral sciences. Ordinal scales are sometimes called rank order scales.

The common practice is to assign numbers on an ordinal scale as follows: Let N represent the number of subjects on whom measures are available. The subject with the highest score or ranking is assigned a rank of 1. The next subject is assigned a rank of 2, and so on. The subject with the lowest score or ranking is assigned a rank equal to N. It is important to note that the original data may be in the form of rank order scores or some other kind of scores. For example, a classroom teacher might rank three students with respect to her impressions of their academic aptitude by assigning them the ranks of 1, 2, and 3. Suppose, however, that IQ scores were available on the three students. Students with IQ scores of 132, 124, and 109 would receive ranks of 1, 2, and 3, respectively. Notice that there is a correspondence between the scores and the ranks, but this correspondence ignores the magnitude of the difference between two adjacent scores. That is, score differences of 8 and 15 are both assigned rank differences of 1. When tie scores are encountered, the ranks are averaged as a general rule. For example, five persons with scores 132, 124, 109, 109, and 100 would be assigned ranks of 1, 2, 3.5, 3.5, and 5. Notice that the scores 109 and 109 received the average of ranks 3 and 4.

Ordinal scales may be either discrete or continuous. However, the statistical procedures designed for use with ordinal data ordinarily assume that the data are continuous. Wherever tie ranks are encountered, the assumption of continuity has been violated. Some statistical procedures designed for use with continuous ordinal data are fairly robust and may be used with relatively continuous data. If the data are markedly discrete, it will ordinarily be necessary to use statistical procedures designed for use with nominal scales.

There is some disagreement among behavioral scientists as to whether ordinal measures should be classified as qualitative or quantitative. The issue appears to revolve around the validity of using the operations of arithmetic (addition, subtraction, multiplication, and division) with the numbers assigned in the ranking process. In the example of the three IQ scores cited earlier, it is immediately apparent that 132 minus 124 is not equal to 124 minus 109. However, when we switch to the ranks assigned to these IQ scores, it appears on the surface at least that 1 minus 2 is equal to 2 minus 3. When the notion of measurement scales (the classification scheme of nominal, ordinal, interval, and ratio scales) was originally introduced, arithmetic operations with rank order scores were regarded as invalid. There now appears to be a growing consensus that the arithmetic operations are performed with the numbers (rather than the measures), and that the operations of arithmetic are valid with the number system. The question remains as to whether the results achieved are useful or meaningful when referred back to the original research. This question cannot be answered with mathematical logic, it must be answered with respect to a given research problem and the data that it yields. It should be noted, however, that the experience of numerous behavioral scientists suggests that arithmetic operations with rank order data usually yield useful and meaningful results.

Statistical procedures which have been developed especially for use with nominal and ordinal measures are often called *nonparametric* statistics. They may be used with data from higher level scales, but this procedure will not usually make full use of the available data.

2.7   INTERVAL MEASUREMENT

An *interval* type of scale is distinguished from an ordinal scale by having equal intervals between the units of measure. This means that a score of 50 points is halfway between scores of 40 and 60 points. The distinct advantage of such a scale is that the measures yielded may be treated with the common arithmetic operations of addition, subtraction, multiplication, and division. The interval scale is a truly quantitative scale.

One notable shortcoming of the interval scale is that it lacks a true zero. This means that one cannot interpret a score of 50 as indicating twice as much of a given trait as a score of 25. The problem arises from the fact that one cannot, as a general rule, meaningfully establish that a person has zero quantity of a particular psychological trait. For example, intelligence tests have been carefully scaled to yield interval measures, but there is no provision in the available tests to identify a person of zero intelligence. Similarly, a student may score zero points on an achievement test, but it has not been demonstrated that he has zero quantity of whatever trait the test is supposed to measure.

Most of the well-standardized psychological tests are intended to yield interval measures. The tests constructed by classroom teachers should probably be regarded as yielding ordinal measures, but no great harm seems to come from treating them as interval measures. While the data from interval scales may be analyzed using procedures designed for nominal and ordinal data, it is usually more profitable to take advantage of the higher level of measurement when it is available.

2.8   RATIO SCALES

Ratio scales have all the properties of interval scales, with the additional property of a true zero. While interval and ordinal scales are more often encountered by the behavioral scientist, he will also make use of such measures as age, height, and weight, which are measured on ratio scales.

With a ratio scale, one can intelligently speak of an individual having zero quantity of a particular trait, or of one individual having three times as much of this trait as another individual. While we may say that a person who weighs 150 pounds has twice the weight of a person who weighs 75 pounds, we cannot say that a person who has an IQ of 150 has twice the intelligence of a person with an IQ of 75.

The data yielded by ratio scales will usually be analyzed in the same fashion as data yielded by interval scales, and the statistical procedures which are specified for use only with interval and ratio data are customarily referred to as *parametric* statistics.

2.9   STATISTICS

Statistics, as a modern academic discipline, provides scientific procedures for collecting, organizing, summarizing, and analyzing quantitative information of the type commonly encountered in the behavioral sciences. First, statistics permits the summarization and presentation of large quantities of this information in such a fashion as to facilitate its communication and interpretation. Second, statistics enables the scientist to extend his research far beyond the restricted setting in which most research is actually conducted. It is this latter application, the use of statistics as a tool for analysis and extension of research findings, with which we are primarily concerned.

Some students and practitioners in the behavioral sciences shy away from quantitative information in general and statistics in particular. Perhaps through ignorance of the discipline, fear of its rigors, or philosophic bias, they have chosen instead to draw their conclusions from tradition, authority, or intuition. Such persons, of course, should not be classed as behavioral scientists. Reliable evidence from empirical research invariably lends itself to quantification and subsequent statistical analysis.

The use of statistical procedures in empirical research is not limited to the behavioral sciences (education, psychology and sociology, for instance). The use of these procedures in research is now universal in such diverse fields as agriculture, biology, chemistry, economics, medicine, and physics. For example, statistical procedures enable the educator to draw conclusions with respect to the efficacy of various instructional methods and materials, the psychologist to determine the precision with which he measures certain human traits, the sociologist to speak with confidence about the incidence of antisocial behavior, the physicist to interpret the activities of subatomic particles, the medical scientist to choose the most effective medicine, and the agricultural engineer the most productive fertilizer.

2.10   DESCRIPTIVE STATISTICS

The techniques of *descriptive statistics* enable us to describe with precision a collection of quantitative information, to do this in a form that is more concise and convenient than the original collection, and to do it in a fashion that makes for ease of interpretation and communication. For instance, an educational researcher might choose to summarize a large number of test scores from a specified population of students by calculating the average score. His description might be further enriched by calculating a measure of dispersion — a single numerical value that reveals valuable information about the manner in which the scores are distributed. Or, he might choose to summarize the collection of scores with a graphical representation. Each of these techniques of descriptive statistics is intended to facilitate the orderly communication and interpretation of an otherwise disorganized mass of raw data.

2.11   STATISTICAL INFERENCE

The techniques of *statistical inference* enable the researcher to draw inferences and generalizations from small groups, called *samples*, to larger groups, called *populations* — from individuals who are direct participants in experimental research to individuals who are not participants — and to do so with a well-defined degree of confidence. A sample is a relatively small group of individuals chosen in a scientific fashion to represent a relatively large group of individuals (the population) which the researcher is interested in studying. The student should take particular note of the fact that the researcher is primarily interested in the population although his research activities are carried on with the sample. This is an essential economy in research — that the scientist be able to extend with confidence his research findings beyond the narrow confines in which the research is conducted. The techniques of statistical inference provide the vehicle for scientific generalizations of this sort.

Statistical inference has been described as a collection of tools for making the best possible decisions in the face of uncertainty — that is, the situation where some, but not all, of the facts are in. Behavioral scientists have long been reconciled to making decisions of this type, and there is little prospect of their being delivered from this uncertainty.

In statistical inference, the term *statistic* is used in a special way. A *parameter* is some numerical property of a population. A *statistic* is a numerical property of a sample that is used to estimate the value of the corresponding population parameter. For example, an investigator might draw a sample from a population and use the average IQ score of the sample as an estimate of the average IQ score of the population. The techniques of statistical inference enable the investigator to do this with a well-defined degree of confidence.

2.12   PREDICTION

The terms *correlation, regression,* and *prediction* are so closely related in statistics that they are often interchangeable. These terms are used only for situations in which the researcher has at least two measures, one related to the other, on each individual. Related measures are those that tend to vary together — when one is larger, the other tends to be systematically larger or smaller. For example, height and weight of human beings tend to be positively correlated; as a general rule, taller persons are heavier persons. Thus, one might predict with a reasonable amount of confidence that a youngster who grows to be six feet tall will be heavier than one who grows to be five feet tall. When two measures are correlated, it is possible, though often with a limited degree of success, to predict one measure from the other. Thus, because IQ and academic success are usually correlated, an educator can predict the possible success of a student whose IQ is known with greater precision than he can predict the success of a student whose IQ is not known. Regression analysis permits prediction with a specified degree of precision.

In recent years, some behavioral scientists have been using regression analysis as a general procedure for experimental analysis. When an electronic computer is available, regression analysis provides an extremely flexible approach to very complex research problems. This text presents an introduction to these techniques along with the more traditional techniques of descriptive statistics and statistical inference.

2.13 PARAMETRIC VERSUS NONPARAMETRIC STATISTICS

The distinction among descriptive, inferential, and correlational statistics has to do with the nature of the problem confronting the investigator and the ultimate use that is to be made of the data. The distinction between parametric and nonparametric statistics has to do with the kind of data available for analysis. For the time being it will be sufficient for the student to distinguish *parametric* statistics as those statistical procedures intended for use with interval and ratio data, and *nonparametric* statistics as those statistical procedures intended for use with nominal and ordinal data.

It should be noted that a statistical procedure designed for use with one kind of data may also be used with any higher order of data. Thus those procedures designed for use with nominal data may be used with any data; those designed for use with ordinal data may also be used with interval and ratio data. Note also that a statistical procedure designed for use with higher order data will ordinarily provide results superior to one designed for use with lower order data. Thus the investigator whose research yields interval or ratio data may be justified in using any statistical procedure in this book; however, the best results with interval or ratio data will ordinarily be achieved with the parametric procedures.

There has been considerable debate in statistical circles with respect to the relevancy of the distinction between ordinal and interval scales. The majority of mathematical statisticians appears to subscribe to the notion that statistical analyses are performed on the numbers yielded by the measures (rather than the measures themselves), and the properties we ascribe to interval scales really belong to the number system. This argument appears to hold as long as the end product of the analysis makes sense. That is, if it makes sense to the investigator to perform arithmetic operations with ordinal data and to interpret the end product in terms of the original measures, it is probably valid to do so. Other statisticians have examined the distinction between ordinal and interval scales empirically. They have conducted large-scale experiments in which various kinds of data have been subjected to various kinds of statistical analysis. Their conclusions also suggest that the distinction between ordinal and interval data is not a particularly relevant one with respect to selecting a method of statistical analysis. Nevertheless, there are circumstances in which the use of the common arithmetic operations with ordinal data will obviously distort the meaning that the investigator attaches to his data. Under these circumstances he must resort to the use of the nonparametric statistics.

The student should be aware that the distinction between parametric and nonparametric statistics presented here is an oversimplification of the real state of affairs. As a matter of fact, statisticians disagree among themselves as to just what is meant by nonparametric statistics. Strictly speaking, the term applies to those statistical procedures used to test hypotheses which do not involve specific values of parameters (such as population means). Sometimes, the term *distribution-free* statistics is encountered. Strictly speaking, this term applies to those statistical procedures in which no assumptions are made about the manner in which variables are distributed. This is in contrast to the parametric statistics, which invariably introduce the assumption that one or more variables are normally distributed (This assumption will be discussed in some detail in later chapters.) It is common practice, however, to use the terms *nonparametric* and *distribution-free* interchangeably. While the distinction between parametric and nonparametric statistics involves all of these things, the student need only be concerned with the distinction made in the opening paragraph of this section for the time being. These other issues will be treated as they arise in the later chapters.

EXERCISES

Note: Answers to most of the exercises will be found in the Appendix.

2.1 Classify each of the following measures as qualitative or quantitative.
  (a) Age in years of students in a certain fourth grade class.
  (b) Age in years of students in a major university.
  (c) Grade levels taught by elementary school teachers.
  (d) Grades recorded on an A, B, C, D, F scale.
  (e) Major fields of study in college.
  (f) Number of problems worked correctly on an arithmetic test.
  (g) Number of runs of a maze completed out of ten trials.
  (h) Number of words spelled correctly on a spelling test.
  (i) Rating of a speech on a one-to-nine scale.
  (j) Religious affiliation (Catholic, Protestant, Jew, other).
  (k) Whether a student is a high school graduate or not.
  (l) Whether a subject is assigned to an experimental or a control group.

2.2 Refer to the measures given in Exercise 2.1. Using strict mathematical standards, seek to classify each of the measures as nominal, ordinal, interval, or ratio.

2.3 Refer to the measures given in Exercise 2.1. To the best of your ability, try to predict whether the measure would behave as continuous, probably continuous, or discrete.

2.4 Given the typical classroom test with 25 multiple-choice questions and with student scores ranging from 13 to 24, should the measures be regarded as

(a) Qualitative or quantitative?
(b) Discrete or continuous?
(c) Nominal, ordinal, interval, or ratio?

2.5 A psychometrist constructs an attitude scale with 20 statements to which the subject responds by choosing one of the following:
      (1) Strongly agree.
      (2) Agree.
      (3) Do not know.
      (4) Disagree.
      (5) Strongly disagree.
(a) If the psychometrist wishes to score each item separately on a one-to-five scale, should he regard the measures as continuous or discrete?
(b) Suppose the psychometrist arrives at a total score by adding the responses to all 20 items, should he regard the measure as continuous or discrete?
(c) Should he regard the scores on the individual items as ordinally or intervally scaled?
(d) Should he regard the total score as ordinally or intervally scaled?

2.6 In a study of factors influencing college student grades, the traditional letter grades are converted to a four-point scale in which A = 4, B = 3, C = 2, D = 1, and F = 0.
(a) If the investigator is seeking to predict final grades in the first semester of Western Civilization, should he regards the grades as discrete or continuous?
(b) Suppose that the investigator were seeking to predict grade-point average for all courses taken during the first semester of the freshman year. Should he regard the grade-point average as discrete or continuous?

2.7 Give the distinguishing characteristics of each of the following:
(a) Descriptive statistics.
(b) Statistical inference.
(c) Correlational statistics.
(d) Nonparametric statistics.
(e) Parametric statistics.

SELECTED REFERENCES

ADAMS, ERNEST W., ROBERT F. FAGOT, and RICHARD E. ROBINSON, "A theory of appropriate statistics," *Psychometrika*, **30,** 99–127 (June 1965).

ANDERSON, NORMAN H., "Scales and statistics: Parametric and nonparametric," *Psychological Bulletin*, **58,** 305–316 (1961).

BAKER, BELA O., CURTIS D. HARDYCK, and LEWIS F. PETRINOVICH, "Weak measurements vs. strong statistics: An empirical critique of S. S. Stevens' proscriptions on statistics," *Educational and Psychological Measurement*, **26,** 291–309 (1966).

BURKE, C. J., "Additive scales and statistics," *Psychological Review*, **60**, 73–75 (1953).

GHISELLI, EDWIN E., *Theory of Psychological Measurement*. New York: McGraw-Hill, 1964.

GUILFORD, J. P., *Psychometric Methods*. New York: McGraw-Hill, 1954.

NUNNALLY, JUM C., *Psychometric Theory*. New York: McGraw-Hill, 1967.

STEVENS, S. S., "On the theory of scales of measurement," *Science*, **103**, 677–680 (June 1946).

3

FREQUENCY DISTRIBUTIONS

3.1 INTRODUCTION

The behavioral scientist frequently encounters large masses of numbers, usually test scores, which communicate little meaning until these are organized or summarized with one of the techniques of descriptive statistics. Table 3.1, for instance, contains scores yielded by an English usage test administered to 120 college students. The scores were recorded in the order in which they were received from the students.

Table 3.1   Random arrangement of 120 scores

13	7	6	16	9	3	12	13	11	14	5	10	4	9	7
9	17	14	13	11	12	19	4	9	7	12	16	7	12	11
4	5	11	17	10	9	10	17	13	16	12	13	18	9	15
16	12	9	7	14	3	16	9	5	10	17	15	11	10	10
6	18	10	15	8	13	14	18	12	8	9	8	12	8	12
11	8	11	12	16	11	7	15	11	14	20	7	6	14	16
5	15	19	14	13	15	8	8	13	11	7	11	12	17	13
14	10	13	8	5	17	11	10	6	12	5	1	11	8	10

It is extremely difficult to make any generalizations about the characteristics of this group of students from the collection of unordered scores. A little more meaning is communicated when the scores are arranged in numerical order, as in Table 3.2.

As indicated earlier, the techniques of descriptive statistics enable the researcher to describe with precision a collection of quantitative information, to do this in a form that is more concise and convenient than the

26

Table 3.2 Ordered arrangement of 120 scores

20	17	16	15	14	13	12	11	11	10	9	8	7	6	5
19	17	16	14	13	13	12	11	11	10	9	8	7	6	5
19	17	16	14	13	13	12	11	11	10	9	8	7	6	4
18	17	15	14	13	12	12	11	11	10	9	8	7	6	4
18	16	15	14	13	12	12	11	10	10	9	8	7	5	4
18	16	15	14	13	12	12	11	10	10	9	8	7	5	3
17	16	15	14	13	12	12	11	10	9	9	8	7	5	3
17	16	15	14	13	12	11	11	10	9	8	8	7	5	1

original collection, and to do it in a fashion that makes for ease of inter-pretation and communication. All of this can be accomplished without loss of information by arranging the data at hand in a *frequency distribution*.

3.2 FREQUENCY DISTRIBUTIONS WITH UNGROUPED DATA

A frequency distribution is a technique for systematically arranging a collection of measures on a given variable to indicate the frequency of occurrence of the different values of the variable. Consider Table 3.3,

Table 3.3 Frequency distribution of 120 ungrouped scores

X	f
20	1
19	2
18	3
17	6
16	7
15	6
14	8
13	10
12	12
11	13
10	10
9	9
8	9
7	8
6	4
5	6
4	3
3	2
2	0
1	1
0	0

Table 3.4 Illustration of the procedure for making a frequency distribution

X	Tally marks
20	/
19	//
18	///
17	//// /
16	//// //

which contains all of the information of Tables 3.1 and 3.2, but in the form of a frequency distribution. The column marked X contains all possible score values, and the column marked f contains the score frequencies. The *frequency* of any score is simply the number of persons who received that score.

It is immediately apparent that the frequency distribution markedly facilitates interpretation of the collection of scores. Notice that the scores of highest frequency stand out clearly. The investigator is able to discern at a glance the general distribution of the scores along the score scale.

The usual procedure for making up a frequency distribution is first to list in order of size all possible score values from the highest to the lowest, and then to read off the scores in the collection in the order in which they were received, making a tally mark for each of these opposite the corresponding score value in the list. Table 3.4 illustrates this procedure for a portion of the data in the previous tables.

3.3   FREQUENCY DISTRIBUTIONS WITH GROUPED DATA

Sometimes the information contained in a frequency distribution will be appreciably less cumbersome and more readily comprehended if the scores are grouped into intervals, each of which contains several score levels. Consider Table 3.5, in which each interval contains two score levels, and Table 3.6, in which each interval contains four score levels. The data are the same as contained in the previous tables.

Notice that Table 3.5 is more compact than any of the previous tables; however, the identity of the individual score is lost, and this loss of information may be of greater importance than the gain in compactness. Table 3.6 is even more compact, but at the expense of a loss of even more information. While Table 3.5 expresses the general form of the distribution quite well, Table 3.6 has lost the form in the grouping of the data.

Table 3.5 Frequency distribution of 120 grouped scores

X	f
19–20	3
17–18	9
15–16	13
13–14	18
11–12	25
9–10	19
7–8	17
5–6	10
3–4	5
1–2	1

Table 3.6 Frequency distribution of 120 grouped scores

X	f
17–20	12
13–16	31
9–12	44
5–8	27
1–4	6

Some older statistics textbooks suggest the grouping of interval or ratio scale measures to simplify calculations with that data. Modern behavioral scientists, however, tend to regard this as a great waste of research information. Generally, the value of the original data is much greater than any possible saving in calculation effort. Normally, the calculations for any substantial research effort will be performed with an electronic computer, and grouping does nothing to facilitate the process.

However, the grouping of data in the fashion described may be a useful technique in descriptive statistics, where the frequency distribution is an end product rather than an intermediate step in a series of calculations. If the grouping of data is undertaken to facilitate interpretation, it is recommended that the number of intervals be between 10 and 25, with 15 to 20 being optimum. Of course, all intervals will normally contain the same number of score levels, with the possible exception of the top and bottom intervals, which the investigator may choose to leave open-ended to accommodate a few extreme scores. Some authors recommend that the interval contain an odd number of score levels so that the middle score is an integer and may be conveniently used as an index value representative of all measures in the interval.

3.4 CLASS LIMITS

Whenever measurements on continuously distributed traits are recorded as integers (whole numbers only, no fractions), a given score must be regarded as being somewhere between half a unit above or below the integer. For example, a youngster getting 12 words right on a spelling test must be regarded as having a score somewhere between $11\frac{1}{2}$ and $12\frac{1}{2}$, since spelling ability is believed to be a continuous rather than a discrete variable. The fact that the score is recorded in whole numbers only is a function of the precision of measurement rather than the discreteness of the variable. For this reason, the *real limits* of continuous data recorded with unit intervals will be regarded as extending one-half unit on either side of the recorded score.

The same principle applies to grouped data. If an interval contains scores ranging from 16 to 20, for instance, the real limits of the interval are $15\frac{1}{2}$ to $20\frac{1}{2}$. Occasionally, the term *integer limits* will be used; in the example quoted, the integer limits are 16 and 20.

The student will want to take note of the idea of real limits, since this is a useful concept which will appear repeatedly in the study of statistics.

3.5 CUMULATIVE AND RELATIVE FREQUENCIES

It is not unusual to report interval frequencies as decimal fractions, or percentages, of the total number of scores in the collection. These are known as *relative frequencies* (*rf*), and they may be simply calculated by the formula

$$rf = \frac{f}{N}$$

where *rf* is the relative frequency, *f* is the frequency of the interval, and *N* is the total number of scores in the collection. The sum of all the relative frequencies in the distribution must equal 1.00 or 100 percent. Relative frequencies are especially important in statistical inference, as they are intimately related to the mathematical idea of probability.

The *cumulative frequency* (*cf*) of a given interval is the frequency of the interval plus the total of the frequencies of all intervals below the given interval. Another way of stating this is to say that the cumulative frequency of any interval is the number of scores in the distribution which fall below the upper real limit of that interval. This is another important idea which will come up again in the study of statistical inference.

The *relative cumulative frequency* (*rcf*) of a given interval is the cumulative frequency divided by the number of measures in the collection:

$$rcf = \frac{cf}{N}$$

Table 3.7 Frequency distribution of 120 scores

X	f	fX	cf	rf	rcf
20	1	20	120	.008	1.000
19	2	38	119	.017	.992
18	3	54	117	.025	.975
17	6	102	114	.050	.950
16	7	112	108	.058	.900
15	6	90	101	.050	.842
14	8	112	95	.067	.792
13	10	130	87	.083	.725
12	12	144	77	.100	.641
11	13	143	65	.108	.541
10	10	100	52	.083	.433
9	9	81	42	.075	.350
8	9	72	33	.075	.275
7	8	56	24	.067	.200
6	4	24	16	.033	.133
5	6	30	12	.050	.100
4	3	12	6	.025	.050
3	2	6	3	.017	.025
2	0	0	1	.000	.008
1	1	1	1	.008	.008
0	0	0	0	.000	.000
	N = 120	T = 1327			

Table 3.7 contains the relative frequencies, cumulative frequencies, and relative cumulative frequencies for the ungrouped frequency distribution considered earlier. It also contains a column for fX, the product of the interval score and frequency, which facilitates the calculation of the sum of the scores, which we will call T.

With respect to the manner of calculating the cumulative and relative cumulative frequencies described here, there is no reason why the cumulative frequency could not be defined as the frequency of the interval plus the total of the frequencies *above* the given interval. Some textbooks give this option in addition to the one given here; however, we shall be consistent in using the definition given earlier in this section.

3.6 BIVARIATE FREQUENCY DISTRIBUTIONS

The concept of frequency distribution may be readily extended to the situation in which two measures are available on each subject, and all of this information may be reported in a single table. Such a table is called a *bivariate frequency table* or *contingency table*.

Suppose that the 120 scores reported in the previous univariate tables were accompanied by information which also identified the student as a freshman, sophomore, junior, or senior. It would be a simple matter to construct four different frequency distributions, one for each class, and then to place them side by side for comparison. All of this can be accomplished, however, with a single bivariate frequency distribution. Table 3.8 makes use of the grouped data of Table 3.6, as well as knowledge of the class of each respondent, in a bivariate frequency table. Notice that the column sums are the same as the frequencies reported in Table 3.6, and that the row sums indicate that the sample was evenly divided among the four classes. These row and column sums are called *marginal frequencies*.

Table 3.8 Bivariate frequency distribution

Scores	1–4	5–8	9–12	13–16	17–20	Totals
Seniors	0	1	6	15	8	30
Juniors	0	3	9	14	4	30
Sophomores	3	7	18	2	0	30
Freshmen	3	16	11	0	0	30
Totals	6	27	44	31	12	120

We will refer to the variable used to determine whether a score is recorded in a given row (the grade level of the student in the example) as the *row variable*. Similarly, we will refer to the variable used to determine whether a score is recorded in a given column (the student's score in the example) as the *column variable*. It should be apparent from examining the example that a bivariate frequency distribution permits the investigator to explore the relationship between the row and the column variables. In the example, it is apparent that higher grade levels are systematically associated with higher scores.

While all of the examples given in this chapter have used quantitative variables, frequency distributions (both univariate and bivariate) may be constructed from qualitative data as well. Of course, the concepts of *fX*, *cf*, and *rcf* are meaningless with qualitative data.

EXERCISES

3.1 Given the following collection of 100 IQ scores:

100	107	85	119	93	111	89	90	113	99
92	108	103	98	86	73	102	100	74	103
112	97	115	72	104	117	98	127	89	107
126	99	71	99	85	95	121	84	96	83
101	66	123	107	97	101	86	106	114	100
85	110	71	102	118	111	103	83	99	126
102	98	96	130	88	114	100	100	92	100
96	81	122	92	80	91	119	81	102	93
118	92	98	88	112	95	110	136	97	98
106	112	76	113	77	100	97	105	101	89

(a) Construct a frequency distribution using unit intervals.
(b) Construct a frequency distribution using intervals of five points. Let 65–69 be the lowest interval.
(c) Construct a frequency distribution using intervals of ten points. Let 60–69 be the lowest interval.
(d) Which of the three frequency distributions communicates the general distribution of the IQ scores most effectively? Which of the three distributions contains the most information?

3.2 Given the following collection of scores on a spelling test: 5, 6, 10, 8, 8, 4, 5, 5, 9, 8, 6, 6, 7, 4, 5, 5, 5, 3, 6, 9
(a) Construct a frequency distribution using unit intervals.
(b) Determine the values of *N* and *T*.
(c) Determine the cumulative frequency for each interval.
(d) Determine the relative frequency for each interval.
(e) Determine the relative cumulative frequency for each interval.

3.3 Complete all columns of the table below except the column labeled PR.

X	f	fX	cf	rf	rcf	PR
8	4					
7	4					
6	11					
5	10					
4	10					
3	4					
2	2					
1	3					
0	2					

3.4 In a recent study of high school student attitudes toward a certain phase of American foreign policy, 22 boys indicated agreement with the policy, 44 indicated disagreement, and 34 had no opinion, while 62 girls indicated agreement, 27 indicated disagreement, and 11 had no opinion. Construct a bivariate frequency distribution with these data. Use the sex of the respondent as the row variable, and the response to the policy as the column variable. Which is the independent variable, and which is the dependent variable? What is the relationship between the two variables?

4

PERCENTILE RANKS

4.1 SCORE TRANSFORMATIONS

The numerical value of a student's score on a test tends to be a highly arbitrary quantity. In addition to student ability, it is also a function of the number of items on the test and of the difficulty of the items. The interpretation of the score may be greatly enriched by expressing it in units which are a function of its relationship to other scores in the distribution. A score has meaning only as it is related to other scores, which is another way of saying that the principal interest of the behavioral sciences is the way in which individuals differ.

When a score in a collection of such scores is expressed as the percentage of scores in the collection that are below this score, the percentage is called a *percentile rank*. A percentile rank is a relative rank, a rank order score based on a scale of 100.

Translating a score into a percentile rank in the manner described is one of several possible *score transformations* that are intended to convey the relationship of a given score to the group. Score transformations may also be used to change the shape of a frequency distribution or to facilitate the comparison of scores from different instruments. When a score is transformed to a new score scale, the original score is referred to as the *raw* score and the new score as the *transformed* score.

To report that a student scored 40 points on an English test and 35 points on a mathematics test has little meaning of itself, but to report that he scored at the 40th percentile in English and at the 60th percentile in mathematics immediately conveys the idea that the student scored slightly below the group average in English and slightly above in mathematics. Such a comparison implies that the two percentile ranks were calculated with reference to a single group or two equivalent groups. Of course, it is important that the group to which the individual is compared be well defined.

4.2   DEFINITIONS

The *percentile rank* of a given raw score is the percentage of scores which is below this particular score. For example, if 50 percent of the students in a distribution are below a given student, his percentile rank is 50. Notice that the percentile rank is a point on the *transformed* score scale. The abbreviation PR will be used to represent percentile rank.

A *centile* (or percentile as it is called by some authors) is a point on the *raw* score scale which corresponds to a given percentile rank. For example, an IQ score of 100 ordinarily corresponds to the 50th centile. The percentile rank is 50, the 50th centile is 100. There are 99 centiles, and these divide a frequency distribution into 100 equal parts. The letter C with a subscript will be used to represent centiles, the subscript indicating which centile. The symbol C_{50}, for instance, represents the 50th centile.

There are nine *deciles*, and these divide a frequency distribution into ten equal parts. Thus, the 1st decile (D_1) is the 10th centile (C_{10}), and the 2nd decile (D_2) is the 20th centile (C_{20}), and so forth.

There are three *quartiles*, and these divide a frequency distribution into four equal parts. Thus, the 1st quartile (Q_1) corresponds to the 25th centile (C_{25}), the 2nd quartile (Q_2) to the 50th centile (C_{50}), and the 3rd quartile (Q_3) to the 75th centile (C_{75}). The 2nd quartile is more often referred to as the *median*. The student is reminded that a quartile is a *point* on a score scale rather than a portion of the distribution.

4.3   CALCULATION OF PERCENTILE RANKS

The percentile rank of any score in a frequency distribution with unit intervals may be calculated by the formula:

$$PR = \frac{100}{N}\left(cf - \frac{f}{2}\right)$$

where f is the frequency of the score, cf is the cumulative frequency, and N is the number of scores in the entire distribution. The expression inside the parentheses is the cumulative frequency at the midpoint of the interval (*cfm*) in which the score appears. Dividing the expression by *N* produces the relative cumulative frequency at the midpoint (*rcfm*), and multiplying by 100 moves the decimal to the right two places to produce a percentage from a proportion. It is conventional to report percentile ranks rounded to the nearest whole number.

Table 4.1 illustrates the calculation of percentile ranks for the frequency distribution that was used for illustrative purposes in Chapter 3. The actual calculations used in determining the percentile rank corresponding to a raw score of 17 are as follows:

$$PR = \frac{100}{120}\left(114 - \frac{6}{2}\right) = \frac{100}{120}(111) = 92.5$$

Table 4.1 Percentile ranks of 120 scores

X	f	cf	cfm	rcfm	PR
20	1	120	119.5	.996	99.6
19	2	119	118	.983	98.3
18	3	117	115.5	.962	96.2
17	6	114	111	.925	92.5
16	7	108	104.5	.871	87.1
15	6	101	98	.817	81.7
14	8	95	91	.758	75.8
13	10	87	82	.683	68.3
12	12	77	71	.592	59.2
11	13	65	58.5	.487	48.7
10	10	52	47	.392	39.2
9	9	42	37.5	.312	31.2
8	9	33	28.5	.237	23.7
7	8	24	20	.167	16.7
6	4	16	14	.117	11.7
5	6	12	9	.075	7.5
4	3	6	4.5	.037	3.7
3	2	3	2	.017	1.7
2	0	1	1	.008	0.8
1	1	1	0.5	.004	0.4

The student should take note of the fact that the raw scores from which percentile ranks are calculated may be ordinal, interval, or ratio data; however, the transformed scores are ordinally scaled. This suggests that adding two percentile ranks or performing similar arithmetic operations with them is mathematically questionable. However, in actual practice, the use of arithmetic operations with percentile ranks is not at all uncommon and appears to work out quite satisfactorily in many circumstances.

Some authors give formulas for calculating percentile ranks from grouped data, but this is likely to yield a transformed score distribution with many gaps in it, and the practice is not recommended. Percentile ranks are most appropriate when calculated from ungrouped raw scores, from a distribution containing at least 200 scores.

4.4   THE DISTRIBUTION OF PERCENTILE RANKS

The distribution of percentile ranks is uniform throughout the transformed score scale. Theoretically, there will be an equal number of scores (one-fifth of the total) below the 20th centile, between the 20th and 40th centiles, between the 40th and 60th, between the 60th and 80th, and above the 80th. For a given collection of scores, this may not be exactly the case, especially if the number of scores is small; however, if the number of scores is several hundred or more, and the raw score scale is relatively continuous (having many levels of performance), the distribution will be almost perfectly uniform. When a graph is drawn of such a distribution using the techniques discussed in Chapter 5, the graph will be rectangular in shape. The distribution of percentile ranks is said to be rectangular or uniform.

4.5   THE CALCULATION OF CENTILES, DECILES, AND QUARTILES

If the student will refer to the PR column in Table 4.1, he will find that none of the raw scores in this distribution corresponds precisely to any centile. For instance, the median (which is also C_{50}, D_5, and Q_2) lies somewhere between a score of 11 and a score of 12. When a centile does not correspond directly with any recorded score, the following procedure is recommended for determining its value:

1. Let $p = $ PR/100, where PR is the percentile rank corresponding to the desired centile. For example, $p = .25$ for Q_1 and .75 for Q_3. Multiply p by N to get Np.
2. If the cumulative frequency of any interval is equal to Np, the desired centile is equal to the upper real limit of the interval. Referring to Table 4.1, one may use this procedure to determine that C_5 is 4.5, C_{10} is 5.5, and C_{20} is 7.5. However, the bulk of the centiles is still indeterminate, and we must proceed to the next step for these.
3. Determine the interval which has the smallest cumulative frequency larger than Np. Then calculate the desired centile by the following formula:

$$C = X_u - \frac{cf - Np}{f}$$

where X_u is the upper real limit, f is the frequency, and cf is the cumulative frequency of the interval.

Now let us illustrate the use of this procedure for determining the values of Q_1, Q_2, and Q_3.

$$Q_1 = 8.5 - \frac{33 - 30}{9} = 8.17$$

$$Q_2 = 11.5 - \frac{65 - 60}{13} = 11.12$$

$$Q_3 = 14.5 - \frac{95 - 90}{8} = 13.88$$

The situation is graphically illustrated for Q_3 in Figure 4.1. It should be noted that for $N = 120$, Q_3 is a point on the score scale which corresponds to $cf = 90$. Another graphic method for determining the values of the various centiles will be considered in Chapter 5.

The procedure outlined here is suitable for calculating centiles from *ungrouped* data. Although the procedure may be modified to accommodate *grouped* data (simply multiply the second term on the right half of the formula by the width of the interval), this author questions the utility of centiles calculated from grouped data.

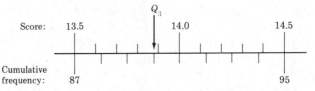

Figure 4.1   Graphic relationship of score and cumulative frequency for a given interval.

EXERCISES

4.1   Refer to the data for Exercise 3.1. Calculate the percentile rank for each score in the distribution. Determine the values of Q_1, Q_2, and Q_3.

4.2   Refer to the data for Exercise 3.2. Calculate the percentile rank for each score in the distribution. Determine the values of D_2, D_4, D_6, and D_8.

4.3   Refer to the data for Exercise 3.3. Complete the PR column.

4.4   Refer to the data of Table 4.1 and to Figure 4.1. Construct a similar figure for Q_1 and another for Q_2.

GRAPHS OF FREQUENCY DISTRIBUTIONS

5.1 INTRODUCTION

One of the techniques which may be employed to organize and summarize a collection of test scores is to represent the data graphically. The graphs which are described in this chapter correspond directly to the frequency distributions of Chapter 3 and often perform the same functions, communicating the same information pictorially. As the student progresses into the study of statistical inference, he will find that graphs not only portray the form of a frequency distribution but also facilitate the solution of many problems in research.

All of the graphs which will be considered here will be constructed with reference to two perpendicular lines, one horizontal and one vertical, which are called the *coordinate axes*. The horizontal axis (or *abscissa*) will represent the score scale, much in the same way that a rule is used. If the data are grouped, the investigator will probably choose to plot the midpoints of the intervals along this axis. Of course, this will be much simpler if the intervals are chosen so that the midpoint is an integer. Frequencies or relative frequencies will be plotted along the vertical axis (or *ordinate*). It is important in the construction of any graph that both axes be clearly labeled so that the reader knows what they represent. The graph itself should also be named or labeled in such a fashion that the reader will know what it is all about.

One of the important ideas associated with graphs of frequency distributions is that the *area* under a graph (or portion of a graph) is representative of the frequency of the corresponding scores. Often, the total area under such a graph will be set to unity, in which case the area above a certain portion of the score scale will be equal to the relative frequency of those scores. This relationship between relative frequency and area is basic to the use of statistics in research.

5.2   THE HISTOGRAM

A *histogram* is essentially a bar graph of a frequency distribution in which the frequencies are represented by areas in the form of vertical bars. Its purpose is to show the frequencies of the various scores graphically. It carries precisely the same information as the original frequency distribution from which it was made, although the graph becomes only an approximation of the frequency distribution if the number of scores is large.

Figures 5.1, 5.2, and 5.3 are histograms of the frequency distributions of Tables 3.3, 3.5, and 3.6, respectively. Notice that the width of each bar stretches from the lower real limit of the score interval to the upper real limit.

In the construction of a histogram, the score scale may be laid out by labeling either the midpoint or the limits of the intervals. The use of the midpoint is the more common practice. If the scores do not range down to

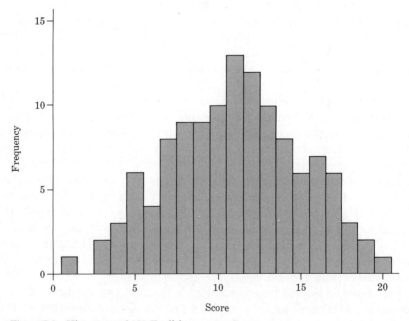

Figure 5.1   Histogram of 120 English usage test scores.

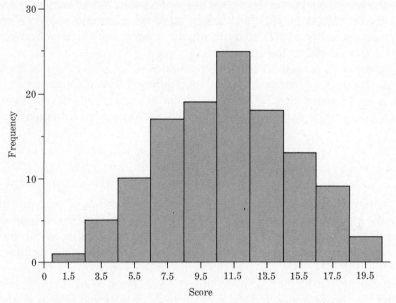

Figure 5.2 Histogram of 120 scores grouped in intervals of two points.

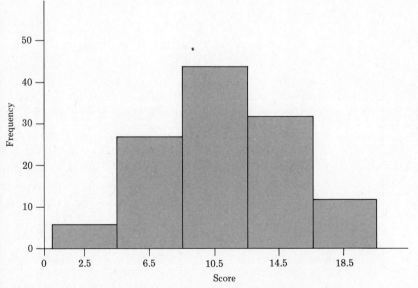

Figure 5.3 Histogram of 120 scores grouped in intervals of four points.

zero, it is permissible to start the score scale at some convenient point above zero which makes provision for the lowest score. Empty score intervals between the highest and lowest score values should be shown on the graph (as in Figure 5.1), and it is a good practice to include one empty interval at each end of the histogram.

The frequency scale must be laid out to provide for all frequencies from zero to the largest frequency if the relationship between frequency and area is to be maintained. The form of the histogram will be the same whether the vertical axis is laid out with frequencies or relative frequencies.

5.3 THE FREQUENCY POLYGON

The frequency distribution may also be portrayed by means of a frequency *polygon*, a graph consisting of straight lines connecting points located above the midpoints of the intervals at heights corresponding to the frequencies. An empty interval is included at each end of the distribution, and the graph is brought down to the horizontal axis (which corresponds to zero frequency) at the midpoints of these two intervals. When the graph is enclosed in this fashion, the total area enclosed is the same as that of the histogram for the same data. The student should note, however, that

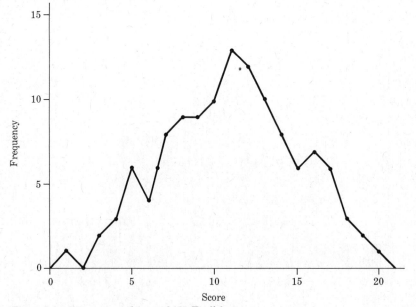

Figure 5.4   Frequency polygon of 120 English usage test scores.

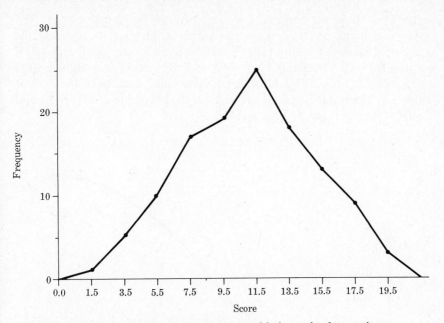

Figure 5.5   Frequency polygon of 120 scores grouped in intervals of two points.

the area of the various intervals is not precisely recorded in the frequency polygon, as it is in the histogram.

Figures 5.4 and 5.5 are frequency polygons constructed from the same data used to construct the histograms in Figures 5.1 and 5.2.

The histogram portrays the frequency distribution as though all scores in each interval were uniformly distributed throughout the interval, whereas the frequency polygon portrays the distribution as though all scores were concentrated at the midpoints of the intervals. The frequency polygon, like the histogram, carries the same information as the frequency distribution from which it was made. Some investigators prefer the frequency polygon when they wish to compare the form of two different frequency distributions. Perhaps the greatest contribution of the frequency polygon is that its form suggests that the trait portrayed is continuously distributed, which is often the case.

5.4 THE OGIVE

An *ogive* resembles a frequency polygon, except that the vertical scale is recorded in cumulative frequencies, relative cumulative frequencies, or percentile ranks. There is another important distinction: The plotting points on the ogive are located above the upper real limits of the intervals rather than above the midpoints, as used with the frequency polygon. The upper real limits are used, since the graph is intended to represent the number of measures falling below specified score values. Such a graph, if constructed on a sufficiently large scale, may be used in estimating centiles and percentile ranks. Figure 5.6 is an ogive constructed with the data of Table 3.3, which was also used to construct the histogram of Figure 5.1 and the frequency polygon of Figure 5.4.

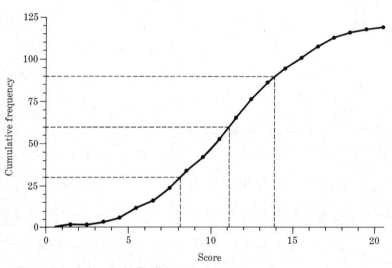

Figure 5.6   Ogive of 120 English usage test scores.

Referring to Figure 5.6, the student may locate the 50th percentile rank on the vertical axis (where $f = 60$) and determine that the value of the median is approximately 11.1 on the horizontal axis. Similarly, he may determine that the value of Q_1 is approximately 8.2 and that Q_3 is approximately 13.9 on the score scale. If the ogive is constructed specifically for the purpose of estimating centiles and percentile ranks, the vertical scale should be divided into 100 equal parts and calibrated in percentile ranks.

5.5   GRAPHS OF CONTINUOUS DISTRIBUTIONS

There is substantial evidence that many of the variables studied by behavioral scientists are continuously distributed even though their measuring instruments yield discrete scores which are only approximations of the supposedly continuous variables. When these measurements are made on relatively small samples drawn from very large populations, it is reasonable to assume that the irregularities in a frequency polygon may be attributed to sampling and measurement errors, rather than being truly characteristic of the distribution of the trait in the parent population. The frequency polygon of a truly continuous variable in a very large population would be a smooth curve, such as that in Figures 5.7, 5.8, 5.9, and 5.10.

The investigator has reproduced the frequency polygon of Figure 5.4 with a dashed line in Figure 5.7 and has also smoothed the original frequency polygon, producing a continuous curve which he hopes will approximate the true distribution of the variable in the parent population. Of course, there is a danger of obscuring the true distribution when a frequency polygon is smoothed in this fashion.

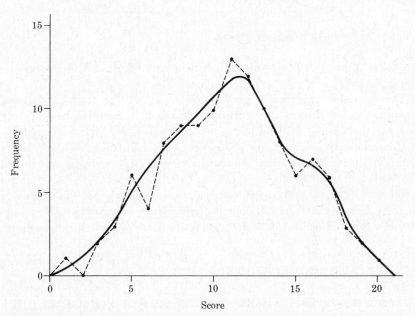

Figure 5.7 Smoothed frequency polygon of 120 English usage test scores.

5.6 THE NORMAL CURVE

Certain direct measures used in the behavioral sciences (for example, the height and weight of adult humans) have been demonstrated to closely approximate a mathematical model called the *normal* distribution. The graph of the normal distribution is a continuous, symmetrical, bell-shaped curve. It is high in the middle, indicating a preponderance of frequencies in the vicinity of the median, and low at the ends, indicating low fre-

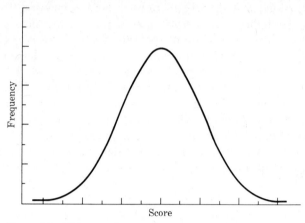

Figure 5.8   The normal curve.

quencies at both extremes of the distribution. Figure 5.8 illustrates the normal curve.

The available evidence suggests that many of the traits underlying psychological measures are normally distributed. For this reason, and because the normal distribution is a useful model for statistical analysis, it is common practice among professional test constructors to adjust the difficulty of the items so that standardized tests yield normally distributed scores.

The normal curve is the foundation of statistical theory. It will be treated in considerable detail in subsequent chapters and will prove to be an extremely useful model for both descriptive statistics and statistical inference.

5.7   SKEWNESS AND KURTOSIS

The term *skewness* refers to the symmetry or asymmetry of a curve. If a curve is symmetrical, such as the normal curve, the right and left halves of

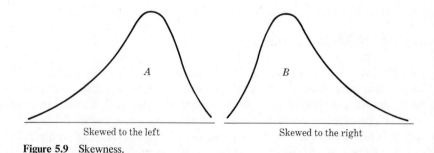

Figure 5.9   Skewness.

the curve are mirror images of each other. A curve that is substantially asymmetrical is said to be skewed. If the scores tend to be concentrated toward the low end of the score scale with the curve trailing off to the right, the curve is said to be positively skewed, or skewed to the right. If the scores tend to be concentrated toward the high end of the score scale, the curve is said to be negatively skewed, or skewed to the left. Both types of skewness are illustrated in Figure 5.9.

The term *kurtosis* refers to the relative peakedness or flatness of a curve in the center of the distribution. If the curve is flatter than a normal curve, it is said to be platykurtic; if the curve is more peaked than a normal curve, it is said to be leptokurtic. These ideas are graphically illustrated in Figure 5.10.

There are measures of skewness and kurtosis, but these are seldom used by the behavioral scientist and will not be discussed in this text.

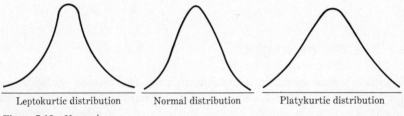

Leptokurtic distribution   Normal distribution   Platykurtic distribution

Figure 5.10   Kurtosis.

EXERCISES

5.1   Refer to the data for Exercise 3.1. Produce a histogram and frequency polygon from the data, using intervals of five points as in (b) of the exercise. Construct an ogive from the data using unit intervals, and estimate the values of Q_1, Q_2, and Q_3 from the ogive.

5.2   Refer to the data of Exercise 3.2. Produce a histogram, frequency polygon, smoothed frequency polygon, and ogive from the data. Estimate the values of D_2, D_4, D_6, and D_8 from the ogive.

5.3   Refer to the data of Exercise 3.3. Produce a histogram, frequency polygon, smoothed frequency polygon, and ogive from the data.

6

STATISTICAL NOTATION

6.1 THE USE OF SYMBOLS

Symbolic notation makes it possible to state and discuss statistical ideas far more precisely and concisely than is ordinarily possible with common words. The student should realize that the use of symbols is intended to *simplify* the use of statistical procedures. The number of different symbols used is relatively few, and the effort needed to master them will prove to be a profitable investment.

The letter N will be used to represent the number of individuals in a collection of individuals being studied. When the collection is divided into subgroups, N will represent the number in the total collection, and n_1 will represent the number in group one, n_2 the number in group two, and so on. If k is the number of groups, then n_k is the number of individuals in the last group. This is a convenient notation for establishing general procedures applicable to the situations in which the number of groups may differ. The symbol n_j may be used to represent the number of individuals in *any* group. This general notation permits the establishment of procedures that apply to any or all groups from group one to group k.

The letter X is the symbol most often used to represent the value of a measure in the collection (a test score, for instance). Through the use of subscripts any score in the collection may be designated. Thus X_1 represents the score of the first individual, X_2 the score of the second, and so on. X_N is the last score in the collection; X_i may be used to represent the score of *any* individual. This general notation permits the establishment of procedures that apply to any or all individuals from the first to the Nth.

When the investigator is confronted with the situation in which there are two measures on each individual, he will probably let Y represent the value of the second measure and use subscripts with it in the same fashion as with X in the previous paragraph.

6.2   THE SUMMATION OPERATOR

Statisticians are so often concerned with summing a collection of measures that a special symbol, the *summation operator*, has been adopted to indicate summation. It is the upper case Greek letter sigma (Σ).

One of the most common uses of the summation operator is to indicate that all of the scores in a distribution are to be added together:

$$\sum_{i=1}^{N} X_i = X_1 + X_2 + X_3 + \cdots + X_N = T$$

The symbols above and below the summation sign indicate the limits of summation, in this case summing all scores from the 1st to the Nth. The idea may be stated in narrative fashion as "taking the sum of the X sub i's as i goes from 1 to N."

When the sum of all measures is to be taken, or when the limits of summation are clearly understood from the context (which is most of the time), statisticians tend to omit the symbols above and below the summation operator and write simply ΣX_i or ΣX. We will also use the upper case letter T (for total) to represent the sum of all scores in a distribution.

The summation operator is used in precisely the same fashion to sum the *number* of individuals in all the subgroups to obtain the total number of individuals:

$$\sum_{j=1}^{k} n_j = n_1 + n_2 + n_3 + \cdots + n_k = N$$

Another common use of the summation operator is to indicate that all of the scores in a collection are to be squared first and then summed:

$$\sum_{i=1}^{N} X_i^2 = X_1^2 + X_2^2 + X_3^2 + \cdots + X_N^2$$

Notice that the operation indicated by the exponent is performed before the operation indicated by the summation symbol.

Still another common use of Σ is when the scores are first summed and then the sum is squared:

$$(\Sigma X_i)^2 = (X_1 + X_2 + X_3 + \cdots + X_N)^2 = T^2$$

Notice the difference between this procedure and the preceding one. Here the operation indicated by the summation symbol is performed before the operation indicated by the exponent. Of course, we can use the symbol T^2 to indicate the square of the summed scores.

Finally, consider the situation in which two measures are available on each individual, and the statistician wishes to take the *product* of each individual's scores and then sum the products for the entire collection:

$$\Sigma X_i Y_i = X_1 Y_1 + X_2 Y_2 + X_3 Y_3 + \cdots + X_N Y_N$$

If the desired operation calls for summing the scores first and then taking the product, the notation would be as follows:

$$(\Sigma X_i)(\Sigma Y_i) = (X_1 + X_2 + \cdots + X_N)(Y_1 + Y_2 + \cdots + Y_N)$$

This operation could also have been indicated by the product $T_x T_y$, where T_x represents the sum of the X-scores and T_y the sum of the Y-scores.

The student should study carefully the operations indicated in Table 6.1 so that he understands fully the manner in which these operations were carried out. The scores indicated under the columns X_i and Y_i are the raw data from which the table was constructed.

Table 6.1   The use of the summation operator

i	X_i	Y_i	X_i^2	Y_i^2	$X_i Y_i$
1	3	2	9	4	6
2	0	1	0	1	0
3	4	3	16	9	12
4	2	2	4	4	4
5	3	5	9	25	15
Sums:	$\Sigma X = 12$	$\Sigma Y = 13$	$\Sigma X^2 = 38$	$\Sigma Y^2 = 43$	$\Sigma XY = 37$

6.3   RULES FOR THE SUMMATION OPERATOR

Rule 1   The summation of an algebraic sum is the algebraic sum of the summations. This is sometimes referred to as distributing the operator. For example:

$$\Sigma (X_i + Y_i - Z_i) = \Sigma X_i + \Sigma Y_i - \Sigma Z_i$$

Rule 2   The summation of the product of a variable and a constant is the product of the constant and the summation of the variable. For example:

$$\Sigma\, CX_i = C\,\Sigma\, X_i$$

Rule 3   The summation of a constant is the constant multiplied by the number of terms in the summation. For example:

$$\Sigma\, C = NC$$

Table 6.2 illustrates arithmetically the use of the three rules for the summation operator. The data were contrived by the author for purposes of illustration.

Table 6.2   Rules for the summation operator

i	X_i	Y_i	Z_i	$X_i + Y_i - Z_i$	C	CX_i
1	3	2	0	5	2	6
2	0	1	1	0	2	0
3	4	3	5	2	2	8
4	2	2	2	2	2	4
5	3	5	2	6	2	6
Totals	12	13	10	15	10	24

With respect to Rule 1:

$$\Sigma\,(X_i + Y_i - Z_i) = 15$$
$$\Sigma\, X_i + \Sigma\, Y_i - \Sigma\, Z_i = 12 + 13 - 10 = 15$$

With respect to Rule 2:

$$\Sigma\, CX_i = 24$$
$$C\,\Sigma\, X_i = 2(12) = 24$$

With respect to Rule 3:

$$\Sigma\, C = 10$$
$$NC = 5(2) = 10$$

Let us now illustrate the use of all three rules with a single statistical expression. Given:

$$\Sigma\,(X_i - M)^2 \qquad \text{where } M \text{ is a constant}$$

First, the operation indicated by the exponent:

$$\Sigma\,(X_i^2 - 2MX_i + M^2)$$

Next, the distribution of the operator according to Rule 1:

$$\Sigma\,X_i^2 - \Sigma\,2MX_i + \Sigma\,M^2$$

Then, the application of Rule 2 to the second term of the expression:

$$\Sigma\,X_i^2 - 2M\,\Sigma\,X_i + \Sigma\,M^2$$

Finally, the application of Rule 3 to the last term:

$$\Sigma\,X_i^2 - 2M\,\Sigma\,X_i + NM^2$$

6.4   DOUBLE SUMMATION

Often the investigator will have a collection of measures that are divided into subgroups — control and experimental groups, for example. Such a collection is subject to double classification (one classification to identify the group, the other to identify the individual in the group) and will probably necessitate the use of double subscripts on some variables and the use of double summation operators. It is generally useful to think of this type of data as organized into rows and columns, with the first subscript indicating the row and the second the column in which the measure is found. Thus, X_{32} represents the score in the third row, second column. If the subgroups are organized into columns (as in Table 6.3), then X_{32} is the score of the third person in the second group. Most of the data presented in this book will be organized in this fashion, with the first subscript indicating the individual and the second the group. There is no reason, however, why the data could not be organized with the groups in rows, in which case X_{32} becomes the second person in the third group. Whichever way the data are organized the first subscript will always indicate the row number and the second the column number.

Table 6.3   Data organized in subgroups

i	Group I	Group II	Group III
1	5	0	2
2	3	1	4
3	2	3	5
4	4	2	3
Totals	14	6	14

Referring to the data in Table 6.3 we immediately see that n_1 is 4, n_2 is 4, n_3 is 4, and N is 12. Let T_1 be the sum of scores in Group I, T_2 the sum of scores in Group II, and T_3 the sum of scores in Group III. Thus T_1 is 14, T_2 is 6, and T_3 is 14. Let T be the sum of scores for *all* groups; in this case T is 34. We will use the expression T_j to represent the sum of scores in *any* group.

The sum of scores in any column is given by the expression:

$$T_j = \sum_{i=1}^{n_j} X_{ij}$$

For instance, summing the scores in Group II:

$$T_2 = \sum_{i=1}^{4} X_{i2} = 0 + 1 + 3 + 2 = 6$$

The sum of all the measures in the entire collection is given by the expression:

$$T = \sum_{j=1}^{k} T_j = \sum_{j=1}^{k} \sum_{i=1}^{n_j} X_{ij}$$

$$= (5 + 3 + 2 + 4) + (0 + 1 + 3 + 2) + (2 + 4 + 5 + 3)$$

$$= 14 + 6 + 14 = 34$$

Whenever more than one summation operator is used in this fashion, the inside operator (the one nearest the variables) will be distributed first.

EXERCISES

6.1 Given the collection of scores, 0, 1, 2, 4, 5, 6, 7, 8, 8, 9, evaluate the following expressions:

(a) ΣX_i

(b) ΣX_i^2

(c) $(\Sigma X_i)^2$

(d) $\Sigma X_i^2 - \dfrac{T^2}{N}$

6.2 Given the following collection of scores in which there are two measures on each of six individuals:

i	X_i	Y_i
1	90	1.2
2	97	2.2
3	104	2.3
4	113	3.9
5	118	3.0
6	125	3.5

Evaluate the following expressions:
(a) ΣX_i
(b) ΣY_i
(c) $\Sigma X_i Y_i$
(d) $(\Sigma X_i)(\Sigma Y_i)$
(e) $\Sigma XY - \dfrac{(\Sigma X)(\Sigma Y)}{N}$

6.3 Given the following collection of scores:

i	Group I	Group II	Group III	Group IV
1	2	1	0	2
2	5	1	3	3
3	4	3	2	0
4	4	4	0	1
5	0	5	4	6
6		2	4	0
7		3	3	5
8		1		2
9		0		
10		5		

Evaluate the following expressions:
(a) Σn_j
(b) ΣX_{i3}
(c) $\Sigma\Sigma X_{ij}$
(d) $\Sigma\Sigma X_{ij}^2$
(e) $(\Sigma\Sigma X_{ij})^2$

7

MEASURES OF
CENTRAL TENDENCY

7.1 INTRODUCTION

One of the most useful ideas in statistics is the representation of a collection of measures by a measure of *central tendency*, that is, a single value chosen in such a fashion as to be *typical* of the collection. Just what constitutes a typical (or average) measure will depend to a large extent on the level of measurement and the manner in which the measures are dispersed throughout the collection. There are three average measures or measures of central tendency in common use: the *mean*, *median*, and *mode*. The student is already familiar with these, although he may not have called them by these names.

When we say that the "average" IQ of a group of students is 105, it is usually understood that the term average refers to the arithmetic mean, which we will call simply the *mean*. The mean of a collection of scores is obtained by adding all of the scores together, then dividing by the number of scores.

When we talk about the "average" teacher's salary, it is usually more meaningful to give the median salary. The median measure is that which divides the collection of measures into two groups of equal size, with half of the measures larger than the median and the other half smaller.

Suppose, however, that a group of people were divided into three groups as follows: 20 percent of them college graduates, 55 percent high school graduates but not college graduates, and 25 percent with less than a high school education. Here the "average" person is a high school graduate, and the term average refers to the mode — the category in which more people are found than any other category.

7.2 THE ARITHMETIC MEAN

By far the most useful of the measures of central tendency is the *arithmetic mean*. As a general rule, when the behavioral scientist uses the term *average* he means the mean. The mean will be represented by either of two symbols: the uppercase letter M, or the variable symbol with a bar across the top, for example, $\bar{X}$ (called "X-bar"). The mean of a collection of measures is simply the sum of the measures divided by the number of measures, and it may be calculated by the formula:

$$M = \bar{X} = \frac{\Sigma X}{N} = \frac{T}{N}$$

If the number of measures is large, it may be desirable to arrange the scores in a frequency distribution and calculate the mean by the formula:

$$M = \bar{X} = \frac{\Sigma f_i X_i}{N} = \frac{T}{N}$$

where X_i is the score of any interval and f_i is the frequency of the same interval.

The mean may be thought of as the score that would be assigned to each individual if the total for the collection were to be evenly divided among all individuals. It is the only one of our three measures of central tendency that is algebraically defined, and this permits its use in a wide range of calculations. Of course, this also implies that the measures from which the mean is calculated are on either an interval or a ratio scale. It is also the only one of the three that is dependent upon the value of every measure in the collection. This latter characteristic is usually a valuable one, but not always, for the mean is much affected by extreme scores. Thus, the mean would not be very typical of a student's work if the teacher makes a practice of awarding zeros when the student is absent or otherwise unproductive. Similarly, it would not be very meaningful to talk about the mean annual income of five school board members if the incomes were $150,000, $6000, $5000, $4500, and $3000. The mean of this collection is $33,700, which is hardly representative of the group.

The mean has other attributes which make it very valuable in statistical inference. Whenever a sample is drawn from a parent population with the intent that a sample statistic provide an estimate of the corresponding population parameter (for example, the sample mean might be used as an estimate of the value of the population mean), it will be highly desirable that the statistic provide a good estimate of the parameter — an estimate in which the investigator may place his confidence. It is important that the student understand that the investigator will ordinarily draw just one sample, calculate the appropriate statistic, and use this statistic as an estimator of the parameter whose value is unknown.

In order to study the relationships between statistics and parameters so that scientists will know which are the best estimators and how much confidence they can place in specified statistics as estimators, statisticians study *sampling distributions*. The statistician proceeds as follows:

1. A very large number of random samples is drawn from a well-defined population, a population for which the parameter of interest is known. Often, the statistican constructs the population so that many of its characteristics are precisely known to him.
2. The statistic corresponding to the parameter of interest is calculated for each sample.
3. A frequency distribution of the statistics (not the scores) is constructed. If the statistics are means, then each sample mean is treated as though it were a single score. Such a frequency distribution is called a *sampling distribution*.

A number of important facts about the relationships between a given statistic and its corresponding parameter may be garnered from the study of such a sampling distribution:

1. If the mean of the sampling distribution is equal to the population parameter, the statistic is said to be *unbiased*. That is, there is no systematic tendency for the statistic to either underestimate or over-estimate the parameter. Some statistics are unbiased, others are biased; the usual situation is one in which the statistic systematically underestimates the parameter, especially with small samples.
2. If the statistics are closely clustered about the value of the parameter, instead of being spread all over the distribution, the statistic is said to be an *efficient* estimator. Some statistics are more efficient than others. Of course, an investigator will probably get a more accurate estimate from a more efficient statistic.
3. The sampling distribution of certain statistics tends to have the same configuration as the *normal* distribution introduced in an earlier chapter. The normal distribution has been extensively studied, and the convenience of working with normally distributed statistics will become increasingly apparent as the student enters into the study of statistical inference.

The student should be aware that the sample mean has all of the characteristics cited above.

It is general statistical practice to use Greek letters to represent parameters and Roman letters to represent statistics. When the behavioral scientist wishes to draw a distinction between the population mean and the sample mean, he will use μ (Greek lowercase mu) to represent the population mean and M or $\bar{X}$ to represent the sample mean. Generally, in statistical inference, the value of the parameter is unknown, and the Greek symbol is used in phrasing hypotheses about it. There are times, in descriptive statistics and some measurement applications, when the distinction between sample and population does not exist, and the symbols may be used interchangeably.

7.3 MATHEMATICAL OPERATIONS WITH THE MEAN

Rule 1   The general mean (M) of a collection of scores that has been divided into several subgroups is the sum of the products of each subgroup mean (M_j or $\bar{X}_j$) multiplied by the number of measures (n_j) in that subgroup and then divided by the total number of scores (N) in all subgroups:

$$M = \frac{\Sigma\, n_j M_j}{N} = \frac{\Sigma\, T_j}{N} = \frac{T}{N} \quad \text{where } n_j M_j = T_j$$

Rule 2   Given a collection of scores that has been divided into several (k) subgroups, each having the same number (n) of measures, the mean of the entire collection is the mean of the subgroup means:

$$M = \frac{n\,\Sigma\, M_j}{N} = \frac{\Sigma\, M_j}{k} \quad \text{where } N = nk$$

Rule 3   When a constant (C) is added to each of a collection of scores, the mean of the new scores ($X + C$) is equal to the mean of the original scores plus the constant ($M_x + C$):

$$M_{x+c} = M_x + C$$

Rule 4   When each of a collection of scores is multiplied by a constant (C), the mean of the new scores (CX) is equal to the mean of the original scores (M) multiplied by the constant (CM):

$$M_{cx} = CM$$

7.4   THE MEDIAN

Statistical textbooks differ as to the definition of the median. The problem derives from the fact that the median is not algebraically defined, and there is seldom true consensus with respect to nonalgebraic definitions. One of the more satisfying definitions is that the median is a point on the score scale below which one-half of the scores fall. This definition makes the median synonymous with the 2nd quartile, 5th decile, and 50th centile.

The median is often popularly defined as the middle score in a collection. If there is an odd number of scores, the value of the median is simply chosen equal to the middle score. If there is an even number of scores, the value of the median will usually be chosen as a point halfway between the two middle scores. This is not a very satisfactory approach to the problem from a mathematical viewpoint, and some problems arise when more than one person has the same score.

Although the median may be used with interval or ratio measures, it is more often used with ordinal measures. The techniques of statistical inference which call for the use of the median often have an underlying assumption that no two persons may have the same score (this is a characteristic of truly continuous data). When this assumption is met, the value of the median may be determined by the method described in the preceding paragraph.

The determination of the value of the median when many persons have equal scores requires a bit more effort. The data should first be arranged in a frequency distribution with unit intervals. If the cumulative frequency of any interval is $N/2$, the median is the upper real limit of that interval. If the cumulative frequency at the midpoint of any interval is $N/2$ (that is, if the interval score has a percentile rank of 50), the median is the same as the interval score. However, if the data do not match either of these criteria, it is necessary to determine the median interval and calculate the median by means of the formula which follows. The median interval is the one with the smallest cumulative frequency equal to or larger than $N/2$:

$$\text{median} = X_u - \frac{(cf - .5N)}{f}$$

where X_u is the upper real limit, f the frequency, and cf the cumulative frequency of the median interval. The student should realize that this formula does not constitute an algebraic definition of the median, but is only a convenient procedure for assigning a numerical value to an indeterminate measure. An example of its use is given in Section 7.6.

The median may be preferred over the mean when the data are on an ordinal scale or when extreme scores are encountered.

7.5 THE MODE

The *mode* of a collection of measures is that measure which occurs most often. Although the mode may be used with ordinal, interval, or ratio measures, it is the only one of the measures of central tendency that may be used with nominal data.

The mode may be highly descriptive of a collection of data, but it is unlikely to enter into statistical calculations, and its use is likely to be restricted to descriptive statistics.

There are many instances in which no single mode can be identified; there may be two or more modes or no identifiable mode at all.

7.6 CALCULATION OF THE MEAN, MEDIAN, AND MODE

The following calculations illustrate the determination of the mean, median, and mode from a frequency distribution. The data are from Table 7.1.

Table 7.1 Frequency distribution of 65 scores

X	f	fX	cf
9	2	18	65
8	6	48	63
7	12	84	57
6	10	60	45
5	12	60	35
4	9	36	23
3	7	21	14
2	4	8	7
1	2	2	3
0	1	0	1
	$N = 65$	$T = 337$	

$$M = \frac{T}{N} = \frac{337}{65} = 5.18$$

$$\text{median} = 5.5 - \frac{35 - 32.5}{12} = 5.29$$

$$\text{mode} = 5 \text{ and } 7 \quad \text{(the distribution is bimodal)}$$

The student should take particular note of the fact that changing the value of a score at either extreme without shifting it to the other side of the median will have no effect on the value of the median; however, it will change the value of the mean.

7.7 SELECTING A MEASURE OF CENTRAL TENDENCY

The first consideration in selecting a measure of central tendency is the level of measurement available. If the data are on a nominal scale, only the mode may be used. If the data are ordinal, the mode or median may be appropriate. With interval data, the investigator is privileged to use the mean, median, or mode. Occasionally, it will be desirable to report two or three different measures of central tendency.

The second consideration in selecting a measure of central tendency is the use to which it will be put. If it is intended to be descriptive, then the measure of central tendency most descriptive of the data should be reported. The important idea here is that the measure of central tendency should be typical of the collection that it represents.

If the investigator wishes to draw inferences from sample to population, his choice of a measure of central tendency will be largely determined by the statistical procedures that are appropriate to his data and hypotheses. The student will have opportunity to acquaint himself with these procedures as he progresses into the study of statistical inference.

The mean has a number of distinct advantages. First, because it is algebraically defined, it lends itself to a variety of mathematical operations. Most of our statistical procedures call for these mathematical operations, and the use of the mean is mandatory. The mean has other qualities of no small importance in drawing inferences from samples to populations. These qualities were discussed in some detail in Section 7.2. Thus, the mean of a sample is more likely to be a good estimate of the corresponding population parameter than the other measures of central tendency. However, the mean may be greatly influenced by the presence of extreme scores and cease to be typical of the collection. This is especially true with small samples, and in such cases the median may be preferred.

Figure 7.1 illustrates the relationship between the mean, median, and mode in symmetrical distributions. Curve A is unimodal and Curve B is bimodal. Figure 7.2 illustrates the relationship between the mean, median, and mode in asymmetrical distributions. Curve A is skewed to the right and Curve B is skewed to the left. Collectively, these curves suggest an alternate definition of the median: It is that point on the score scale such that a line perpendicular at the point will divide the area under the curve into two equal parts.

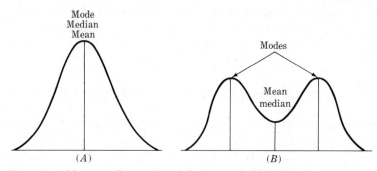

Figure 7.1 Mean, median, and mode in symmetrical distributions.

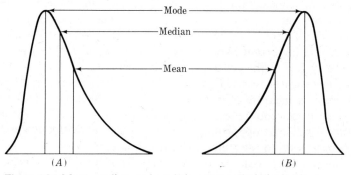

Figure 7.2 Mean, median, and mode in asymmetrical distributions.

EXERCISES

7.1 Given the following collection of scores: 9, 8, 8, 7, 6, 5, 5, 5, 4, 3,
  (a) Calculate the arithmetic mean.
  (b) Determine the value of the median.
  (c) Determine the mode of the distribution.
  (d) If each score in the distribution were multiplied by 5, what would the value of the mean of the new scores be?
  (e) If 10 were added to each score in the distribution, what would the value of the mean of the new scores be?
  (f) If 6 were subtracted from each score in the distribution, what would the value of the mean of the new scores be?

7.2 Given the following distribution of scores:

X_j	f_j
20	1
18	1
17	3
16	2
14	4
12	5
11	5
10	6
9	4
7	3

    (a) Calculate the arithmetic mean.
    (b) Determine the value of the median.
    (c) Determine the mode of the distribution.

7.3 Refer to the data of Exercise 3.3.
    (a) Calculate the arithmetic mean.
    (b) Determine the value of the median.
    (c) Determine the mode of the distribution.

7.4 Refer to the data of Exercise 6.3.
    (a) Calculate the mean for each of the subgroups.
    (b) Using Rule 1 of Section 7.3, calculate the general mean for the entire collection.

7.5 One class of 24 students has a mean of 45.7 on a certain test. A second class of 30 students has a mean of 41.6 on the same test. Calculate the general mean for the two classes.

8

MEASURES OF DISPERSION

8.1 INTRODUCTION

The essential characteristic of any variable is that it varies — that it yields different measures for different individuals. The populations that the behavioral scientist studies will always vary on one or more dimensions, and he will be interested in an index which accurately describes the manner in which they vary. It is customary to refer to this index as a *measure of dispersion*. Often, a fairly adequate summary of a collection of measures may consist simply of a measure of central tendency and a measure of dispersion. These will usually be the mean and an index which describes the dispersion of the measures about the mean.

Six measures of dispersion will be discussed. None of these is appropriate for use with nominal variables. The dispersion of nominal measures is probably adequately treated by reporting the proportion of the measures in each classification or by drawing a bar graph that pictorially expresses the dispersion of the measures. Two of the six measures of dispersion, the range and the semi-interquartile range, are suitable for use with ordinal variables. The other four assume measurement on at least an interval scale.

8.2 THE RANGE

The range is the simplest measure of dispersion, and it is defined as the difference between the largest and the smallest scores in the collection. This may be expressed in algebraic form by $R = X_h - X_l$, where X_h is

64

the highest score and X_l the lowest. Some statisticians choose to define the range in such a fashion as to include both the highest and lowest scores, in which case it is one larger than the range yielded by our definition. We will be consistent in using the first definition, but the student should be aware that the other definition may appear in the literature.

Because the range uses only two scores out of the entire collection, it may not be very descriptive of the total population. As a general rule, good statistical practice demands the use of as much data as possible. There is also a tendency for the range to vary with sample size — larger samples tending to yield more extreme values. Because the range is totally determined by the two extreme scores, which tend to vary considerably from sample to sample, it is not a very stable statistic. Because the size of the range is a function of sample size, it is said to be a *biased* statistic. These latter characteristics are of importance only when a sample is used to represent a population.

There are times, however, when the range is a highly descriptive measure of dispersion and should be reported. There are also a few nonparametric procedures that use the range and that are used to draw inferences from samples to populations.

8.3   THE SEMI-INTERQUARTILE RANGE

It has been suggested that the range may not be very descriptive of a distribution because its value is totally determined by the two extreme scores. For the same reason it is not a very stable statistic, and it tends to be a function of sample size. These shortcomings are to some extent overcome by selecting the semi-interquartile range as a measure of dispersion.

The term *interquartile range* refers to the interval $Q_3 - Q_1$, which contains half the scores in the distribution. The value of this statistic is not at all influenced by the values of the extreme scores. Dividing the interquartile range by two yields the semi-interquartile range, as indicated by the formula:

$$Q = \frac{Q_3 - Q_1}{2}$$

Consider the two distributions of scores in Table 8.1. The measures are from an attitude scale dealing with race relations. Group I consists of 20 white male union members, and Group II consists of the wives of these men. With a little effort one may quickly determine that the median is 5 and the range is 8 for both groups, which suggests that the two groups do not differ in their attitudes on race relations. With a little more effort one may determine that the semi-interquartile range is 1.5 for Group I and 0.5 for Group II, which suggests that the attitudes of the females are not as widely dispersed as those of the males. The behavioral scientist would probably interpret this finding as suggesting that the females are less inclined to choose extreme positions — that they are more homogeneous than the males on the trait being measured.

Table 8.1   Attitudes on race relations

Group I (Males)		Group I (Females)	
X_i	f_i	X_i	f_i
9	1	9	1
8	1	8	1
7	3	7	0
6	3	6	3
5	4	5	10
4	3	4	2
3	2	3	2
2	2	2	0
1	1	1	1

8.4   THE MEAN DEVIATION

Behavioral scientists make a great deal of use of *deviation scores* — scores calculated as the difference between a raw score and the mean. A deviation score is usually represented by the lowercase form of the symbol used to represent the raw score. The formula for a deviation score is:

$$x_i = X_i - M$$

Persons who score below the mean receive a negative deviation score, and persons who score above the mean receive a positive deviation score. The algebraic sum of all the deviation scores in a collection is equal to zero:

$$\Sigma \, x_i = 0$$

The *mean deviation* is a measure of dispersion which is equal to the mean of the *absolute* values of the deviation scores. An absolute value is one in which the algebraic sign of the variable is disregarded (all values are treated as positive values). The mean deviation is expressed symbolically as follows:

$$MD = \frac{\Sigma \, |X_i - M|}{N}$$

Absolute values are used, since the mean of the algebraic values is zero. The use of absolute values in statistical calculations is to be avoided, as it precludes the possibility of additional calculations. (This is a mathematical restriction upon the use of absolute values.) For this reason, the mean deviation is used primarily to introduce the measures of dispersion which follow.

8.5   THE SUM OF SQUARES

One way to avoid working with absolute values is to square the deviations before summing. This is common statistical practice, and the student can expect to perform this operation many times. The simplest measure of dispersion using this procedure is the sum of the squared deviations from the mean, which statisticians abbreviate as *sum of squares*.

There are several mathematically equivalent expressions for the sum of squares, and the student must be thoroughly familiar with all of them:

$$SS = \Sigma x_i^2 = \Sigma (X_i - M)^2$$

$$SS = \Sigma X_i^2 - \frac{(\Sigma X_i)^2}{N} = \Sigma X_i^2 - \frac{T^2}{N} = \Sigma X_i^2 - NM^2$$

Statisticians sometimes refer to the quantity ΣX_i^2 as the uncorrected sum of squares and the quantity SS or Σx_i^2 as the corrected sum of squares. The arithmetic difference between the uncorrected sum of squares and the corrected sum of squares (T^2/N or NM^2) is often referred to as the correction for the mean, or simply the correction factor.

The several formulas for the sum of squares all get the same results, and the choice of formula is usually determined by the nature of the data and the availability of calculating equipment. Consider the data in Table 8.2. Because these data are simple, the various formulas for the sum of squares are about equally convenient. When an electronic computer or desk calculator is available, it will usually be most convenient to calculate the sum of squares in this fashion:

$$SS = \Sigma X_i^2 - \frac{T^2}{N} = 340 - \frac{2500}{10} = 90$$

Precisely the same result is achieved when we calculate the sum of squares by the formula:

$$SS = \Sigma X_i^2 - NM^2 = 340 - 10(25) = 90$$

Table 8.2   Calculation of the sum of squares

i	X_i	X_i^2	x_i	x_i^2
1	0	0	−5	25
2	1	1	−4	16
3	2	4	−3	9
4	4	16	−1	1
5	5	25	0	0
6	6	36	1	1
7	7	49	2	4
8	8	64	3	9
9	8	64	3	9
10	9	81	4	16
$N = 10$	$T = 50$	$\Sigma X^2 = 340$	$\Sigma x = 0$	$\Sigma x^2 = 90$

Notice in Table 8.2 that it was convenient to calculate Σx_i^2 directly, but only because the scores and the mean were integers. While it is not at all uncommon for the scores to be integers, it is much more likely that the mean will not be an integer, and this greatly complicates the calculation of the squared deviation scores. However, there is a computational scheme which can be used to simplify the process. For example, consider the data in Table 8.3, which have a mean of 3.2. Let C be an integer constant approximately equal to the mean. In the example, we have let C equal 3. Then, let w_i be a deviation score, the deviation of X_i from C. Thus, $w_i = X_i - 3$. The sum of squares may now be calculated by the formula:

$$SS = \Sigma w_i^2 - \frac{(\Sigma w_i)^2}{N} = 7 - \frac{1}{5} = 6.8$$

Table 8.3   Calculation of the sum of squares

i	X_i	X_i^2	w_i	w_i^2
1	2	4	−1	1
2	2	4	−1	1
3	3	9	0	0
4	4	16	1	1
5	5	25	2	4
$N = 5$	$T = 16$	$\Sigma X^2 = 58$	$\Sigma w = 1$	$\Sigma w^2 = 7$

That this is in fact the sum of squares may be demonstrated by calculating the sum of squares with one of the other formulas:

$$SS = \Sigma X_i^2 - \frac{T^2}{N} = 58 - \frac{256}{5} = 6.8$$

The computational scheme just described will probably be preferred when the sum of squares must be calculated without the use of computational aids such as a desk calculator or electronic computer.

Although the sum of squares enters into many statistical calculations, including the measures of dispersion which follow, it is not usually reported as a descriptive index. The variance and standard deviation which follow are the most important measures of dispersion in both descriptive and inferential statistics.

8.6   THE VARIANCE

The variance is defined as the mean of the squared deviations from the mean. The population variance is represented by σ^2 (σ is the lowercase Greek letter sigma), and it is obtained by dividing the sum of squares by the number of subjects:

$$\sigma^2 = \frac{SS}{N} \qquad \text{where } SS = \Sigma (X_i - \mu)^2$$

This definition of the variance holds only for the situation in which the size of the population (the value of N) is finite. There are important theoretical sampling distributions (such as the normal and other continuous distributions tabled in the Appendix) in which a population of infinite size is assumed. These require a more general definition of the variance which can only be appreciated by persons possessing a knowledge of the calculus and of moment-generating functions. No such background is assumed for the student using this text; however, he should be aware that mathematical statisticians often use a more general definition of the variance. This limitation will not prove to be a handicap for the student interested in behavioral research.

In statistical inference where the sample variance is used to estimate the population variance, it is not sufficient simply to substitute the sample mean for the population mean in the formula above. Such a procedure would result in a biased statistic, one which is systematically influenced by sample size, tending to underestimate the population variance with small sample sizes. In order for the sample variance to be used as an estimate of the population variance, the systematic bias introduced by small sample size must be corrected for by adjusting the formula, that is, substituting $N - 1$ for N. It can be demonstrated both mathematically and empirically that this correction provides an unbiased estimate of the population variance. We will use the symbol S^2 to represent a sample variance which has been calculated in this fashion; thus the formula becomes:

$$S^2 = \frac{SS}{N-1} \qquad \text{where } SS = \Sigma\,(X_i - M)^2$$

The sample variance formula with the adjustment to provide an unbiased estimate of the population variance will always be used in statistical inference. However, there is some disagreement as to which formula should be used in descriptive statistics and those measurement applications where the collection of measures is not necessarily a sample intended to represent some larger population. The author is partial to the use of the population formula because it is consistent with mathematical procedures to be considered in subsequent chapters. He also recommends the calculation of the sum of squares as a first step to be completed before the calculation of the variance, regardless of which formula is used.

8.7   THE STANDARD DEVIATION

The standard deviation is simply the square root of the variance. It will be represented by the symbols σ and S.

The standard deviation has an advantage over the variance as a descriptive measure of dispersion in that it is expressed in the same units as the original measures. For example, modern intelligence tests are commonly scaled with mean of 100 and standard deviation of 15. A person with a score of 115 is one standard deviation above the mean. The variance in this case is 225, which is not a particularly descriptive piece of information. However, both the variance and the standard deviation will be used a great deal in statistical inference.

The student is advised to become so familiar with the various formulas for the sum of squares, variance, and standard deviation that he recognizes them instantly when they enter into a series of calculations and thus be able to reproduce them at will.

8.8   MATHEMATICAL OPERATIONS WITH THE VARIANCE
AND STANDARD DEVIATION

Rule 1   When a constant (C) is added to each of a collection of scores, the sum of squares, variance, and standard deviation of the new scores ($X + C$) remains the same as the sum of squares, variance, and standard deviation of the original scores:

$$\sigma_{x+c} = \sigma_x$$

Rule 2   When each score in a collection is multiplied by a constant (C), the standard deviation of the new scores (CX) is equal to the constant multiplied by the standard deviation of the original scores:

$$\sigma_{cx} = C\sigma_x$$

Rule 3   When each score in a collection is multiplied by a constant (C), the variance of the new scores (CX) is equal to the constant squared (C^2) multiplied by the variance of the original scores:

$$\sigma^2_{cx} = C^2\sigma^2_x$$

The student is reminded that subtraction is a special case of addition (the addition of a negative number) and that division is a special case of multiplication (multiplication by a reciprocal). The rules above apply to those situations where a constant is subtracted from each score or where each score is divided by a constant.

8.9   HOW TO EXTRACT THE SQUARE ROOT OF ANY POSITIVE NUMBER

The square roots of the whole numbers from 1 to 1000 are given in Table 2 of the Appendix. The use of the table may be extended to other numbers as follows:

1. If we shift the decimal place in the square two places (so that a number within the range of the table is produced), the decimal place in the root is shifted one place in the same direction.
2. For example, from the table we ascertain that the square root of 83 is 9.11 and that the square root of 830 is 28.81.
3. Using the rule in (1) and the information in (2), we determine that:

The square root of 0.0830 is 0.02881.
The square root of 0.83 is 0.911.
The square root of 8.30 is 2.881.
The square root of 83 is 9.11.
The square root of 830 is 28.81.
The square root of 8300 is 91.1.
The square root of 83000 is 288.1.

This strategy will often require that the user round the number whose square root is desired to two or three significant places, but this will be adequate for many situations in which the standard deviation is to be calculated.

In the event that a procedure for calculating the square root of any number to any desired accuracy is called for, the following strategy is recommended:

1. Given a number whose square root is desired, make an educated guess as to the square root. For example, suppose we wish to take the square root of 6.33. We know that the square root of 4 is 2 and that the square root of 9 is 3. Obviously, the desired root is somewhere between 2 and 3, so we will guess 2.40.
2. Divide the number whose square root is desired by the guess. In our example, 6.33 divided by 2.40 is 2.64.
3. Take the average of the guess and the quotient from step (2). Thus the average of 2.40 and 2.64 is 2.52.
4. This average is the first approximation of the square root. It may be checked for accuracy by squaring it and comparing the square to the original number whose square root was desired. In the example, the square of 2.52 is 6.35. This is very close to the original 6.33.
5. If greater accuracy is desired, repeat the whole procedure, substituting the average (the first approximation) for the guess in step (1). This procedure may be repeated as many times as necessary.

The whole procedure may be summarized in a single expression:

$$\text{root} \approx \frac{\dfrac{\text{square}}{\text{guess}} + \text{guess}}{2}$$

A process which may be repeated in this fashion to achieve a desired degree of accuracy is called an *iterative* process. An iterative process is sometimes more efficient than a direct assault upon a problem, and the technique is often used with modern electronic computers.

EXERCISES

8.1 Given the following collection of scores: 2, 3, 5, 6, 6, 8,
   (a) Calculate the range of the scores.
   (b) Calculate the sum of squares.
   (c) Calculate the variance.
   (d) Calculate the standard deviation.
   (e) Suppose that 3 points were added to each of the scores in the collection. What would be the value of the standard deviation for the new collection of scores?

(f) Suppose that each score in the original collection were multipled by 10. What would be the value of the standard deviation of the new collection of scores?

(g) Suppose that each score in the original collection were multiplied by 2. What would be the value of the variance of the new collection of scores?

8.2 Given the following collection of scores: 0, 1, 2, 4, 5, 6, 7, 8, 8, 9,

(a) Calculate the range.

(b) Calculate the semi-interquartile range.

(c) Calculate the mean deviation.

(d) Calculate the sum of squares.

(e) Calculate the variance.

(f) Calculate the standard deviation.

8.3 Given the following collection of scores: 9, 11, 12, 12, 14, 15, 16, 16, 17, 17, 17, 19,

(a) Calculate the sum of squares.

(b) Calculate the variance.

(c) Calculate the standard deviation.

8.4 Given the following collection of scores: 2, 3, 3, 4, 5, 5, 5, 7, 8, 8

(a) Calculate the sum of squares.

(b) Calculate the variance.

(c) Calculate the standard deviation.

(d) Assume that the collection of scores is a random sample from some population. Calculate an unbiased estimate of the population variance.

8.5 Refer to the data of Table 3.7.

(a) Calculate the sum of squares.

(b) Calculate the variance.

(c) Calculate the standard deviation.

SELECTED REFERENCES

GAMES, PAUL A., "Further comments on N vs. $N - 1$," *American Educational Research Journal*, **8**, 582–584 (May 1971).

HUBERT, LAWRENCE, "A further comment on N vs. $N - 1$," *American Educational Research Journal*, **9**, 323–325 (Spring 1972).

KNAPP, THOMAS R., "Notes and comments on N vs. $N - 1$," *American Educational Research Journal*, **7**, 625–626 (November 1970).

LANDRUM, WM. L., "A second comment on N vs. $N - 1$," *American Educational Research Journal*, **8**, 581 (May 1971).

9

STANDARD SCORES

9.1 THE IDEA OF A STANDARD SCORE

The numerical value of the score a student achieves on an examination is usually highly arbitrary, having meaning only as it is related to the other scores achieved on the test. One way of relating meaning to this numerical score was discussed in a previous chapter: the use of percentile ranks to indicate the placement of the score in the distribution by stating the percentage of scores in the distribution which are smaller. Another approach commonly used calls for indicating the placement of the score with respect to the mean of the distribution. In the latter procedure the usual alternatives are to state the score in terms of standard deviations above or below the mean, or to change the mean and standard deviation of collection to standard values and report the scores as values which conveniently communicate the idea of deviation from the mean.

Score distributions that have been altered to have means and standard deviations of standard value are called *standard scores*. The mathematical operation by which a distribution of scores receives a new mean and standard deviation is called a *score transformation*. The score transformations considered in this chapter are *linear* transformations, which means that the shape of the distribution of the standard scores will be precisely the same as that of the original or raw scores. The student should take particular note of this, for there is a common misconception that standard scores are normally distributed. In order for standard scores to be normally distributed, it is necessary that the original score distribution be normal, or that a nonlinear transformation be used to produce a new distribution that is normal. Normally distributed standard scores play a special role in statistical inference.

In constrast to the percentile ranks considered earlier, standard scores are algebraically defined. Percentile ranks are ordinally scaled and may be derived from raw scores which are on ordinal, interval, or ratio scales. Standard scores produced by linear transformations are intervally scaled and may be derived from raw scores which are on interval or ratio scales. Although a score of zero may play a special role in standard scores, it should not be interpreted as an absolute zero of the type encountered with ratio scales.

9.2   STANDARD SCORES WITH ZERO MEAN AND UNIT STANDARD DEVIATION

One of the most useful and certainly the most common linear score transformations is that which produces a score distribution with mean of zero and standard deviation of one. A score reported on such a scale is essentially reported as so many standard deviations above or below the mean.

Consider the distribution of the deviation scores ($x_i = X_i - M$) from a specified collection of raw scores (X_i). From our studies of mathematical operations with the mean, we know that when a constant (in this case, $-M$) is added to each score in a collection, the mean of the new score distribution is equal to the mean of the original distribution plus the constant (in this case, $M - M$), so that the distribution of the deviation scores has a mean of zero. The standard deviation of this distribution is the same as that of the raw scores, since the standard deviation of a distribution is unaffected by adding or subtracting a constant from each score.

If all of the scores in a distribution are multiplied by a constant, the standard deviation of the new distribution will be equal to the original standard deviation multiplied by the constant. Thus the standard deviation of any distribution may be changed to unity by simply multiplying each score by the reciprocal of the standard deviation ($1/\sigma$). In the case under discussion, the mean of the distribution is unaffected, since a constant multiplied by zero is still zero.

This discussion leads to the mathematical definition of a standard score (z) having mean of zero and standard deviation of one:

$$z_i = \frac{X_i - M}{\sigma} = \frac{x_i}{\sigma}$$

For example, given a score distribution with mean of 100 and standard deviation of 15, the z-score of an individual with a raw score of 120 may be calculated as follows:

$$z = \frac{120 - 100}{15} = \frac{20}{15} = 1.33$$

This is equivalent to saying that the individual scored 1.33 standard deviations above the mean. Consider another example from the same collection, where the individual scored 90 on the original score scale:

$$z = \frac{90 - 100}{15} = \frac{-10}{15} = -0.67$$

This is equivalent to saying that this individual scored 0.67 standard deviations below the mean. Of course, an individual from this group with a raw score of 100 would have a standard score of zero.

9.3 PROPERTIES OF z-SCORE DISTRIBUTIONS

When the mean and standard deviation are calculated for a given collection of raw scores, and all of the raw scores are transformed into z-scores, the resultant z-score distribution has the following properties:

1. The mean of the z-score distribution is zero. This is true not only for z-scores but also for deviation scores (where $x_i = X_i - M$).
2. Raw scores below the mean become negative z-scores, and raw scores above the mean become positive z-scores. This property also applies to deviation scores.
3. Both the variance and the standard deviation of a z-score distribution are one.
4. The unit for z-scores is the standard deviation. For example, when z equals two, the score is two standard deviations above the mean.
5. When z-scores are calculated for random samples, the range of the z-scores is a function of sample size. Typically, large samples tend to range from -3.00 to $+3.00$, but smaller samples tend to have a smaller range.
6. The sum of the z-scores is zero. This property also applies to deviation scores.
7. The sum of the squares of the z-scores is equal to the number of scores in the collection ($\Sigma z_i^2 = N$). This is true only if the standard deviation is calculated using N in the denominator (see Sections 8.6 and 8.7).
8. The shape of the z-score distribution is precisely the same as that of the original score distribution.

9.4   OTHER LINEAR SCORE TRANSFORMATIONS

The investigator will often regard the use of a score distribution with negative score values as unnecessarily awkward. It is a simple matter to convert z-scores to a new distribution having any convenient mean and standard deviation, eliminating the negative scores in the process. It is customary to refer to such scores by the capital letter Z, as in the following formula:

$$Z_i = M' + z_i S'$$

M' is the mean and S' the standard deviation desired in the new distribution.

Suppose the investigator desires a score distribution with mean of 50 and standard deviation of 10, a common choice that yields a score distribution with all positive values, ranging from approximately 20 to 80, while providing for extreme scores ranging from 0 to 100. For example, the individual with a z-score of -0.67 has a Z-score as follows:

$$Z = 50 + 10z = 50 - 6.7 = 43.3$$

Another commonly used standard score distribution has a mean of 500 and standard deviation of 100. For the example above:

$$Z = 500 + 100z = 500 - 67 = 433$$

Modern intelligence tests are commonly scaled in this fashion with a mean of 100 and standard deviation of 15, which is a vast improvement over the old intelligence quotient (MA/CA) approach to IQ.

9.5   COMBINING STANDARD SCORES

The scores earned by students on examinations are typically recorded as the number of questions answered correctly, the number of words spelled properly, or the number of problems solved. In addition to having different score units, one might expect the difficulty of the test items to vary from examination to examination. For these and other reasons the raw scores received on social studies, spelling, and arithmetic tests are not directly comparable, and a composite score for a given student covering several such examinations is highly questionable. In fact, if the tests have markedly different standard deviations, a composite of several arithmetic tests might be questioned.

Standard scores of the type under discussion, however, are pure or abstract numbers having no unit. Furthermore, when calculated for the same group of students, standard scores on different examinations have directly comparable means and standard deviations. For this reason there is usually no objection to the calculation of a composite score using several standard scores.

Often, the teacher or psychometrist will prefer to assign different weights to different scores when making up a composite test score. Suppose that an instructor gave three regular tests and one final examination during a course of study, and that he wished one of the regular tests to contribute only one-fourth as much as the other two to the course grade, and the final examination to count one and one-half times as much. The composite score might be computed as follows:

$$\text{total score} = 0.25z_1 + z_2 + z_3 + 1.5z_4$$

Of course, the instructor might seek an average rather than a total score: in the example above the total score would be divided by 3.75. This would not produce an average z-score, however, since means vary less than the scores from which they are calculated and the standard deviation of these average scores would be less than one.

EXERCISES

9.1 Given the following collection of scores: 1, 4, 4, 5, 5, 6, 7, 8,
    (a) Calculate the mean and standard deviation.
    (b) Calculate a z-score for each measure in the collection.
    (c) Transform all of the scores in the collection to produce a new distribution with mean of 50 and standard deviation of 10.

9.2 Given the following collection of scores: 7, 7, 8, 9, 9, 10, 10, 11, 11, 12, 13, 13,
    (a) Calculate the mean and standard deviation.
    (b) Calculate a z-score for each measure in the collection.
    (c) Transform all of the scores in the collection to produce a new distribution with mean of 100 and standard deviation of 15.

9.3 Refer to the data of Table 4.1.
    (a) Calculate the mean and standard deviation.
    (b) Calculate a z-score for each measure in the collection.

10

NORMAL DISTRIBUTIONS

10.1 THE NORMAL CURVE AS A MODEL

Many years ago, learned men developed the *law of errors* from their observations of events such as archery contests. At first the law of errors was simply a group of generalizations that were found to hold true for many different kinds of phenomena. These generalizations are: (1) small errors occur more often than large errors; (2) very large errors very seldom occur; (3) errors become progressively less numerous as they become larger; (4) negative errors (errors to the left) are about as numerous as positive errors (errors to the right). Over 200 years ago Abraham DeMoivre defined the law of errors mathematically when he derived the equation of the normal curve.

Consider the situation in which a true-false examination with six questions is administered to a group of students who have absolutely no knowledge of the materials on which they are being tested. Suppose also that the examination is very well written and that no clues are given to assist the individual in arriving at a correct answer without knowledge of the material. In other words, the student can only guess at the answers, and the chance of getting a right answer to any question is one in two. If we can assume that the responses are completely random, we can arrive at certain conclusions with respect to the distribution of the scores from this test. First, the average score for the group of students would be three answers right and three answers wrong, and this score would also be the mode. Statisticians would call this the *expected value*. However, not all students would get this score; some would score higher and others lower. Quite a few students would be expected to score four right, and about the same number would get only two right. Somewhat fewer students would be expected to get either five right and one wrong *or* five wrong and one right. And very few students would be expected to get all of the questions right or all wrong. This is a practical application of the law of errors.

We would expect the frequency distribution of these scores to be unimodal and symmetrical, with the concentration of the scores in the vicinity of the mean (the expected value) of the distribution. Actually, the chances of a student getting any particular score on such an examination are readily calculable from basic probability theory, which will be considered in a later chapter. Figure 10.1 shows both the histogram and frequency polygon for the expected distribution of these scores.

If the number of questions in the examination were increased to ten, the expected distribution would be as shown in the histogram and frequency polygon of Figure 10.2. The shape of the frequency polygon suggests that the distribution of the trait being measured is continuous, and that if the polygon were smoothed it would closely approximate a normal curve. As a matter of fact, it can be demonstrated mathematically and empirically that the distribution of these scores would approach a normal distribution if the number of questions and the number of students were both very large.

The normal curve is essentially a mathematical model for chance errors. The frequency distributions of many events found in nature closely approximate the normal distribution, and it has proved to be an extremely useful model for the study of these events. The distribution of scores yielded by psychological tests often closely approximates the normal distribution, and the available evidence suggests that the traits underlying these measures are in fact normally distributed. For this reason and because the normal distribution is a useful model for statistical analysis,

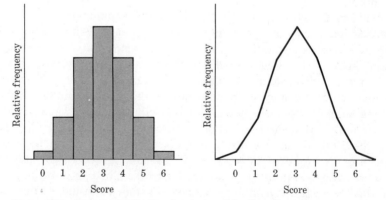

Figure 10.1 Chance distribution for six true-false items.

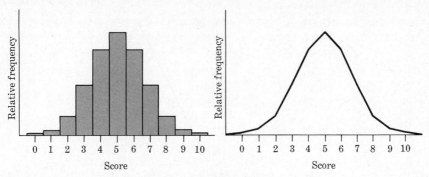

Figure 10.2 Chance distribution for ten true-false items.

there is often a conscious effort on the part of the test constructor to produce an instrument that yields normally distributed scores. This may be accomplished by adjusting the difficulty of the items or by using a nonlinear score distribution which produces normally distributed standard scores. The task is made somewhat easier by the fact that the tests tend to yield scores with rather large chance error components.

10.2 THE EQUATION OF THE NORMAL CURVE

The study of relationships between variables is central to any scientific endeavor, and every scientific equation is an expression of such a relationship. Whenever two variables are related in such a fashion that the value of one variable is determined by the value of the other, the two are said to be *functions* of each other; an equation can be written to express this relationship. Such a relationship is indicated by the general equation $Y = f(X)$, which says that Y is a function of X. Whenever the values of two variables are paired in this fashion, a graph of the relationship may be drawn. The X-variable will be plotted along the horizontal axis, and the Y-variable along the vertical axis, with a single point on the graph representing each pair of scores. When all the points are connected, a curve is produced that is a graphic representation of the equation. Mathematicians call this curve the *locus* of the equation. Figure 10.3 is a graph of the equation $Y = 2X - 3$. Notice that negative values are as readily plotted on the graph as positive ones. The equation in this case is a simple linear equation, and its locus is a straight line. This is characteristic of equations with two variables, neither of which has an exponent.

The normal curve is a graphic plot of a rather complicated mathematical function. This equation defines a family of curves (infinite in number), any

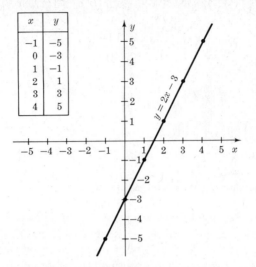

Figure 10.3   Graph of a linear equation.

one of which may be completely determined by specifying the values of the mean and standard deviation:

$$Y = \frac{1}{\sigma\sqrt{2\pi}} \, e^{-(X-\mu)^2/2\sigma^2}$$

where Y = height of the curve corresponding to a given score
$\quad\;\; X$ = a score value corresponding to a given height
$\quad\;\; \mu$ = the mean of the X-variable
$\quad\;\; \sigma$ = the standard deviation of the X-variable
$\quad\;\; \pi$ = a constant approximately equal to 3.1416
$\quad\;\; e$ = a constant approximately equal to 2.7183

The role of the mean and standard deviation in determining the particular normal curve is illustrated in Figure 10.4. Curves A and B have the same mean but different standard deviations. Curves B and C have the same standard deviation but different means.

The equation of the normal curve may be somewhat simplified by specifying that the mean be zero and the standard deviation unity:

$$Y = \frac{1}{\sqrt{2\pi}} \, e^{-z^2/2}$$

This equation in standard score form is sometimes referred to as the *standard normal curve*. Figure 10.5 is a graph of the standard normal curve.

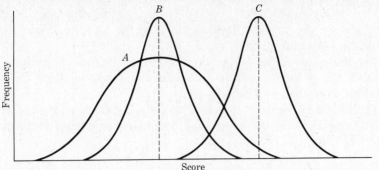

Figure 10.4 Mean and standard deviation of normal curves.

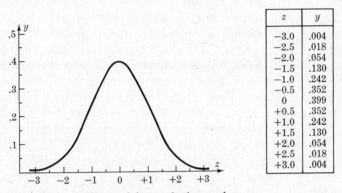

z	y
−3.0	.004
−2.5	.018
−2.0	.054
−1.5	.130
−1.0	.242
−0.5	.352
0	.399
+0.5	.352
+1.0	.242
+1.5	.130
+2.0	.054
+2.5	.018
+3.0	.004

Figure 10.5 Ordinates of the standard normal curve.

10.3 SUMMARY OF THE PROPERTIES OF NORMAL CURVES

1. A normal curve is a graph of a particular mathematical function — a model for events whose outcome is left to chance.
2. There is an infinite number of different normal curves, each one determined by the values of the mean and standard deviation. The standard normal curve has a mean of zero and a standard deviation of unity.
3. A normal curve is symmetrical and bell-shaped, with its maximum height at the mean.
4. A normal curve is continuous; there is a value of Y for every value of X, including fractional values. Its use as a model for frequency distributions assumes the X-variable is continuously distributed.
5. The value of Y is positive for all values of X. While scores may be positive or negative, there is no such thing as a negative frequency.

6. The curve approaches but never touches the X-axis (mathematicians say it is *asymptotic* to the X-axis). The score scale ranges from minus infinity to plus infinity; however, the frequency of scores beyond ± 3 standard deviations is near zero.
7. The inflection points of the curve occur one standard deviation on either side of the mean. Inflection points are points at which the bend in the curve changes direction.
8. The total area between the curve and score scale is unity, and areas under portions of the curve may be treated as relative frequencies.

10.4   AREAS UNDER THE NORMAL CURVE

In Chapter 5, "Graphs of Frequency Distributions," it was established that the frequencies of the various scores are represented by the areas of the vertical bars in a histogram. This also holds true for graphs of continuous distributions if vertical lines are drawn from the real limits on the score scale to the curve. Of course, the determination of the area of a bar having a curved top is not quite as simple as the determination of the area of a simple rectangle, as encountered in the histogram. Ordinarily, the calculation of the area underneath a portion of a curve requires the use of integral calculus. In the case of the normal curve this is complicated by the fact that the curve has no simple integral, and the determination of the frequency of a given score or range of scores would be a laborious process even for a mathematician. Fortunately, we are spared the task of these involved calculations, since the areas under the normal curve have been extensively tabled.

Figure 10.6 illustrates the relationship between scores (expressed as standard deviations from the mean) and relative frequencies (expressed as proportions of the area under the curve). Notice that the proportion of the area between the mean and any point on the z-score scale is the same for all normal distributions. For example, 34.13 percent of the scores in a normal distribution fall between the mean and minus one standard deviation. With a little effort one may readily ascertain that 15.87 percent of the area under the curve is to the *left* of a score of minus one. This means that approximately 16 percent of the normally distributed scores are below a score of minus one standard deviation, and such a score corresponds roughly to the 16th centile. Similarly, one may determine that a score of plus two standard deviations roughly corresponds to the 98th centile.

Table 3 in the Appendix gives the proportion of the total area under a normal curve between the mean and any score expressed in standard deviations from the mean. Of course, any score expressed in standard deviations from the mean is a z-score; in this case we are concerned only with the z-scores that are known to be normally distributed. Because the

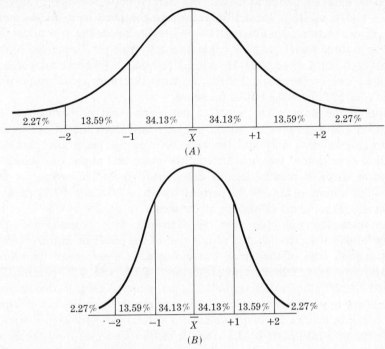

Figure 10.6 Areas under the normal curve.

normal curve is symmetrical, the areas from the mean are the same for both positive and negative z-scores, and only positive scores are recorded in the table. To illustrate the use of Table 3, the student should verify that 44.06 percent of the area under the curve is between the mean and $z = 1.56$ and that 49.87 percent of the area is between the mean and $z = -3.02$. The areas are recorded as proportions in the table and must be multiplied by 100 to be reported as percentages.

Now, let us illustrate the use of the table for translating any normally distributed z-score into a percentile rank. If the score is positive, one need only add .5000 (the area to the left of the mean) to the tabled value and multiply by 100 to get the percentile rank. For example, a z-score of .05 corresponds to a percentile rank of 52, a z-score of 1.13 to a percentile rank of 87, and a z-score of 2.06 to a percentile rank of 98. The author has taken the liberty of rounding the percentile ranks to whole numbers. If the score is negative, one need only subtract the tabled value from .5000 and multiply by 100 to get the percentile rank. For example, a z-score of $-.88$ corresponds to a percentile rank of 19 in any normally distributed collection of scores. The important idea is that percentile rank is represented by the area to the *left* of a score. The student may find it easier to understand this idea if he makes a freehand sketch of a normal curve for each of the examples above, carefully locates the mean and the given score on the score scale, draws vertical lines at these two points that divide the area under the curve into three parts, and uses Table 3 to determine the proportion of the area in each of the three parts. He is reminded that .5000 of the total area is on either side of the mean. Finally, he must determine what proportion of the total area is to the left of the score.

Suppose the investigator wants to determine what proportion of the scores falls between plus or minus one standard deviation of the mean. To do this he need only add the area between the mean and plus one standard deviation to the area between the mean and minus one standard deviation. Thus, it may be readily determined that 68.26 percent of the scores fall within ±1.00, 95.44 percent within ±2.00, and 99.74 percent within ±3.00 standard deviations of the mean.

The researcher will often want to determine what proportion of the scores falls in the extreme area at one end of the curve or in the extreme areas at both ends of the curve. For example, 2.28 percent of the scores are above +2.00, 2.28 percent are below −2.00, and 4.56 percent are beyond ±2.00 standard deviations of the mean. Again, if the student experiences any difficulty in determining where these values come from, he is urged to make a freehand sketch of a normal curve for each example, divide the area into parts by drawing vertical lines upward from the mean and from the given score or scores, and determine the area in each part. Many investigators find this is a useful technique when working with areas under the normal curve.

EXERCISES

10.1 Given a score point located 1.58 standard deviations above the mean in a normal distribution, determine the following:
(a) The area between the mean and the score.
(b) The total area to the left of the score.
(c) The total area to the right of the score.
(d) The percentile rank of the score.

10.2 Given a score point located 1.58 standard deviations below the mean in a normal distribution, determine the following:
(a) The area between the mean and the score.
(b) The total area to the left of the score.
(c) The total area to the right of the score.
(d) The percentile rank of the score.

10.3 Given a score point located 2.81 standard deviations above the mean in a normal distribution, determine the following:
   (a) The area between the mean and the score.
   (b) The total area to the left of the score.
   (c) The total area to the right of the score.
   (d) The percentile rank of the score.

10.4 Given a score point located 2.81 standard deviations below the mean in a normal distribution, determine the following:
   (a) The area between the mean and the score.
   (b) The total area to the left of the score.
   (c) The total area to the right of the score.
   (d) The percentile rank of the score.

10.5 Given a percentile rank of 22 in a normal distribution, determine the corresponding z-score.

10.6 Given a percentile rank of 91 in a normal distribution, determine the corresponding z-score.

10.7 Given a percentile rank of 50 in a normal distribution, determine the corresponding z-score

10.8 What proportion of the scores in a normal distribution is higher than $+3.00$? What proportion is lower?

10.9 Suppose an investigator had a z-score on each person in a group known to be normally distributed and that he wished to identify any person in the top 5 percent of the score distribution. What z-score would be needed to place a person in the top 5 percent?

10.10 Suppose an investigator had a z-score on each person in a group known to be normally distributed and that he wished to identify any person in the middle 50 percent of the group. What z-score range would include the middle 50 percent of the distribution?

11

NORMALIZED STANDARD SCORE DISTRIBUTIONS

11.1 INTRODUCTION

Two kinds of score transformations have been discussed in previous portions (Chapters 4 and 9) of this text. The first, transformation of raw scores to percentile ranks, was a nonlinear score transformation that altered the distribution of the scores. No matter what form the distribution of the raw scores might have, the distribution of the percentile ranks was always rectangular. The second, transformation of raw scores to z-scores or Z-scores, was a linear transformation that did not alter the form of the distribution of the scores. No matter what form the distribution of the raw scores might have, the distribution of the standard scores was the same. In this chapter, nonlinear score transformations which produce normally distributed scores will be considered.

There are several reasons why a behavioral scientist might desire to transform a collection of raw scores from whatever distribution form they might have to a normal distribution. He might be convinced that the underlying trait which is being assessed indirectly by a psychological instrument is in reality normally distributed and more accurately represented by normally distributed scores. Or, he might desire to make comparisons of measures on several different dimensions of the same population, and this is more readily justified if the measures are all distributed in the same fashion. Finally, most of the techniques of statistical inference are specifically intended for use with normally distributed measures, and the use of these techniques is more readily defended if the score distribution is normal.

There are two commonly used procedures for achieving normally distributed measures. The test constructor may seek to build an instrument that yields normally distributed raw scores, and this is accomplished by manipulating the difficulty of the items. If the underlying trait is indeed normal, this should not be too difficult to do; however, it may require considerable research and revision. As an alternative, the investigator may choose to use a nonlinear transformation that yields normally distributed standard scores — a procedure that usually entails the determination of percentile ranks as an intermediate step.

Several different normal standard score systems are used by behavioral scientists, and the relationships between the most popular ones are illustrated in Figure 11.1.

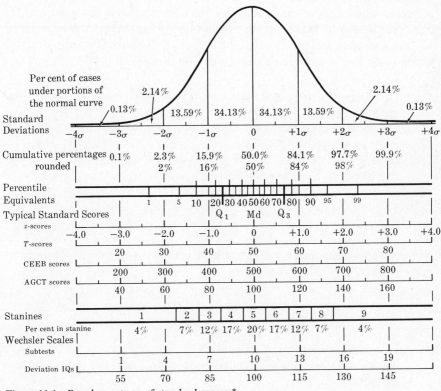

Figure 11.1 Popular systems of standard scores.*

*Figure 11.1 is reprinted from *Test Service Bulletin*, **48** (1955), The Psychological Corporation, New York.

11.2   NORMALIZED z-SCORES

To convert raw scores of any form to normalized z-scores, it is first necessary to transform the raw scores to percentile ranks. Once this is accomplished, the investigator may refer to a table of areas under the normal curve and translate the percentile ranks into normal z-scores.

For example, suppose a percentile rank of 20 is to be transformed into the equivalent normal z-score. A percentile rank of 20 has 20 percent of the distribution below or to the left of it and 30 percent of the distribution between the score and the mean. Turning to Table 3 in the Appendix, we find that a z-score of $\pm.84$ has .2995 of the total area under the curve between the score and the mean. The normal z-score equivalent to a percentile rank of 20 must be $-.84$, since percentile ranks less than 50 correspond to negative z-scores. For a second example, let us transform a percentile rank of 60 to the equivalent normal z-score. This will be a positive z-score with .1000 of the area under the curve between the score and the mean. Turning to Table 3 again, we find that .25 is the appropriate z-score.

A convenient strategy for converting percentile ranks from any distribution to normally distributed z-scores is outlined as follows:

1. Let A_i be the area in Table 3 of the Appendix corresponding to a specified z_i.
2. To convert percentile ranks from any data to normally distributed z-scores, proceed as follows:
    (a) If $PR_i = 50$, $z_i = 0$.
    (b) If $PR_i > 50$, z_i is positive.

$$A_i = \frac{PR_i - 50}{100}$$

   z_i corresponds to A_i in Table 3.
    (c) If $PR_i < 50$, z_i is negative.

$$A_i = \frac{50 - PR_i}{100}$$

   z_i corresponds to A_i in Table 3.
3. To convert normally distributed z-scores to percentile ranks, proceed as follows:
    (a) If $z_i = 0$, $PR_i = 50$.
    (b) If z_i is positive, $PR_i = 50 + 100A_i$, where A_i corresponds to z_i in Table 3.

(c) If z_i is negative, $PR_i = 50 - 100A_i$, where A_i corresponds to z_i in Table 3.

Suppose, for example, that a percentile rank of 20 is to be transformed into the equivalent normal z-score:

$$A_i = \frac{50 - 20}{100} = 0.30$$

Referring this value of A_i to Table 3, we find it corresponds to $z_i = \pm 0.84$; because PR_i is less than 50, we conclude $z_i = -0.84$.

For a second example, suppose that a percentile rank of 60 is to be transformed to the equivalent normal z-score:

$$A_i = \frac{60 - 50}{100} = 0.10$$

Referring this value of A_i to Table 3, we find it corresponds to $z_i = \pm 0.25$; because PR_i is greater than 50, we conclude that $z_i = +0.25$.

Of course, z-scores arrived at using this procedure are not only normally distributed but also have mean of zero and standard deviation of one. A linear transformation (as in Section 9.4) may be used to produce any desired mean and standard deviation. For example:

$$IQ = 15z_i + 100$$

11.3   *T*-SCORES

T-scores are normally distributed scores with mean of 50 and standard deviation of 10. They may be calculated from normally distributed z-scores by the linear equation:

$$T_i = 50 + 10z_i$$

T-scores are especially convenient to use, since all score values are positive, and the range of the scores is usually about 20 to 80. *T*-scores are customarily rounded to integer values and should be carefully identified to avoid their being mistaken for percentile ranks.

Several similar systems are used with standardized tests. The College Entrance Examination Board, for example, uses a mean of 500 and standard deviation of 100. Deviation IQ scores are calculated using a mean of 100 and standard deviation of 15.

11.4 STANINES

Stanines are normally distributed integer scores with mean of 5, standard deviation of 2, and range from 1 to 9. It is probably most appropriate to think of each stanine as covering a range of scores; a stanine of 5, for example, corresponds to all z-scores in the range $-.25$ to $+.25$. A stanine of 6 corresponds to z-scores in the range $+.25$ to $+.75$, and a stanine of 7 to z-scores of $+.75$ to $+1.25$. Other values may be easily determined by referring to Figure 11.1. Stanines are a very crude form of normally distributed scores, but they are especially convenient when large masses of data must be stored on computer punchcards. Each variable recorded in stanine form requires only one column on a punchcard.

EXERCISES

11.1 Complete the table, filling in with normally distributed z-scores, T-scores, and stanines.

PR	z-score	T-score	Stanine
90			
80			
70			
60			
50			
40			
30			
20			
10			

11.2 Refer to the data of Table 4.1. Produce a normalized z-score distribution for this collection of measures.

11.3 Refer to the data of Exercise 3.1.
  (a) Produce normalized z-scores for the entire collection.
  (b) Produce T-scores for the entire distribution.
  (c) Produce stanines for the entire distribution.

PEARSON PRODUCT MOMENT CORRELATION

12.1 THE STUDY OF RELATED VARIABLES

The author likes to think of scientific research as the systematic study of relationships among variables. The first eleven chapters of this book have been concerned primarily with the analysis of a single variable. We have occupied ourselves with the fundamental characteristics of the variable, some techniques for facilitating its manipulation, and some procedures for examining its distribution in the group under study. We now progress to the situation in which there are two or more observations available on each member of the group; here we are particularly concerned with the possible relationships among these variables. For example, it is commonly accepted that students with high IQ scores tend to earn higher grades in school than students with low IQ's. We are interested in determining whether or not such relationships exist, the degree of relationship, and the practical implications of any such relationship. A prerequisite to any study of this type is that an observation on one variable be paired with an observation on a second variable for each individual in the group under study.

The term *correlation* refers to the degree of correspondence or relationship between two variables. Correlated variables are those which tend to vary together — when one is larger, the other tends to be systematically larger or smaller. Consider the situation in which an achievement test is administered to a group of students at the beginning of the school year and again at the end of the year. Generally, there will be a tendency for students who scored above the group average at the beginning also to score above the average at the end. Similarly, most of the students who scored below the average at the beginning will probably score below the average at the end. The two measures are said to be *positively correlated*. Occasionally, the behavioral scientist will encounter a situation in which a high score on one variable tends to be paired with a low score on another variable. These two variables are said to be *negatively correlated*. Often, whether a correlation is positive or negative is a completely arbitrary thing. A variable that correlates positively with extroversion correlates negatively with introversion. But extroversion and introversion are ordinarily regarded as two different ways of looking at the same variable. The important idea is that there is *concomitant variation* of two correlated variables.

A *correlation coefficient* is an index of relationship between two variables. There are several different kinds of correlation coefficients but most of these have certain characteristics in common. When two variables are perfectly positively correlated, a direct relationship exists such that a higher score on one variable is always associated with a higher score on the other variable. This perfect positive correlation will have a coefficient of one. When two variables are perfectly negatively correlated, an inverse relationship exists such that a higher score on one variable is always associated with a lower score on the other. This perfect negative correlation will have a coefficient of minus one. When there is no relationship between the two variables, that is, when there is no tendency for a higher score on one to be associated with either a higher or lower score on the other, the variables are said to be uncorrelated, and the coefficient is zero. While perfect correlations and zero correlations are sometimes encountered by the physical scientist, most of the variables studied by the behavioral scientist will be correlated, but imperfectly so, yielding a correlation coefficient somewhere between zero and plus or minus one.

This chapter is concerned with the most commonly used measure of correlation, the Pearson product moment correlation coefficient, which was named for its originator, pioneer behavioral scientist Karl Pearson. Several other correlation coefficients, some of which are adaptations of the Pearson, will be considered in Chapter 13.

12.2 GRAPHING THE RELATIONSHIP BETWEEN TWO VARIABLES

When two measures are available on each of a group of individuals, the first measure will be designated as the *X*-variable and the second as the *Y*-variable. As before, the relationship between the two variables will be graphed with the *X*-variable plotted along the horizontal axis and the *Y*-variable along the vertical axis. A single point will be located on the graph for each pair of scores. Figures 12.1 to 12.5 are graphs constructed in this fashion. A complete listing of the scores from which the graphs were constructed is included with each graph.

Figure 12.1 is a graph illustrating perfect positive correlation for six pairs of scores and Figure 12.2 is a similar graph for perfect negative correlation. Figure 12.3 illustrates positive — but less than perfect — correlation and Figure 12.4 negative — but less than perfect — correlation. Figure 12.5 illustrates the case of two variables with no discernible correlation.

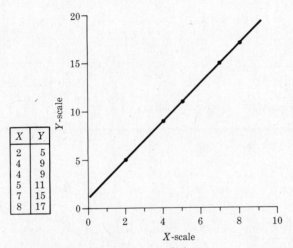

X	Y
2	5
4	9
4	9
5	11
7	15
8	17

Figure 12.1   Perfect positive correlation.

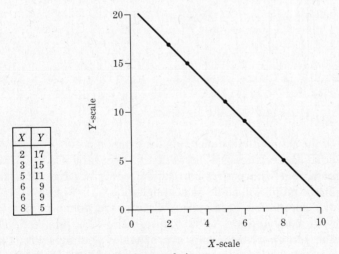

X	Y
2	17
3	15
5	11
6	9
6	9
8	5

Figure 12.2   Perfect negative correlation.

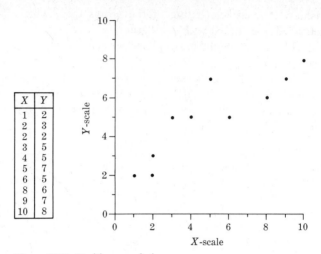

X	Y
1	2
2	3
2	2
3	5
4	5
5	7
6	5
8	6
9	7
10	8

Figure 12.3   Positive correlation.

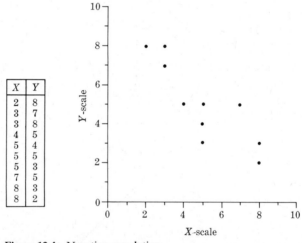

X	Y
2	8
3	7
3	8
4	5
5	4
5	5
5	3
7	5
8	3
8	2

Figure 12.4   Negative correlation.

One of the assumptions underlying the Pearson product moment correlation is that the relationship between the two variables is a *linear* one. This means that with perfect correlation, either negative or positive, the points on the graph will all lie on a straight line, as in Figures 12.1 and 12.2. When the correlation is less than perfect, the points depart from the straight-line configuration, as in Figures 12.3, 12.4, and 12.5. The more nearly the pattern of the points approximates a straight line, the higher the degree of correlation between the two variables. A graph constructed

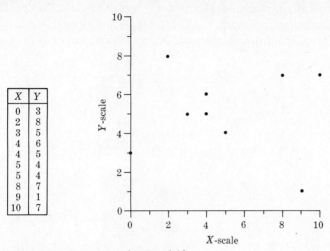

X	Y
0	3
2	8
3	5
4	6
4	5
5	4
5	4
8	7
9	1
10	7

Figure 12.5   Two uncorrelated variables.

in this fashion is called a *scatter diagram*, or simply *scattergram*. Just as a histogram or frequency polygon is a graphic illustration of the manner in which a single variable varies, a scattergram is a graphic illustration of the manner in which two variables vary together.

12.3   CALCULATION OF THE PEARSON CORRELATION COEFFICIENT

By far the most popular of the several available coefficients of correlation is the Pearson correlation coefficient, which may be defined as the mean of the z-score products of two paired variables. It is represented by r and may be calculated from the formula:

$$r = \frac{\Sigma\, z_x z_y}{N}$$

The student is reminded that the sum of the squares of the z-scores in a given distribution is equal to N (from Section 9.3). Thus, if in a given collection of paired scores every z_x is equal to the corresponding z_y, the scores are perfectly positively correlated, and the sum of the z-score products is the same as the sum of the squared z-scores for either variable. When the sum of the z-score products equals N (as it would in this case), the correlation coefficient is unity. Similarly, if every z_x is paired with a z_y of equal magnitude but opposite in sign (every positive score is paired with a negative score), the sum of products will be equal to minus N, and the correlation coefficient will be equal to minus unity.

Although the formula in z-score form serves to illustrate the definition of the correlation coefficient, it will usually be more convenient to calculate the coefficient from this mathematically equivalent formula:

$$r = \frac{SP}{\sqrt{SS_x SS_y}}$$

where

$$SP = \Sigma\, xy = \Sigma\, XY - \frac{(\Sigma\, X)(\Sigma\, Y)}{N}$$

$$SS_x = \Sigma\, x^2 = \Sigma\, X^2 - \frac{(\Sigma\, X)^2}{N}$$

$$SS_y = \Sigma\, y^2 = \Sigma\, Y^2 - \frac{(\Sigma\, Y)^2}{N}$$

The symbols SP and $\Sigma\, xy$ are used to represent the sum of products of the deviation scores, which we will abbreviate to *sum of products*. Notice the close resemblance of the sum of products to the sum of squares, the distinction between the two being the substitution of a product of two variables for the square of a single variable (which is the product of a variable with itself). There is another distinction worthy of note: Although the sum of squares is always positive, the sum of products may be either negative or positive.

Occasionally, it will be convenient to calculate the correlation coefficient from other formulas which are mathematically equivalent:

$$r = \frac{\Sigma\, xy}{N\sigma_x\sigma_y} \quad \text{or} \quad r = \frac{\Sigma\, xy}{(N - 1)S_x S_y}$$

The calculation of the sum of products will be illustrated with the data of Figures 12.1 and 12.2.

For Figure 12.1:

$$SP = 378 - \frac{(30)(66)}{6} = 48$$

For Figure 12.2:

$$SP = 282 - \frac{(30)(66)}{6} = -48$$

The student will probably profit from completing these calculations for himself from the raw data given with the two graphs.

The calculation of the correlation coefficient will be illustrated with the data of Figures 12.1, 12.2, 12.3, and 12.4.

For Figure 12.1:

$$r = \frac{48}{\sqrt{(24)(96)}} = 1.00$$

For Figure 12.2:

$$r = \frac{-48}{\sqrt{(24)(96)}} = -1.00$$

For Figure 12.3:

$$r = \frac{53}{\sqrt{(90)(40)}} = .883$$

For Figure 12.4:

$$r = \frac{-34}{\sqrt{(40)(40)}} = -.85$$

The student should complete one or two of these calculations to familiarize himself with the procedure. That the sum of products is zero (a necessary condition for the correlation to be zero) for the data of Figure 12.5 is an exercise which is left to the student.

12.4   FACTORS WHICH INFLUENCE THE VALUE OF THE COEFFICIENT

The magnitude of the correlation coefficient is unaffected by adding a constant to either or both of the two variables. In fact, some computation effort may be saved by subtracting a constant from each score in one or both of the distributions when the scores have large numerical values.

The magnitude of the correlation coefficient is a function of the variability of the measures. A correlation coefficient calculated from a group having a wide range of talent will be larger than that from a group which is quite homogeneous on one or both of the variables. For example, the correlation between IQ score and an achievement score may be quite high for a group of students ranging in ability from borderline mental retardates to the intellectually gifted. The correlation between the same two variables for a group consisting of only exceptionally bright students is likely to be very low. This suggests that reporting the correlation of two variables is meaningful only when the group under study is specified.

The correlation coefficient assumes that the relationship between the two variables is a linear one. Consider the scattergram of Figure 12.6. There is obviously perfect correspondence between the two variables, but it is not a linear one, and the degree of relationship between the two variables will be grossly underestimated by the calculation of a Pearson correlation coefficient. Curvilinear correlation will be discussed in Chapter 42. As a general rule, whenever the relationship between two variables is sufficiently nonlinear to invalidate the use of the Pearson correlation coefficient it will be readily apparent from an inspection of the scattergram.

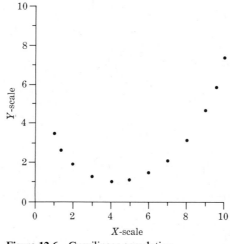

Figure 12.6   Curvilinear correlation.

In order for the Pearson correlation coefficient to reach its extreme values of plus and minus one, it is also required that the distributions of the two variables have the same shape. For example, if one of the variables is continuous and the other is a binary variable, the coefficient will always be somewhat less than unity. Similarly, if one of the variables is skewed to the left and the other is skewed to the right, the coefficient will always be less than unity.

12.5 INTERPRETATION OF THE PEARSON CORRELATION COEFFICIENT

The student is cautioned against making any absolute interpretation of the correlation coefficient. It is an abstract number, a convenient index of relationship which has been defined in such a fashion as to lie between the limits minus one to plus one. It is invariably expressed as a decimal fraction, but this fraction should not be interpreted as a proportion or percent, and it should not be interpreted in terms of the original score units, since it is independent of the unit and the magnitude of the scores. In fact, there is no need for the two variables to be expressed in the same score units.

Occasionally, one reads that correlation coefficients on the order of .30 to .70 suggest a moderate relationship, while coefficients less than these indicate a low relationship and coefficients larger than these a high relationship. This is a fallacious argument in statistical inference, since the significance of a correlation coefficient is a function of sample size; large correlations achieved with small samples may be completely meaningless. It is also a fallacious argument in descriptive statistics, since the nature of the group and the variables being studied plus the use to which the coefficient is to be put will determine whether a particular coefficient is large or small. For example, a correlation of .70 between scholastic aptitude as measured in the first grade and grade point average in college would be phenomenal. However, a coefficient of .70 between two supposedly equivalent forms of an achievement test would be so low as to suggest major revision is in order. It is important to note also that the degree of relationship is independent of the sign of the correlation coefficient; a coefficient of $-.70$ suggests the same degree of relationship as a coefficient of $+.70$.

Another problem arises in the interpretation of the correlation coefficient because one may add a constant to each score of one of the variables without changing the value of the coefficient. Suppose that the investigator wants to determine the relationship between scores from the same test administered to the same group of subjects on two different occasions. A high correlation coefficient might lead the investigator to believe that the two sets of scores are essentially equivalent when, in reality, the scores of the group are significantly higher or lower on the second administration. Similarly, a high correlation between the scores for a group of school children on tests of reading ability and numerical ability does not indicate that the two traits are about equally developed. The correlation coefficient is an index of the concomitant variation of two variables, but it says nothing about the magnitude of the variables.

Possibly the most valid and useful interpretation of the correlation coefficient is achieved by squaring it. With a little effort (and a knowledge of regression), one may demonstrate that r^2 is the ratio of two variances. The denominator of this ratio is the total variance of one of the variables, and the numerator is that part of this variance which can be predicted from or attributed to the other variable. Thus, r^2 is that proportion of the variance of one measure which is accounted for by the other. Consider the situation in which a pretest is administered to a group of students prior to a learning experience, and a posttest is administered to the same group following the learning experience. If a correlation coefficient is calculated from the two variables, r^2 may be interpreted as that proportion of the posttest variance which is accounted for by the pretest. This proportion of the variance of the posttest existed prior to the learning experience and cannot be attributed to the learning which took place between the two test administrations. Suppose that the correlation between an intelligence test score and an achievement test score is .50: then one might conclude that 25 percent of the variance of the achievement scores is accounted for by intelligence, as measured by the intelligence test. The quantity r^2 is sometimes called the *coefficient of determination*, because it is that proportion of the variance of a variable which may be predicted by or determined by another variable. The quantity $1 - r^2$ is sometimes called the *coefficient of nondetermination*, because it is that proportion of the variance which cannot be predicted by or determined by the other variable.

Possibly the most serious mistake that an investigator can make in the interpretation of a correlation coefficient is to regard a significant correlation as proof of a cause-and-effect relationship. For example, one might investigate the study habits of college students and determine that there is a negative correlation between the amount of time spent in study and course grades. This does not establish that higher grades result from less study, but suggests that brighter students do not study as much as the dull ones. However, the correlation of IQ scores of parents and their children has been reported to be on the order of .50, and the investigator is probably entitled to interpret this as evidence of a causal relationship. The observed correlation between two variables is sometimes due to a cause-and-effect

Table 12.1   Typical values of r

Descriptions of variables		Nature of subjects	Typical value of r
X	Y		
Iowa Test of Educational Development, Grade 9	Freshman grade-point average in college	Over 600 college students	.58
The Stanford-Binet Intelligence Test IQ	The same test given one week later	Elementary-school pupils	.90
Verbal reasoning ability (as measured by the Differential Aptitude Test)	Nonverbal reasoning ability	High-school juniors	.65
Height	Achievement in college physics	Male college seniors	.00

relationship, and this is science at its best; but a significant correlation is not of itself sufficient evidence to establish a causal relationship.

Interesting examples of correlation coefficients as encountered in behavioral research have been compiled by Glass and Stanley* and are reported in Tables 12.1 and 12.2.

Table 12.2   Correlations between IQ's of related or unrelated children as a function of genetic similarity and similarity of environment

Nature of pairing	Typical value of r
Identical twins, reared together	.88
Identical twins, reared apart	.75
Fraternal twins of same sex	.53
Fraternal twins of opposite sex	.53
Siblings, reared together	.49
Siblings, reared apart	.46
Parent with own child	.52
Foster parent with child	.19
Unrelated, reared together	.16

12.6   THE CORRELATION MATRIX

Very often in behavioral research, measures on several variables will be available on each subject in the group being studied, and the investigator will be interested in reporting the correlation between each pair of variables. The most convenient form for reporting a large number of correlations for the same group of subjects is the correlation matrix. A matrix is a rectangular array of numbers arranged in rows and columns, the rows being numbered from top to bottom and the columns from left to right. In the case of the correlation matrix the rows and columns represent the variables, and the numbers entered in the matrix are the correlation coefficients. Thus, the number entered in the second row and third column is the coefficient of correlation between the second and third variables. Table 12.3 is a correlation matrix reporting all possible correlations for five variables. Notice that the correlation matrix is a symmetric matrix, that the entry in row one and column three is the same as that in row three and column one. Notice also that the correlation of each variable with itself is 1.00 and this is the value entered in the principal diagonal.

*Tables 12.1 and 12.2 are reprinted from *Statistical Methods in Education and Psychology* by Gene V. Glass and Julian C. Stanley, published by Prentice-Hall, Inc., Englewood Cliffs, New Jersey, with the permission of the publisher.

Table 12.3   Correlation matrix for five variables

Variable	1	2	3	4	5
1	1.00	.29	.68	.05	.17
2	.29	1.00	.44	.22	.03
3	.68	.44	1.00	.39	.12
4	.05	.22	.39	1.00	.41
5	.17	.03	.12	.41	1.00

EXERCISES

12.1   Given the following collection of paired measures:

X: 5   10   2   3   1   2   4   8   6   9
Y: 9   8   6   7   3   3   6   4   6   8

(a)   Construct a scattergram from the data.
(b)   Compute the Pearson correlation coefficient.

12.2   Given the following collection of paired measures:

X: 1   4   4   5   5   5   6   7   8
Y: 9   7   8   6   1   6   2   1

(a)   Construct a scattergram from the data.
(b)   Compute the Pearson correlation coefficient.

12.3   A social studies teacher administered an attitude scale to a group of high school students and recorded the scores of the boys separately from those of the girls. The scale purports to measure student attitudes toward "going steady," and a high score is indicative of a favorable attitude toward this activity. The scores are as follows:

Boys: 4   9   1   2   5   8   9   5   3   4
Girls: 11   8   6   7   10   8   5   11   6   8

(a)   Record the sex of each student as a binary variable, awarding a score of one to each boy and a score of zero to each girl. Call this variable X, and call the score on the attitude scale Y. You should then have 20 pairs of measures, each pair having a sex "score" and an attitude score. Calculate the Pearson correlation coefficient from the paired variables.
(b)   What significance should be attached to the sign of the correlation coefficient?
(c)   What proportion of the attitude score variance is accounted for by the sex of the respondents?

12.4 Given the following data in which X and Y are paired scores, z_x and z_y are linear transformations of X and Y, respectively, and $X + 2$ is the sum of the X-score plus a constant:

X	Y	z_x	z_y	$X + 2$
2	2	−1.5	−1.5	4
4	6	−0.5	+0.5	6
5	5	0.0	0.0	7
6	4	+0.5	−0.5	8
8	8	+1.5	+1.5	10

Compute:
(a) The Pearson correlation coefficient for X and Y.
(b) The Pearson correlation coefficient for z_x and z_y.
(c) The Pearson correlation coefficient for X and $X + 2$.
(d) The Pearson correlation coefficient for Y and $X + 2$.
(e) The Pearson correlation coefficient for z_y and Y.
(f) The Pearson correlation coefficient for z_y and X.

13

OTHER CORRELATION COEFFICIENTS

13.1 THE SPEARMAN RANK CORRELATION COEFFICIENT

Occasionally, it will be desirable to compute a correlation coefficient when one or both variables is on an ordinal scale. There is also the situation in which the score distribution is extremely heterogeneous, and the numerical values of the scores are less meaningful than the ranks of the scores. This is the same situation as that encountered earlier when the mean was less meaningful than the median. The Spearman rank correlation coefficient is an adaptation of the Pearson product moment correlation coefficient for use with ordinal data. It may be calculated by substituting the ranks of the individuals for their scores in the Pearson formula. However, by placing the restrictions on the data that each variable will have N ranks and that these will range from 1 to N, the formula may be simplified to:

$$r_s = 1 - \frac{6 \Sigma d_i^2}{N^3 - N}$$

The author has chosen to represent the Spearman coefficient by r_s rather than by ρ, the Greek letter rho, which is sometimes used, since he believes that this is consistent with the good statistical practice of reserving Greek letters for parameters. The symbol d_i represents the difference in the two ranks assigned the ith individual.

The calculation of the Spearman rank correlation coefficient is first illustrated with a sample of size five and with the scores limited to the ranks 1 through 5. The reader's attention is directed to the data recorded in Table 13.1. Only the first four columns are utilized in the calculation of the Spearman coefficient; the remaining three columns will be used to demonstrate the equivalence of the Spearman and Pearson coefficients with data of the sort described.

Table 13.1   Equivalence of the Spearman and Pearson coefficients

X_i	Y_i	d_i	d_i^2	X_i^2	Y_i^2	$X_i Y_i$
1	3	-2	4	1	9	3
2	1	1	1	4	1	2
3	4	-1	1	9	16	12
4	2	2	4	16	4	8
5	5	0	0	25	25	25
15	15	0	10	55	55	50

First, the calculation of the Spearman coefficient:

$$r_s = 1 - \frac{6(10)}{125 - 5} = \frac{60}{120} = .050$$

Second, the calculation of the Pearson coefficient, as in Section 12.3:

$$r = \frac{5}{\sqrt{(10)(10)}} = \frac{5}{10} = .050$$

The calculation of the Spearman coefficient from scores which have been ranked will be illustrated with the data of Figure 12.3, where .883 has already been determined to be the value of the Pearson coefficient. To rank the data, the scores are first placed in an ordered series; it does not matter whether the highest score or the lowest score is placed first. The scores are then numbered (ranked) from 1 to N. Where tie scores are encountered, the ranks are summed and divided equally among the tied scores so that scores having the same numerical value also have the same rank. The X-variable is ranked in this fashion:

X-score: 1   2   2   3   4   5   6   8   9   10

X-rank: 1   2.5   2.5   4   5   6   7   8   9   10

The Y-variable is ranked separately, but in the same fashion:

Y-score: 2   2   3   5   5   5   6   7   7   8

Y-rank: 1.5   1.5   3   5   5   5   7   8.5   8.5   10

After the scores have been ranked, the ranks are paired in precisely the same pairings as the original scores, and the difference between the ranks is determined for each pair. This procedure is illustrated in Table 13.2.

Table 13.2 Calculation of the Spearman rank coefficient

X-score	Y-score	X-rank	Y-rank	d_i	d_i^2
1	2	1.0	1.5	.5	.25
2	3	2.5	3.0	.5	.25
2	2	2.5	1.5	−1.0	1.00
3	5	4.0	5.0	1.0	1.00
4	5	5.0	5.0	.0	.00
5	7	6.0	8.5	2.5	6.25
6	5	7.0	5.0	−2.0	4.00
8	6	8.0	7.0	−1.0	1.00
9	7	9.0	8.5	−.5	.25
10	8	10.0	10.0	.0	.00

$$\Sigma\, d_i^2 = 14.00$$

Once the differences have been squared and summed, the calculation of the Spearman coefficient is quite simple:

$$r = 1 - \frac{6(14)}{1000 - 10} = 1 - \frac{84}{990} = 1 - .085 = 0.915$$

This value is the Pearson correlation coefficient calculated from the ranks. It is slightly different from the earlier value of the Pearson coefficient calculated from the scores ($r = 0.883$). Ordinarily, with interval data, the calculation of the correlation coefficient from scores would be preferred. However, for the individual who does not have immediate access to a calculating machine, the Spearman formula provides a good approximation of the Pearson formula with considerably less computational effort. The approximation improves with larger samples, and the Spearman coefficient is probably completely appropriate for use by the classroom teacher in the study of relationships among test scores. In the research setting, however, the Pearson coefficient will be chosen by the investigator for the study of relationships between test scores.

An alternative formula is computationally more laborious, but has the advantage that it may be corrected for tied ranks. When the scores are limited to the ranks 1 through N, the sum of squares and the sum of products may be calculated as follows:

$$SS_x = SS_y = \frac{N^3 - N}{12}$$

$$SP = \frac{1}{2}(SS_x + SS_y - \Sigma\, d^2)$$

Then:

$$r_s = \frac{SP}{\sqrt{SS_x SS_y}}$$

In the second example given:

$$SS_x = SS_y = \frac{990}{12} = 82.5$$

$$SP = \frac{1}{2}(82.5 + 82.5 - 14.0) = 75.5$$

$$r_s = \frac{75.5}{82.5} = 0.915$$

which is precisely the same answer as before.

The effect of tied ranks is to reduce the value of the sum of squares, a phenomenon which is not reflected in the formulas given above. In the event that ties are present in one or both variables, a tie correction factor may be calculated for each variable as follows:

1. For each rank level where ties exist,

$$T = \frac{f^3 - f}{12}$$

where f is the number of observations at that rank. For example, if two observations are tied at the same rank, $T = (8 - 2)/12 = 0.5$ for that rank level.
2. These are summed over all the ranks for each variable, yielding ΣT_x and ΣT_y.
3. The formulas for the sums of squares are adjusted thus:

$$SS_x = \frac{N^3 - N}{12} - \Sigma T_x \qquad SS_y = \frac{N^3 - N}{12} - \Sigma T_y$$

In the example,

$$\Sigma T_x = 0.5 \quad \text{and} \quad \Sigma T_y = 1.0$$

Then:

$$SS_x = 82.5 - 0.5 = 82.0$$
$$SS_y = 82.5 - 1.0 = 81.5$$

$$SP = \frac{1}{2}(82.0 + 81.5 - 14.0) = 74.75$$

$$r_s = \frac{74.75}{\sqrt{(82.0)(81.5)}} = \frac{74.75}{81.75} = 0.914$$

The effect of the correction factor on the example is negligible, but it does move the value of the Spearman coefficient in the direction of the value of the Pearson coefficient.

The effect of the tie correction tends to be miniscule when the number of ties is few, as in the example. The effect of the tie correction may be appreciable, however, when the number of ties is large. Some experimentation on the part of the user may be desirable to determine whether the tie correction factor has an appreciable effect with his data.

The Spearman rank correlation coefficient is a nonparametric statistic with certain underlying assumptions. One of these is that the measures are truly continuously distributed and that no two persons have precisely the same quantity of a continuously distributed trait. If the precision of measurement is so gross that large numbers of tie scores are awarded, then use of the Spearman coefficient is highly questionable. If the proportion of ties is not large, its effect on the value of the coefficient is negligible.

13.2 THE KENDALL TAU RANK CORRELATION COEFFICIENT

The Kendall coefficient is an alternative to the Spearman rank correlation coefficient. Although the Spearman coefficient is an adaptation of the Pearson coefficient for use with ordinal data, the rationale for the Kendall coefficient is quite different. The strategy consists of (1) determining the number of consistencies minus the number of discrepancies in a series of paired ranks; and (2) forming the ratio of this first quantity to the number of consistencies which would be present if there were perfect positive correspondence between the two sets of ranks. A four-step computational procedure is suggested, and it is illustrated with the same data used to illustrate the calculation of the Spearman coefficient and reported in Table 13.1.

1. Arrange the X-ranks in order from 1 to N and pair with the corresponding Y-ranks. For the example:

$$X\text{-ranks:} \quad 1 \quad 2 \quad 3 \quad 4 \quad 5$$

$$Y\text{-ranks:} \quad 3 \quad 1 \quad 4 \quad 2 \quad 5$$

2. Beginning with the first pair (that for which $X = 1$), compare the corresponding value of Y to each succeeding value of Y. Assign a value of $+1$ to each Y-rank, which is larger than the first, and a value of -1 to each Y-rank, which is smaller than the first. Then take the algebraic sum.

For the first pair $(X = 1)$: $\quad -1 \quad +1 \quad -1 \quad +1 = 0$
For the second pair $(X = 2)$: $\quad +1 \quad +1 \quad +1 = +3$
For the third pair $(X = 3)$: $\quad -1 \quad +1 = 0$
For the fourth pair $(X = 4)$: $\quad +1$

Finally, take the sum over all ranks to get the value of S. For the example, $S = 0 + 3 + 0 + 1 = +4$.

3. Calculate the optimum value of S, that value which would be achieved if there were perfect positive correspondence between the two sets of ranks. This can be demonstrated to be equal to $\frac{1}{2}N(N - 1)$.

4. Form the ratio:

$$\tau = \frac{S}{\frac{1}{2}N(N - 1)} = \frac{2S}{N^2 - N}$$

For the example:

$$\tau = \frac{+8}{25 - 5} = +0.40$$

This compares with the $+0.50$ using the Spearman coefficient and the same data. The Kendall coefficient has the advantage that it has been generalized to a partial correlation coefficient, which is not true for the Spearman. Partial correlation coefficients are beyond the scope of this discussion. A tie correction formula has been derived for the Kendall coefficient, but it does not appear to appreciably affect the value of the coefficient and will be omitted here.

13.3 THE GOODMAN-KRUSKAL GAMMA COEFFICIENT

The Goodman-Kruskal gamma (γ) coefficient is an index of relationship for ordinal data arranged in a bivariate frequency table. It is the only correlation coefficient in this text designed specifically for use with discrete ordinal data. The gamma coefficient has been popular with sociologists, but has received little attention from researchers in education and psychology. The rationale for the statistic has much in common with that for the Kendall coefficient, and the student should develop some feeling for it as we develop the computational strategy.

The procedure will be illustrated with the data recorded in Table 13.3. The letters H, M, and L represent high, medium, and low scores in the example.

Table 13.3 Discrete ordinal data in a bivariate frequency table

	H	M	L
H	3	2	1
M	1	2	1
L	2	1	2

1. Arrange the data in a bivariate frequency table with the frequency of the highest scores on both variables in the upper left-hand cell and the frequency of the lowest scores on both variables in the lower right-hand cell (see Table 13.3, for example).

2. Develop a "P" value as follows: Take the sum of all the cell frequencies that are below and to the right of the upper left-hand cell ($cell_{11}$), and multiply this sum by the frequency of this cell. Repeat this procedure for every cell in the table which has other cells to the right and below it. For the example:

$$Cell_{11}: \quad (2 + 1 + 1 + 2)\,(3) = 18$$

$$Cell_{12}: \quad (1 + 2)\,(2) = 6$$

$$Cell_{21}: \quad (1 + 2)\,(1) = 3$$

$$Cell_{22}: \quad (2)\,(2) = 4$$

Finally, take the sum of all these quantities and call it P. For the example, $P = 18 + 6 + 3 + 4 = 31$.

3. Develop a "Q" value as follows: Take the sum of all cell frequencies which are below and to the left of the upper right-hand cell, and multiply this sum by the frequency of this cell. Repeat this procedure for every cell in the table which has cells to the left and below it. For the example:

$$Cell_{13}: \quad (1 + 2 + 2 + 1)\,(1) = 6$$

$$Cell_{12}: \quad (1 + 2)\,(2) = 6$$

$$Cell_{23}: \quad (2 + 1)\,(1) = 3$$

$$Cell_{22}: \quad (2)\,(2) = 4$$

Finally, take the sum of all these quantities and call it Q. For the example, $Q = 6 + 6 + 3 + 4 = 19$.

4. Calculate the gamma coefficient from:

$$\gamma = \frac{P - Q}{P + Q}$$

For the example:

$$\gamma = \frac{31 - 19}{31 + 19} = \frac{12}{50} = 0.24$$

With a little effort, it can be demonstrated that the Pearson coefficient for this same data is 0.19. Many authors would question the use of the Pearson coefficient with discrete ordinal data of this sort.

For a two-by-two table in which the cell frequencies are represented by the letters a, b, c, and d,

a	b
c	d

the formula for the Goodman-Kruskal coefficient reduces to

$$\gamma = \frac{ad - bc}{ad + bc}$$

13.4   THE POINT BISERIAL CORRELATION COEFFICIENT

The point biserial correlation coefficient is simply a Pearson correlation coefficient with one continuous variable and one dichotomous variable. The dichotomous variable is assumed to be discrete, with a uniform distribution in each of the two categories. To classify students as successful or unsuccessful (or as graduates and nongraduates) and use this as a dichotomous variable implies that all of the unsuccessful students are equally unsuccessful and that all of the successful students are equally successful. The point biserial is often used in test item analysis, in which a correlation is calculated between the scores on a single test item and scores on the total test. The idea is to determine whether or not the individual test item is consistent with the total test. This implies that the students who answer the item incorrectly are equally ignorant of the quality measured by the item and that the students who answer the item correctly are equally well-informed. Most statisticians agree that the point biserial is mathematically defensible, but some authorities insist that a true dichotomy must be established before the procedure is valid. The author is inclined to feel that the problem is one of precision of measurement, and the fact that the measurement is crude (that is, limited to two levels) does not invalidate the measure or its use in the calculation of a correlation coefficient.

When the point biserial is calculated with the established Pearson formulas, the dichotomized variable may be assigned any two numerical values—one and zero, or 13 and 27, for example. The value of the coefficient is unaffected by the exact values assigned to the dichotomy; therefore the common practice is to assign values of one and zero, since this greatly reduces the computational effort. The use of a binary variable of this sort permits the derivation of a computational form which is somewhat simpler to use than the Pearson formulas, but which is mathematically equivalent:

$$r_{pbi} = \frac{M_1 - M_0}{\sigma_x} \sqrt{pq}$$

where X = the continuous variable

σ_x = the standard deviation of the continuous variable

M_1 = the X-mean of the group scoring one on the dichotomy

M_0 = the X-mean of the group scoring zero on the dichotomy

p = the proportion of the total group scoring one on the dichotomy

q = the proportion of the total group scoring zero on the dichotomy

The student's attention is directed to the fact that this formula is mathematically equivalent to the Pearson formulas given earlier only if σ_x is calculated using N rather than $N - 1$ in the denominator.

The equivalence of the Pearson and the point biserial correlation coefficients is demonstrated with the data of Table 13.4:

Table 13.4   Correlation of continuous and dichotomous variables

X	X^2	Y	Y^2	XY
1	1	1	1	1
1	1	1	1	1
2	4	0	0	0
6	36	1	1	6
6	36	1	1	6
7	49	0	0	0
8	64	0	0	0
9	81	0	0	0
40	272	4	4	14

$$r = \frac{SP}{\sqrt{SS_x SS_y}} = \frac{-6}{\sqrt{(72)(2)}} = -0.50$$

$$r_{pbi} = \frac{M_1 - M_0}{\sigma_x} \sqrt{pq} = \frac{3.5 - 6.5}{3} \sqrt{\left(\frac{1}{2}\right)\left(\frac{1}{2}\right)} = -0.50$$

The student is encouraged to follow through with the calculations from beginning to end to demonstrate that the author's calculations are correct and to gain experience with the formulas.

It is completely appropriate to use nominal variables in the dichotomy, assigning a one for male and a zero for female, for example. In this case, if the females score higher on the continuous variable than the males, a negative coefficient will be obtained. Of course, whether the coefficient is negative or positive is completely arbitrary with nominal variables.

It should be noted that the magnitude of the point biserial is a function of the sizes of p and q, and it can only be as large as -1 or $+1$ when p and q are one half. When the subjects are not evenly divided in the dichotomy, unity correlation cannot be achieved.

13.5 THE PHI COEFFICIENT

The phi coefficient is an extension of the point biserial (and the Pearson) coefficient to the situation in which both variables are dichotomized. Consider the data of Table 13.5 and the calculation of the Pearson coefficient from the following data:

Table 13.5 Correlation of two dichotomized variables

X	X^2	Y	Y^2	XY
1	1	1	1	1
1	1	1	1	1
1	1	1	1	1
1	1	1	1	1
1	1	0	0	0
0	0	1	1	0
0	0	1	1	0
0	0	0	0	0
0	0	0	0	0
0	0	0	0	0
5	5	6	6	4

$$r = \frac{SP}{\sqrt{SS_x SS_y}} = \frac{1.0}{\sqrt{(2.5)(2.4)}} = \frac{1.0}{2.45} = 0.41$$

Data of this sort are quite often organized into a two-by-two bivariate frequency table, with the letters a, b, c, and d representing the frequencies of the four cells, as shown in Table 13.6.

Table 13.6 Bivariate frequency table for phi coefficient

	$X = 0$	$X = 1$
$Y = 1$	$a = 2$	$b = 4$
$Y = 0$	$c = 3$	$d = 1$

Using this notational scheme, we may calculate the phi coefficient from the cell frequencies of the bivariate frequency table by the formula:

$$\phi = \frac{bc - ad}{\sqrt{(a + b)(c + d)(a + c)(b + d)}}$$

which, in the example under consideration, is:

$$\phi = \frac{12 - 2}{\sqrt{(6)(4)(5)(5)}} = \frac{10}{24.5} = 0.41$$

The same assumptions that apply to the single dichotomy of the point biserial also apply to both dichotomies of the phi coefficient. Unity correlation can only be achieved when the subjects are evenly divided between the two categories of both dichotomies, that is, when $a + b = c + d$ and $a + c = b + d$.

13.6   THE BISERIAL CORRELATION COEFFICIENT

The biserial correlation coefficient is an index of relationship between two continuously distributed variables, one of which is treated as a dichotomy. The most common example of a continuously distributed variable which is treated as a dichotomy is the academic pass-fail variable. While academic ability is certainly a continuous variable (and probably normally distributed), pass-fail (or graduation-nongraduation) is a common dichotomous interpretation of this trait. The biserial correlation coefficient is calculated on the assumption that the dichotomized variable is in fact normally distributed. It has no direct relationship to any of the previous correlation coefficients, although it usually ranges from minus one to plus one.

The biserial correlation coefficient may be calculated from the formula:

$$r_{bi} = \frac{M_p - M_q}{\sigma_x} \left(\frac{pq}{y} \right)$$

where X = the variable recorded as continuous

σ_x = the standard deviation of the X-variable

M_p = the X-mean of the successful group

M_q = the X-mean of the unsuccessful group

p = the proportion of the subjects in the successful group

q = the proportion of the subjects in the unsuccessful group

y = the ordinate of the normal curve which divides it into two parts with p, the proportion of the total area in one part, and q, the proportion of the area in the other part

Calculated in this fashion, the biserial coefficient will be positive if the successful group has the higher X-mean and negative if the unsuccessful group has the higher mean. Table 4 in the Appendix gives the height of the normal curve for various values of p and also gives the value of pq/y.

Suppose that 60 percent of the students enrolled in a certain graduate school complete the M.A. degree and 40 percent do not complete the degree. Suppose also that the mean IQ of the students who graduate is 120, the mean of those who do not is 110, and the standard deviation is 15. The biserial correlation between IQ score and whether or not a student graduates is:

$$r_{bi} = \frac{120 - 110}{15}(.621) = .41$$

The biserial correlation coefficient has been used in almost every situation in which the point biserial is used. It may be used with discriminant analysis to determine the probability of success for a student with a given score on the continuously recorded variable. This is a special application of prediction.

Some authors question the utility of the biserial coefficient, especially in those situations in which the normality of the dichotomized variable is not established. The biserial coefficient tends to behave peculiarly when this assumption is violated. Odd-shaped distributions may produce coefficients larger than unity.

13.7 THE TETRACHORIC CORRELATION COEFFICIENT

The tetrachoric correlation coefficient is an extension of the biserial to the situation in which two normally distributed variables are dichotomized. Its computation is quite involved; however, a good approximation of the coefficient is achieved if the phi coefficient is first calculated from the data, then transformed by the following mathematical expression:

$$r_{tet} = \text{sine}\ (\phi\ 90°)$$

Table 5 in the Appendix was compiled using this transformation and is convenient for determining the value of the tetrachoric coefficient from a knowledge of the phi coefficient. The data used to illustrate the calculation of the phi coefficient in Section 13.5 yielded a coefficient of 0.41. When this is referred to in Table 5 in the Appendix, the value of the tetrachoric coefficient is seen to be 0.60.

This approach provides a good approximation only if both variables are dichotomized in the vicinity of the median. For more generalized procedures for determining the value of the tetrachoric correlation coefficient, the reader is directed to the Selected References at the end of the chapter.

13.8   THE CONTINGENCY COEFFICIENT

All of the correlation coefficients considered have required measurement on at least an ordinal scale or have permitted the use of nominal measurement limited to a dichotomous variable. The contingency coefficient is intended for use with nominal data recorded in bivariate frequency tables. There is no restriction upon the number of categories, provided the number of measures is quite large. The use of the contingency coefficient requires a knowledge of the chi-square statistic and will be considered later.

EXERCISES

13.1   Two art judges ranked five oil paintings as follows:

Judge I	Judge II
1 *African Sunset*	1 *African Sunset*
2 *Blue Mist*	2 *Lily of the Valley*
3 *Lily of the Valley*	3 *The Old Fisherman*
4 *The Old Fisherman*	4 *Blue Mist*
5 *Country Scene*	5 *Country Scene*

Calculate the Spearman rank correlation coefficient from the data. Are the judges in substantial agreement?

13.2   Refer to the data of Exercise 12.1. Rank the data and calculate the Spearman rank correlation coefficient. How does your answer compare with the Pearson coefficient calculated earlier with these data?

13.3   Refer to the data of Table 13.3. Calculate the Spearman rank correlation coefficient with and without the tie correction factor. How do your answers compare with the Goodman-Kruskal gamma coefficient and the Pearson product moment correlation coefficient reported in the text?

13.4   Refer to the data of Exercise 13.1. Calculate the Kendall tau coefficient. How does your answer compare with that achieved using the Spearman rank coefficient?

13.5 Refer to the data of Exercises 12.1 and 13.2. Calculate the Kendall tau coefficient. How does your answer compare with that achieved using the Pearson and Spearman coefficients?

13.6 Refer to the data reported in Table 13.6. Calculate the Goodman-Kruskal gamma coefficient. How does your answer compare with the answer for the phi coefficient reported in the text?

13.7 Refer to the data of Exercise 12.3. Calculate the point biserial correlation coefficient for the data. How does your answer compare with the Pearson coefficient calculated earlier with these data?

13.8 Repeat Exercise 13.7 using the biserial correlation coefficient. How does this answer compare with the point biserial coefficient? Is this an appropriate use of the biserial coefficient?

13.9 The political preference of each of a group of subjects was obtained along with an indication of whether the subject intended to vote for or against a proposed constitutional amendment. Of the group, 22 Republicans and 10 Democrats indicated they would vote for the amendment, and 8 Republicans and 16 Democrats indicated they would vote against it. Arrange the data in a bivariate frequency table and calculate the phi coefficient.

13.10 Repeat Exercise 13.9 using the tetrachoric correlation coefficient. How does your answer compare with the phi coefficient? Is this an appropriate use of the tetrachoric coefficient?

SELECTED REFERENCES

CARROLL, JOHN B., "The nature of the data, or how to choose a correlation coefficient," *Psychometrika*, **26**, 347–372 (1961).

HAYES, S. P., "Diagrams for computing tetrachoric correlation coefficients from percentage differences," *Psychometrika*, **11**, 163–172 (1946).

JENKINS, WILLIAM L., "An improved method for tetrachoric *r*," *Psychometrika*, **20**, 253–258 (1955).

SLAICHERT, WILLIAM M., "Techniques for estimating the coefficient of correlation from a fourfold table," unpublished Ph.D. dissertation, Iowa State College, 1951.

WEISS, ROBERT S., *Statistics in Social Research*. New York: John Wiley, 1968.

14

PRINCIPLES OF PREDICTION

14.1 THE IDEA OF REGRESSION

The terms *correlation*, *regression*, and *prediction* are so closely related in statistics that they are often used interchangeably. The term *correlation* refers to the degree of correspondence or relationship between two variables. Whenever two variables are correlated, it is possible, though often with a limited degree of success, to predict the score on one variable from a knowledge of the score on the other. *Regression* refers to a phenomenon characteristic of prediction with less than perfect correlation.

In 1885 Francis Galton published a paper entitled *Regression towards Mediocrity in Hereditary Stature*. It was already well established that many human characteristics are inherited—that tall people tend to have tall offspring, and that short people tend to have short offspring, and so on. But Galton was interested in predicting the physical characteristics of the progeny from a knowledge of the physical characteristics of the parents. He observed that very tall parents do indeed tend to have tall offspring — but not as tall as the parents. Similarly, very short people tend to have short offspring—but not as short as the parents. This phenomenon is popularly known as *regression toward the mean*.

Consider the situation in which a student's grade point average in college is known to be correlated with rank in his high school graduating class, and the investigator is interested in predicting college grades from high school rank. Let X represent the predictor variable (high school rank) and Y the predicted variable (college grades). Given an individual whose score on the X-variable is not known, the best predicted score for

this individual is the Y-mean. That is, in the long run of many such predictions to be made for many different individuals for whom there is no
knowledge of the X-variable, predicting the Y-mean for each of them will
give less prediction error than any other Y-score. Even when the X-vari-
able is known, the investigator will not be able to perfectly predict the
Y-variable, since the relationship between the two is imperfect. The investigator is now predicting out of partial ignorance rather than the total
ignorance he experienced when he predicted the Y-score without any
knowledge of the X-score. As a result, his prediction will be influenced by
both his knowledge and his ignorance, and the predicted Y-score will be
closer to the mean than the corresponding X-score. If the two variables
are expressed as standard scores with zero mean, the absolute value of
the predicted score (z_y) will be smaller than the absolute value of the
predictor score (z_x). Although regression toward the mean often appears to
be a natural phenomenon, the modern statistician prefers to regard it as
a mathematical assumption built into the statistics of prediction.

14.2 EQUATIONS AND GRAPHS OF LINEAR RELATIONSHIPS

In order for the behavioral scientist to predict scores on one variable
from a knowledge of scores on another, he must also have a knowledge
of the relationship between the two. This means he must first study a
group of subjects on whom both measures are available before he can
extend his findings to subjects on whom only one measure is available.
The bulk of this chapter will be devoted to the group on whom both
measures are available, although we are primarily interested in being able
to predict the second variable from a knowledge of the first.

One very important assumption underlying the prediction process considered here is that the relationship between the two variables is linear
and can be represented by a straight line drawn on a scattergram. The
relationship between two linearly correlated variables may also be represented by the equation of a straight line, which is in the general form of
$Y = bX + c$.

Figure 14.1 is a graph of the particular linear equation, $Y = 2X - 3$.
The b in the general form of the equation has been replaced by the constant 2 in this particular equation, and c has been replaced by -3. There
is a value of Y for every possible value of X in this equation, which is
another way of saying that there is an infinity of solutions to the equation.
All of these solutions are represented by the line in Figure 14.1.

The constant b is known as the *slope of the line*, which is the rate at
which Y changes with change in X. Its value may be determined from the
graph of the line by selecting any two values of X (call them X_1 and X_2)

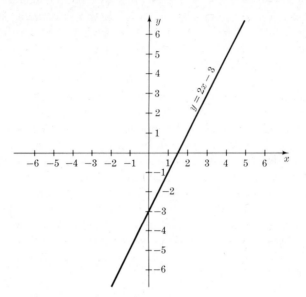

Figure 14.1   Graph of a linear equation.

plus the corresponding values of Y (call them Y_1 and Y_2), then calculating the slope from:

$$b = \frac{Y_2 - Y_1}{X_2 - X_1}$$

The slope of the line also happens to be the tangent of the angle which the line makes with the horizontal axis. The constant c is called the Y-intercept, and it may be determined from the graph as that point at which the line crosses the vertical axis. It is the value of Y that corresponds to $X = 0$.

14.3   THE LEAST SQUARES CRITERION

If two variables are perfectly correlated, a scattergram may be constructed and a straight line drawn through *all* of the points on the graph. When predicting from this situation, one need only identify a point on the line which is directly above the X-score on the horizontal axis and identify the corresponding Y-score on the vertical axis. More often, the investigator will determine the equation of the line and solve for Y by substituting the value of X in the equation.

Of course, it is highly unlikely that the behavioral scientist will ever encounter perfect correlation in a practical prediction situation. He can anticipate that the points on the scattergram will not all fall on a straight line, and he must seek to identify a straight line which best expresses the relationship between the two variables. He is interested in arriving at a linear prediction equation which gives the least amount of error when predicting the second measure from the first, that is, for that group of individuals on whom both measures are known. The assumption is made that this equation will also give the least amount of error when predicting the second measure for individuals on whom only the first measure is known.

Let X represent the first score and Y the second score of the group on whom we have both scores. We will seek a linear equation with which we may predict Y from X with the least amount of error. The predicted scores will lie on the straight line, but many of the Y-scores previously recorded from actual measurements will not lie on this line. We must make a distinction between the Y-scores previously recorded and the Y-scores predicted from the equation. Let $\hat{Y}$ represent the predicted scores (those which all lie on the prediction line). To the extent that these differ from the recorded Y-scores, they contain an error component. The difference between the predicted score ($\hat{Y}$) and the recorded score (Y) for a given predictor score (X) is error and will be represented by the lower case letter e:

$$e_i = Y_i - \hat{Y}_i$$

Thus, if the recorded Y-score for a given individual is smaller than the predicted score, the error is negative; if the recorded Y-score is larger than the predicted score, the error is positive. The situation is illustrated in Figure 14.2. The data were borrowed from Figure 12.3, for which we have already calculated the sum of squares, sum of products, and the correlation coefficient. The author has also determined the appropriate prediction equation and has drawn the prediction line. The predicted score for an individual whose X-score is 5 is also 5 in this particular case. Notice, however, that the one individual who scored 5 on the X-variable has a score of 7 on the Y-variable. The prediction error for this individual is 2. Of course, in the typical research setting, where the number of subjects is much larger, several subjects would probably have an X-score of 5, but there is no reason to expect that each would have the same recorded Y-score. Each, however, would receive the same predicted Y-score ($\hat{Y}$), and the error would vary from person to person with some errors being positive and some negative.

Now, how does one determine the prediction equation that gives the least error? There is an infinite number of prediction equations for a given set of data in which the algebraic sum of the errors is zero. In fact, if the point on the scattergram that corresponds to both the X- and the Y-mean is located, any straight line drawn through this point will be a prediction

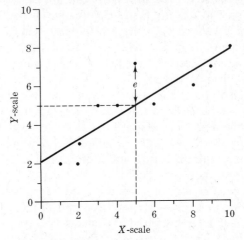

Figure 14.2   Linear prediction error.

line for which the sum of the errors is zero. Obviously, the solution does not lie in having a prediction equation with large amounts of negative error canceled by equally large amounts of positive error. Another solution might call for the use of the absolute values of the errors, but this is a mathematical dead end; thus we seek a solution that permits additional manipulation of the findings. The answer is quite simple—it is the one we encountered in selecting the variance over the mean deviation as a measure of dispersion. We shall square the errors and choose as our prediction equation that equation for which the sum of the squared errors ($\Sigma\ e_i^2$) is smaller than that for any other equation. This is called the *least squares criterion*, and its derivation requires the use of differential calculus; however, the problem has already been solved, and we shall be able to determine the prediction equation through the use of simple algebra. This prediction equation is more often referred to as a *regression equation* and its locus as a *regression line*.

14.4   REGRESSION EQUATIONS FOR STANDARD SCORES

One of the consequences of regression toward the mean is that the regression line will pass through a single point corresponding to the means of the Y-, $\hat{Y}$-, and X-distributions. Whenever scores are expressed as deviations from the mean (as in the case of $x_i = X_i - M$ or $z_i = x_i/\sigma$), the mean of the distribution of the deviation scores is zero. If the regression equation is calculated with deviation scores, there will be a point on the regression line at which both x and y (or z_x and z_y) equal zero. In other words, the y-intercept will be zero and will drop out of the equation.

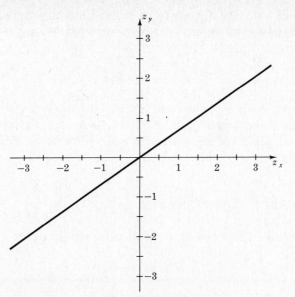

Figure 14.3   Regression with standard scores.

The least squares solution yields a regression equation for use with z-scores having a slope equal to the correlation between the two variables, and the equation becomes:

$$\hat{z}_y = r_{xy}z_x$$

If the two sets of z-scores are perfectly linearly correlated, r_{xy} will equal unity, and the regression line will lie at a 45-degree angle to the horizontal axis. If there is no relationship between the two variables, r_{xy} will be zero, and the regression line will coincide with the horizontal axis, which is also the y-mean, and the predicted score for all subjects is the y-mean. If the two variables are correlated, but imperfectly, the regression line will intercept the horizontal axis at some angle between 0 and 45 degrees.

Figure 14.3 shows the prediction line for the situation in which the correlation is .67 between two sets of z-scores. Notice the regression toward the y-mean; an individual with $z_x = +3$ is awarded $\hat{z}_y = +2$, while an individual with $z_x = -3$ is awarded $\hat{z}_y = -2$.

14.5   REGRESSION EQUATIONS FOR DEVIATION SCORES

The general form of the regression equation for deviation scores is:

$$\hat{y} = bx \qquad \text{where } x = X - M$$

The value of b may be derived by substituting into the regression equation for z-scores and by the application of some algebra:

$$b = \frac{r_{xy}\sigma_y}{\sigma_x} = \frac{SP}{SS_x}$$

The use of deviation scores will simplify the problems of regression analysis when a single variable is predicted from a knowledge of several others. This is called *multiple regression* and will be considered in one of the later chapters.

14.6 REGRESSION EQUATIONS FOR RAW SCORES

The general form of the regression equation for raw scores is:

$$\hat{Y} = bX + c$$

The only distinction between deviation scores and raw scores is the placement on the score scale (the shifting of the mean). Thus, the value of the slope is the same as for the deviation scores:

$$b = \frac{r_{xy}\sigma_y}{\sigma_x} = \frac{SP}{SS_x}$$

The value of the Y-intercept is chosen so that the Y-mean corresponds to the X-mean (so that they fall at the same point on the regression line):

$$c = \bar{Y} - b\bar{X}$$

Consider the data of Figure 14.2, for which the following results were previously determined: $\bar{X} = 5$, $\bar{Y} = 5$, $SS_x = 90$, $SS_y = 40$, and $SP = 53$. The regression equation may be determined as follows:

$$b = \tfrac{53}{90} = .59$$
$$c = 5 - .59(5) = 2.05$$
$$\hat{Y} = .59X + 2.05$$

The regression line of Figure 14.2 was constructed with a knowledge of the regression equation. This equation may now be used to predict the $\hat{Y}$-score for any individual for whom the X-score is known. For example, for an individual whose X-score is 2:

$$\hat{Y} = .59(2) + 2.05 = 3.23$$

The student should refer to Figure 14.2 and verify that this is in fact the $\hat{Y}$-score that corresponds to $X = 2$ on the regression line. He should also note that neither of the two persons in the original distribution who scored 2 on the X-variable has this Y-score.

The problem of errors of prediction will be considered next.

When one variable is predicted from another, it is customary to refer to the predictor variable as the *independent* variable and the predicted variable as the *dependent* variable.

14.7   THE STANDARD ERROR OF ESTIMATE

Let us return for a moment to the situation in which the investigator has no knowledge of the independent variable. Here the predicted score will be the Y-mean, and 68 percent of the subjects will fall within one standard deviation of the Y-mean if the population is normally distributed. Thus the investigator may predict the Y-mean for all subjects and have the assurance that 68 percent of his predictions will be accurate within plus or minus one standard deviation. This is a perfectly worthless piece of information, but it does illustrate the important idea of the standard deviation as an estimate of error.

The error of prediction for a given subject was previously defined as the difference between his actual score on the dependent variable and his predicted score on the dependent variable; that is, $e_i = Y_i - \hat{Y}_i$.

Of course, knowledge of the magnitude of this error after the final score is in is not a very useful piece of information in the practical prediction situation. What is needed is an estimate of this error at the time of prediction—when the score on the dependent variable has not yet been awarded to this particular subject. The investigator will require an accurate estimate of the amount of error involved in any given prediction situation. One such estimate is the *standard error of estimate*, which may be defined as the standard deviation of the errors of prediction:

$$SE_{\text{est}} = \sqrt{\frac{\Sigma(Y - \hat{Y})^2}{N - 2}} = \sqrt{\frac{\Sigma e^2}{N - 2}}$$

As a general rule, it will be considerably more convenient to calculate the standard error of estimate from the following mathematically equivalent formula:

$$SE_{\text{est}} = S_y \sqrt{1 - r^2} \qquad \text{where } S_y = \sqrt{\frac{SS_y}{N - 2}}$$

The standard error of estimate, of course, must be calculated from that group on whom both the independent and dependent measures are available, but it is intended for use with those individuals on whom the dependent variable is not yet available. The student has now entered into the study of statistical inference. The individuals from whom the regression equation was calculated are the sample; the population consists of those individuals for whom the dependent variable might be predicted using the regression equation. The investigator is interested in extending his findings from the sample to the population and in having an expression of the confidence with which he can do this. Notice that the formula for the standard deviation of the errors is calculated with $N - 2$ in the denominator. This term, $N - 2$, is the degrees of freedom of the statistic, a rather involved topic which will be considered in greater detail in the chapters which follow. Its use, as with the use of $N - 1$ before, is to insure an unbiased estimate of the population parameter.

The calculation of the standard error of estimate is illustrated with the data of Figure 14.2:

$$SE_{est} = \sqrt{\frac{40}{8}} \sqrt{1 - .78} = 1.05$$

This suggests that 68 percent of the individuals for whom a given predicted score is calculated will in fact score within plus or minus 1.05 score points of the predicted score. Referring to Figure 14.2, we note that seven of the ten subjects from whom the regression equation was calculated did in fact score within 1.05 points of their predicted scores. There are certain assumptions underlying the use of the standard error of estimate in this fashion. First, the group from whom the regression equation was calculated must be genuinely representative of the group with whom the regression equation is used for purposes of prediction. Second, the errors of prediction are assumed to be normally distributed. Third, the errors of prediction are equally distributed at all points on the regression line. This last condition is known as *homoscedasticity*, and its violation usually increases the errors of prediction for extreme scores. This is not usually a problem in practical prediction, since the possible success or failure of the extreme individuals is more easily predicted than that of the individuals in the middle of the distribution. In other words, there may be considerable error in the predicted scores for the extreme cases, but this is likely to be of little practical consequence to the behavioral scientist and should not prohibit the use of the standard error of estimate.

EXERCISES

14.1 Given the following collection of scores, in which the X-variable is IQ score and the Y-variable is the grade received in a certain course of study:

i	X_i	Y_i	X_i^2	Y_i^2	X_iY_i
1	120	4	14400	16	480
2	110	3	12100	9	330
3	100	2	10000	4	200
4	90	0	8100	0	0
5	80	1	6400	1	80
	500	10	51000	30	1090

(a) Make a scattergram of the scores recorded above.
(b) Calculate the correlation between the two variables.
(c) Determine the slope of the regression line.
(d) Determine the Y-intercept of the regression line.
(e) Write the regression equation.
(f) Draw the regression line on the scattergram.
(g) Calculate the predicted score for each of the X-scores recorded above.
(h) Calculate $\Sigma\, e^2$.
(i) Notice that three of the points on the scattergram lie on a straight line. The equation of this line is $Y = .1X - 8$. Draw the line on the scattergram and calculate $\Sigma\, e^2$ for this prediction line.
(j) Compare the answers to (h) and (i).
(k) Determine the standard error of estimate.
(l) What is the predicted grade for an individual with an IQ of 95?

14.2 Refer to the data of Figure 12.4.
(a) Determine the slope of the regression line.
(b) Determine the Y-intercept of the regression line.
(c) What is the predicted score corresponding to $X = 3$?
(d) What is the predicted score corresponding to $X = 8$?
(e) Calculate the standard error of estimate.

15

STATISTICAL PROCEDURES IN MEASUREMENT

15.1 INTRODUCTION

Measurement and research in the behavioral sciences are intimately related—good research is invariably dependent upon measurement. By the same token, certain research is essential to the development of a good measurement device. Most of the measurement devices used by behavioral scientists can be classified as tests, and there are three essential attributes of a good test: *standardization, validity,* and *reliability.*

Standardization means that the test has been administered to a well-defined group and that careful records have been kept with respect to their performance. These records are called *norms* and the well-defined group is the *norm group*. The selection of an appropriate norm group is important, because it serves as the basis for the determination of test validity and reliability, and because it will often be the point of reference for interpretation of measures derived from other subjects. To facilitate the process of comparison, standardization will generally include the development of one or more kinds of standard scores.

A test is valid when it measures what it is intended to measure. The important idea here is the use which is to be made of the test scores. A test may be valid for one use, but this is no assurance that it is valid for other uses. For example, an IQ test might be a valid measure of scholastic aptitude, but it might be completely invalid as a measure of executive ability.

Reliability refers to the precision of measurement. It is an essential ingredient in validity; a test cannot be very valid if it is not very reliable.

However, a test may be reliable without being valid. The professional test constructor will seek to express reliability and validity in mathematical terms. Summarizing, validity is an expression of the extent to which a test measures what it is supposed to measure, reliability of how well it measures whatever it measures.

The personal judgment of the particular individual who scores a test should not be a factor affecting the score. To the extent that several qualified persons can score a test for a given individual and award the same score, a test is said to be objective; to the extent that several different scorers do not award the same score, there is an error component in the score. *Objectivity* is an essential ingredient in reliability. Because a lack of objectivity is reflected in a lack of reliability, it will not usually be necessary to seek a mathematical definition of objectivity.

15.2   THEORY OF RELIABILITY

Man has never been able to demonstrate that any measurement on a continuous scale is without error, and all of the available evidence suggests that every such measurement has some error in it. The logical conclusion is that every measurement on a continuous scale consists of two components, the true value plus some measurement error. This idea may be expressed mathematically by:

$$x_i = t_i + e_i$$

where x_i = an attempt to measure something (the observed value)

$\quad t_i$ = the true value

$\quad e_i$ = the error component in the measurement

If we can assume that an error of measurement is as likely to be positive as it is to be negative, it is logical to conclude that the algebraic sum and the mean of the errors would both be zero if an infinite number of attempts to measure the same thing were made. The definition of the true score, then, is the mean of an infinite number of attempts to measure the same thing.

This idea may be extended to the situation in which one measure is taken on each of several subjects, so that the true score is a variable. Then the following relationship may be defined:

$$\sigma_x^2 = \sigma_t^2 + \sigma_e^2$$

Reliability may be defined as the ratio of the true score variance to the observed score variance. The symbol r_{xx} will be used to represent the coefficient of reliability, a quantity which exists only in theory and which is sometimes called the *self-correlation:*

$$r_{xx} = \frac{\sigma_t^2}{\sigma_x^2} = 1 - \frac{\sigma_e^2}{\sigma_x^2}$$

There are several different ways to estimate the coefficient of reliability, and they are distinguished from one another by defining error in slightly different ways.

It can be demonstrated that the coefficient of reliability is a function of the number of items in a test — the greater the number of items in a test, the more reliable the test tends to be. Thus a shortened form of an established test can be expected to be less reliable than the original. The so-called Spearman-Brown prophecy formula is an expression of the relationship between test length and reliability:

$$r_e = \frac{kr_{xx}}{1 + (k - 1)r_{xx}}$$

where r_{xx} = is the reliability of the original test

r_e = is the expected reliability of the test with the number of items changed

k = is the ratio of the number of items in the changed test to the number in the original

When the length of the test is doubled, the formula reduces to:

$$r_e = \frac{2r_{xx}}{1 + r_{xx}}$$

15.3   TEST-RETEST RELIABILITY

The most obvious method for determining reliability of a test calls for administering it to the same sample on two different occasions, then defining reliability as the Pearson product moment correlation between the two sets of scores. This method assumes that it is practical and valid to administer the test to the same group of persons twice in a relatively short period of time. To some persons, this is the only true reliability — to others, it is only one way of getting at reliability, and perhaps not the best way for a given instrument.

Test-retest reliability is sometimes called a *coefficient of stability* and carries with it the idea that the trait being measured is fairly stable through time — at least through the period of time that separates the two administrations. It also implies that the second score is not greatly affected by the double exposure.

It is not a very practical procedure for measurements of traits that are believed to be in a state of flux or traits that are being subjected to undue influence during the time between administrations. These factors would tend to decrease the estimate of reliability. And it is not a very practical approach to reliability where remembering an answer or calculation or procedure might affect the second score. These factors would tend to give a spurious increase in the estimated reliability.

Error in test-retest reliability is anything that leads a person to get different scores on the two administrations.

15.4 EQUIVALENT FORMS RELIABILITY

One way of solving some of the problems associated with test-retest reliability is to have two equivalent forms of the same instrument. Both forms are administered to the same group of subjects, and the estimated reliability is simply the Pearson product moment correlation between the two sets of scores. The two administrations may take place quite close in time, and there is little danger of error introduced by change through time or due to practice effects.

However, to the extent that the two forms are not equivalent, a new source of error is introduced. As a general rule, this approach to reliability will yield a somewhat lower estimate of reliability than test-retest. There is also the problem of generating two equivalent forms of the same instrument, a task that may not be economically or otherwise feasible.

Error in this procedure is also anything that leads a person to get different scores on the two administrations, but this is not exactly the same error as in the test-retest procedure and may not lead to the same coefficient of reliability.

15.5 SPLIT-HALF RELIABILITY

One of the approaches to reliability most popular with the test constructor is the split-half technique. A single administration of the instrument is made, the test split into two halves which are scored separately, and a Pearson correlation coefficient between the two scores is calculated. Then the Spearman-Brown prophecy formula is used to compensate for the fact that the reliability was estimated from a test one-half the length of the final form. For example, suppose the correlation between the two halves were .80, then the expected reliability of the full test would be:

$$r_e = \frac{2(.80)}{1 + .80} = .89$$

While a number of ways of carrying out a split-half reliability have been devised, the best is probably the odd-even approach in which the odd-numbered test items are treated as one-half of the test and the even-numbered as the alternate half. This insures that approximately the same amount of time will be devoted to both halves and helps to insure some equivalence of the two forms if there is a tendency for the items to become more difficult near the end of the test.

This type of reliability is one form of *internal consistency* reliability. It is not very meaningful in speeded tests, and to the extent that the test is speeded, the coefficient of reliability is inflated. Whenever a test has a time limit, and some of the subjects do not complete all items, the score is as much a function of the number of items completed as it is of the number of items answered correctly. There is a tendency in this situation for the items in the first part of the test to be answered correctly and those in the last part to be answered not at all (which is interpreted as answering incorrectly). The correlation between the odd and even items of such a test will be very high.

There are many desirable characteristics in split-half reliability: the conditions of administration are near identical for both halves, there is little or no time lag between administrations; and the attitudes or response sets of the respondents should be near identical for the two administrations. But some researchers suggest that this is not reliability at all, and it certainly does tend to yield a higher estimate of the reliability than test-retest. There is no dispute that split-half reliability is completely inappropriate where a substantial speed component is measured.

15.6 RELIABILITY BASED ON ITEM STATISTICS

Kuder and Richardson developed their approaches to reliability based on item statistics in an attempt to overcome some of the deficiencies of split-half reliability. Their formulas split a test into as many parts as there are items in the test. Assumptions are made that all of the items measure a single trait and that each respondent attempts every item. The basic formula is Kuder-Richardson formula 20:

$$r_{xx} = \left(\frac{n}{n-1}\right)\left(\frac{\sigma^2 - \Sigma\,pq}{\sigma^2}\right)$$

where n = the number of items in the test
$\quad \sigma^2$ = the variance of the test scores
$\quad p$ = the proportion of the subjects answering a given question right
$\quad q = 1 - p$

Some computational effort can be saved through the use of Kuder-Richardson formula 21, if one can assume that all of the items are of equal difficulty:

$$r_{xx} = \frac{n\sigma^2 - M(n - M)}{(n - 1)\sigma^2}$$

Notice that this reliability estimate may be calculated with only a knowledge of the mean and variance of the scores plus the number of items on the test.

To the extent that a test measures on more than one dimension (and this is often desirable to get maximum validity), the Kuder-Richardson formulas will underestimate the reliability. To the extent that the test is speeded, they will overestimate the reliability.

15.7 THE STANDARD ERROR OF MEASUREMENT

Beginning with the mathematical definition of reliability given earlier, one may derive a formula for the standard deviation of the errors of measurement. This is more often referred to as the *standard error of measurement:*

$$SE_m = \sigma_e = \sigma_x \sqrt{1 - r_{xx}}$$

This is usually interpreted as an indication of the probable extent of error in the score of a given individual. Suppose an IQ test has a reliability of .96 and standard deviation of 15:

$$SE_m = 15 \sqrt{1 - .96} = 3$$

Thus, if one can assume that the errors of measurement are normally distributed about a given score and equally distributed throughout the score range, 68 percent of the subjects will be regarded as having a true IQ score within 3 points of their IQ scores as determined from the test.

15.8 THE PROBLEM OF VALIDITY

In the physical sciences both direct and indirect measurement are encountered. The weight of an object is measured directly by placing it on a balance and comparing it to a standard. Temperature, however, is measured indirectly. When the physical scientist says he is measuring temperature, it is more probable that he is measuring the length of a column of mercury and inferring from it the temperature. He does not measure heat directly — he measures not what temperature is, but what it does.

The same idea holds true with respect to most measurement by the behavioral scientist. First, he must carefully assign a definition to the trait under study — not in terms of what it is, but what it does. He asks the question, how does one behave if one possesses this trait? Or, what behavior distinguishes an individual with much of this trait from one with little of it? For example, he might define intelligence as the ability to solve certain kinds of problems. This is called an *operational definition;* the trait itself is not observable, but it is defined in terms of observable phenomena. The behavioral scientist then measures the observable phenomena and draws conclusions with respect to the unobservable trait. Of course, there is always the danger that he is not measuring what he proposes to measure at all. This is the problem of validity — to what extent is he measuring what he intends to measure?

15.9 CONTENT VALIDITY

Content validity is demonstrated by showing how well the content of a test samples the situations about which conclusions are to be drawn. It is especially important for achievement and proficiency measures.

This is sometimes referred to as logical or sampling validity, or validity by definition. It usually implies the use of some expertise to define a universe of interest, the careful drawing of a representative sample of ideas from this universe, and the preparation of test items that match these ideas. Some empirical method will be used to demonstrate the reliability of the instrument, but empirical methods may not necessarily be required to establish validity.

Statistical methods may be applied to the problem of content validity. For example, the test constructor may perform an analysis of the test items which includes a point biserial (or biserial) correlation between the score on each item and the score on the total test. The assumption is that the necessary expertise is available to identify intelligently what the total test is all about. Test items that are not consistent with the total are either revised or eliminated.

15.10 CRITERION-RELATED VALIDITY

Criterion-related validity is demonstrated by correlating test scores with one or more external criteria. Since few direct measures are available to the behavioral scientist, the big problem is to find a criterion. Consider, however, a test intended to measure scholastic aptitude; in this case, the criterion would be the degree of success in a scholastic setting. Such an approach to validity is essentially a matter of prediction — something which behavioral scientists are able to do with only a limited degree of success. Nonetheless, this is usually regarded as the best approach to validity when a defensible criterion can be identified. Criterion-related validity is sometimes called *empirical* validity. The coefficient of validity is the correlation between the test and the external criterion; we should not expect it to be as high as the coefficient of reliability.

Concurrent validity is a cross-sectional approach to the problem of prediction. Instead of studying the future success of the subject, scores of already successful or unsuccessful people are used as a standard of comparison. Thus, one might validate a test of mechanical aptitude by securing scores of mechanics and nonmechanics to demonstrate that the test can discriminate between the two. This is sometimes called *discriminant* validity; it is not as highly regarded as predictive validity. In the final analysis, the most useful test is one that can be used to predict human behavior, and concurrent validity is not direct evidence of this quality.

15.11 CONSTRUCT VALIDITY

Construct validity is not a simple idea. It includes both logical and empirical approaches to validity, as do all other approaches, but it seeks a unique combination of the two. The approach is as follows, in a general sense, but the exact techniques will vary greatly:

1. The investigator sets forth a proposition that a test measures on a certain dimension, or designs a test to measure that dimension.

2. He then inserts that proposition into current theory about that dimension.
3. He hypothesizes empirical relationships based on the theory.
4. Then a study of the relationships is carried out using the test to measure the dimension.

If the findings are consistent, some support is lent to the measurement and the theoretical framework within which it is presented. Gor example, the author constructed an instrument that purports to measure certain personal values. Another investigator hypothesized that psychologists who subscribe to different psychological theories are people who differ on certain personal values. The investigator's conclusions were borne out by research, lending support to both the validity of the instrument and the theory of the investigator.

BASIC PROBABILITY THEORY

16.1 DEFINITION OF PROBABILITY

By way of introduction to probability theory, let us first review the terms *population* and *sample*. A *population* is a collection of objects, events, or individuals having some common characteristic which the researcher is interested in studying. A *sample* is a smaller group chosen from the population to participate directly in the research as a representative of the population. Invariably, the use of probability theory to interpret the research will prescribe the use of a *random* sample. This is a sample chosen in a fashion which insures that every object, event, or individual in the population has equal chance of being drawn for the sample. The important idea is that the selection of each object in the sample is a chance event.

It is very difficult to arrive at a definition of the term *probability* that is both completely acceptable to the mathematician and readily communicable to the nonmathematician. However, the following definition is relatively simple and quite adequate for the use of probability theory in the behavioral sciences: The probability that an event (object or person) will occur is equal to the relative frequency of occurrence of that event (object or person) in the population under consideration.

For example, suppose the population consists of 100 persons who have previously been determined to include 30 females and 70 males. The probability of selecting a male in random drawings of a single individual from this group is 70/100. Similarly, the probability of drawing a female is 30/100. Turning to a different kind of sampling, let us suppose that when a student guesses the answer to an objective question on an examination his answer is a completely random one. Then the probability of guessing the right answer to a true-false question is 1/2, and the probability of guessing the right answer to a multiple-choice question with five responses is 1/5. In these last examples, the probability of the event is its relative frequency in a hypothetical frequency distribution, or *long-run* frequency distribution as it is sometimes called. When statisticians say the probability of heads in the flip of a coin is one-half, they mean the relative frequency of heads is 0.50 in a distribution consisting of a very large number of flips of the coin.

16.2 SEVEN FUNDAMENTAL RULES OF PROBABILITY

Rule 1 The probability of occurrence of any one of a set of equally likely events is one divided by the number of events. This may be stated symbolically by $P(e_i) = 1/N$.

For example, given a group of six boys, whose names are Joe, John, Robert, Jim, Elmer, and George, the probability of selecting Elmer in a single random draw is 1/6.

Rule 2 If a frequency distribution exists, and if there is more than one object or event in a given class, the probability of drawing an object of this particular class is equal to the class frequency divided by the number of objects; that is, $P(e_i) = f_i/N$.

For example, given a frequency distribution consisting of 20 Englishmen, 30 Canadians, and 5 Australians, the probability of selecting an Australian in a single random draw from this population is 5/55, which is 1/11.

Rule 3 The probability of any event cannot be less than zero (there are no negative probabilities) or greater than one (one represents a sure thing). Stated mathematically, $0 \leq P(e_i) \leq 1$.

Rule 4 The sum of the probabilities of all of the possible events in a population must equal one. This may be stated symbolically by $\Sigma P(e_i) = 1$.

Rule 5 The probability of an event occurring plus the probability of that event not occurring is equal to one. This is usually stated symbolically by $q = 1 - p$, where p represents the probability of occurrence and q represents the probability of nonoccurrence.

This rule suggests that any population may be dichotomized and treated with statistical procedures appropriate for use with dichotomous populations. (The population considered above could be regarded as consisting of Canadians and non-Canadians, for example.)

Rule 6   The addition rule of probability   The probability that any one of a set of mutually exclusive events will occur is the sum of the probabilities of the separate events.

When two events are mutually exclusive, if one occurs, the other cannot occur. For example, in an objective type of examination, the student's answer to a question is either right or wrong; it cannot be both.

Refer to the population containing 20 Englishmen, 30 Canadians, and 5 Australians. The probability of an Englishman is 20/55, and the probability of a Canadian is 30/55. The probability that a single random selection will produce either an Englishman or a Canadian is 50/55, the sum of the probabilities of the separate events.

Rule 7   The multiplication rule of probability   The probability that a combination of independent events will occur is the product of the separate probabilities of the events.

When two events are independent, the occurrence or nonoccurrence of one has no effect on the other.

For example, the probability that a student will guess the correct answer to the first question on a true-false examination is 1/2, and the probability that he will guess the correct answer to the second question is also 1/2. But before he answers any questions, the probability that he will guess the correct answers to the first two questions is 1/4, the product of the separate probabilities. Similarly, the probability that he will guess the correct answers to the first three questions is 1/8, and to the first four questions 1/16. Again, we are assuming that his answers are completely random.

16.3 SAMPLING FROM A FINITE POPULATION
WITHOUT REPLACEMENT

A special problem arises when sampling from populations sufficiently small that the drawing of a sample appreciably changes the distribution of the population. Suppose a population consists of six men and three women. The probability of drawing a male from this population is 6/9 or 2/3. However, if a male is drawn and not replaced, the population becomes five men and three women, and the probability of drawing a second male is 5/8. Thus, the probability of drawing two males in a row without replacing the first is 5/12, the product of the separate probabilities. Similarly, the probability of drawing three males in a row without replacement from a population consisting of six males and three females is: 6/9 × 5/8 × 4/7, which is 5/21. The behavioral scientist is unlikely to draw samples of this sort, but he should be aware of the implications if he encounters such a procedure.

16.4   SAMPLING FROM A FINITE POPULATION WITH REPLACEMENT

The problem of sampling from a small population is appreciably simplified if the sampling is done with replacement. As before, the probability of drawing a male from a population containing six males and three females is 2/3. If a male is drawn and replaced, then the probability of drawing a male the second time is also 2/3, and the probability of drawing two males in a row with replacement is 4/9. The probability of drawing three males in a row with replacement is 8/27.

16.5   SAMPLING FROM VERY LARGE POPULATIONS

Most of the sampling procedures considered in this text assume that the researcher is drawing samples from a population so large that sampling does not appreciably change the distribution of the population. This is equivalent to sampling with replacement. As a general rule, behavioral scientists will draw their samples from very large populations, and these populations will be regarded as approaching infinite size.

16.6   FACTORIAL NOTATION

Consider the situation in which a high school principal is asked to rank five teachers in order of their teaching abilities. He may choose the first teacher from any one of the five, but has only four from which to choose the second, three from which to choose the third, two from which to choose the fourth, and one for the fifth. This is a problem in sampling without replacement, and if the principal's selection is a purely random one, the probability of any particular order of the five teachers is 1/5 × 1/4 × 1/3 × 1/2 × 1/1, which is 1/120.

Calculating the product of the number of choices available to the principal yields 5 × 4 × 3 × 2 × 1, which is 120, the number of possible ways in which the five teachers could be ranked. If the selection were completely random, each of these arrangements would be equally probable, each having a probability of 1/120.

Calculations like these occur 'often enough in probability problems to warrant a special type of notation, called *factorial* notation: *n*! (read "*n*-factorial") is the product of all of the integers from *n* to 1. For example 3! is 3 × 2 × 1, which is 6, which is the number of ways three different objects can be arranged in order. Suppose the three objects were represented by the letters *A*, *B*, and *C*. They can be arranged in order six ways, as follows: *ABC*, *ACB*, *BAC*, *BCA*, *CAB*, and *CBA*. In the same fashion, one may determine the number of ways six things can be arranged in order: 6! = 6 × 5 × 4 × 3 × 2 × 1 = 720.

It is customary to let both 1! and 0! equal one. The logic of this will become clear to the student as he studies the following sections and works some examples.

16.7 PERMUTATIONS

Suppose a principal had been asked to choose and rank the best three teachers out of a group of five. He would have had five from which to choose the first, four from which to choose the second, and three from which to choose the third, yielding a total of 60 ways in which three teachers can be chosen and ranked from a group of five. A group of objects chosen and ordered in this fashion is called a *permutation*. The number of permutations of *n* objects taken *r* at a time is:

$$P(n, r) = \frac{n!}{(n - r)!}$$

In the example quoted above, the value of *n* was 5 and the value of *r* was 3:

$$P(5, 3) = \frac{5!}{(5 - 3)!} = \frac{5!}{2!} = \frac{5 \times 4 \times 3 \times 2 \times 1}{2 \times 1} = 60$$

In a completely random selection, each of the possible permutations is equally probable.

16.8 COMBINATIONS

Suppose the principal were requested simply to choose the three best teachers from a group of five without ranking the three. The possible ways of ranking three objects is 3!, which is 6. This suggests that the number of combinations of five things taken three at a time (without respect to order) is equal to the number of permutations divided by 3!, that is, 60/6, which is 10. The general equation for determining the number of combinations possible from n objects taken r at a time is:

$$C(n, r) = \frac{n!}{r!(n - r)!}$$

In a completely random situation, each of these possible combinations is equally probable.

The symbol $\binom{n}{r}$ is also used to represent the number of combinations of n objects taken r at a time.

EXERCISES

16.1 A group of students is known to contain 15 students of superior ability, 40 students of average ability, and 12 students of below average ability.
   (a) What is the probability of a random choice yielding a student of superior ability?
   (b) What is the probability of a random choice yielding either a student of superior ability or a student of below average ability?

16.2 A true-false examination contains eight questions.
   (a) What is the probability of a student guessing the right answer to any given question? Assume a completely random selection.
   (b) What is the probability of a student getting all of the questions right simply by guessing?
   (c) What is the probability of a student getting all of the questions wrong simply by guessing?

16.3 A multiple-choice examination contains five questions, each with three responses.
   (a) What is the probability of a student guessing the right answer to any given question?
   (b) What is the probability of a student guessing a wrong answer to any given question?
   (c) What is the probability of a student getting all of the questions right simply by guessing?

16.4 A college student claims he can identify four different brands of cigarettes by taste. An experiment is set up to test his ability, he is blindfolded and given four cigarettes (one of each brand), one at a time, and asked to identify them by brand name.

(a) How many different ways can four different cigarettes be presented to him one at a time?

(b) What is the probability that he will correctly identify all four brands simply by guessing?

16.5 A graduate student is given the choice of answering any four questions from a group of seven on an essay examination. How many different options does he have; that is, how many groups of four questions can be selected from a collection of seven?

16.6 Eight coeds will compete in a beauty contest, and the judges will award trophies to the first-, second-, and third-place winners.

(a) How many ways can three winners be ranked out of a field of eight competitors?

(b) Suppose some college boys place the names of the eight girls in a hat and attempt to duplicate the judges' selection by drawing names from the hat. What is the probability that the first draw from the hat will be the first-place winner? What is the probability that three names drawn from the hat would be the first-, second-, and third-place winners, in that order?

17

THE BINOMIAL PROBABILITY DISTRIBUTION

**17.1 APPLICATION OF THE FUNDAMENTAL RULES
OF PROBABILITY**

Consider the situation in which a student who has not studied his lessons is confronted with a "pop-quiz" containing five multiple-choice questions each having three possible responses. Let us assume that his answers are random guesses, so that the probability of his guessing the correct answer to any given question is $1/3$ and the probability of his guessing a wrong answer to any given question is $2/3$.

The probability that the student will guess all five correct answers is determined by applying the multiplication rule of probability: $1/3 \times 1/3 \times 1/3 \times 1/3 \times 1/3 = (1/3)^5$, which is $1/243$.

The probability that the student will get four right answers and one wrong is a bit more complicated. First, the number of combinations of five things taken four at a time is five. This should be verified from the combinatorial formula given in Section 16.8. Each of these five possible combinations of four right answers and one wrong is equally probable. Now, let us consider any one of these equally probable combinations. The probability that he will get a particular set of four questions right is $1/3 \times 1/3 \times 1/3 \times 1/3$, and the probability that he will get the other question wrong is $2/3$. Applying the multiplication and addition rules of probability, we see that the probability that the student will get four answers right and one wrong is $5(1/3)^4(2/3)$, which is $10/243$. (The five equally probable combinations are added together by multiplying by five.)

Now consider the situation in which the student gets three answers right and two wrong. There are ten equally probable combinations of three right and two wrong answers to five questions. Applying the multiplication and addition rules as before, we find that the probability that the student will get three right and two wrong is $10(1/3)^3(2/3)^2$, which is $40/243$.

Similarly, the probability that the student will get two right and three wrong may be determined to be $10(1/3)^2(2/3)^3$, which is $80/243$. And the probability he will get one right and four wrong is $5(1/3)^1(2/3)^4$, which also happens to be $80/243$. Finally, the probability of getting all five questions wrong is $1(1/3)^0(2/3)^5$, which is $32/243$. Now, consider the total picture:

$$P(5 \text{ right, } 0 \text{ wrong}) = 1(1/3)^5(2/3)^0 = 1/243$$
$$P(4 \text{ right, } 1 \text{ wrong}) = 5(1/3)^4(2/3)^1 = 10/243$$
$$P(3 \text{ right, } 2 \text{ wrong}) = 10(1/3)^3(2/3)^2 = 40/243$$
$$P(2 \text{ right, } 3 \text{ wrong}) = 10(1/3)^2(2/3)^3 = 80/243$$
$$P(1 \text{ right, } 4 \text{ wrong}) = 5(1/3)^1(2/3)^4 = 80/243$$
$$P(0 \text{ right, } 5 \text{ wrong}) = 1(1/3)^0(2/3)^5 = 32/243$$

$$\text{Sum of the probabilities} = 243/243 = 1$$

Let us take note of some of the relationships evident in these calculations. The first multiplier, which we will refer to later as the coefficient, is the number of combinations of right answers possible for a given score. It may be calculated using the combinatorial formula, where n is the number of questions and r is the number of right answers. Notice that the coefficients are symmetrical, that the first and last coefficients in the distribution are equal to one, that the second and next to last are equal to n. The quantity in the first parentheses is the probability (p) of getting a right answer, and the exponent is the number of right answers (r). Similarly, the quantity in the second parentheses (q, or $1 - p$) is the probability of getting a wrong answer, and the exponent is the number of wrong answers ($n - r$). Calculations of this sort are greatly facilitated through the use of the binomial probability distribution equation.

Finally, take note of the fact that the sum of all of the probabilities in the distribution is equal to one. This is an example of the use of the addition rule of probability, which may be extended to any combination of probabilities in the distribution. For instance, the probability of a student getting four or more questions right is equal to the probability of getting four right plus the probability of getting five right, which is $11/243$.

17.2   DICHOTOMOUS POPULATIONS

Many dichotomous variables are encountered in the behavioral sciences: for example, male versus female sex, right versus wrong answers to test questions. Many other nominal variables are usefully dichotomized by classifying all objects in a given population as either belonging or not belonging to a given classification: for example, psychologists and non-psychologists, delinquents and nondelinquents. The reader is reminded that these categories must be mutually exclusive and collectively exhaustive. One cannot dichotomize a population by classifying people as Republicans and Democrats, discarding any individuals who do not fit in one of these two classes. Occasionally, it will be desirable to reduce a continuous variable to a dichotomy: for example, classifying students as above or below average ability.

Populations that are split into two mutually exclusive and collectively exhaustive categories are called *dichotomous* populations. The binomial probability distribution is a useful device for studying dichotomous populations.

17.3   THE GENERAL EQUATION OF THE BINOMIAL PROBABILITY DISTRIBUTION

The general equation of the binomial probability distribution (BPD) is given by:

$$P(r \text{ in } n) = \binom{n}{r} p^r q^{n-r}$$

where n = the size of a random sample of events

r = the number of events in the first category

$n - r$ = the number of events in the second category

p = the probability of any single event falling in the first category

$q = 1 - p$, which is the probability of any single event falling in the second category.

For a given problem, only r is a variable, and r may vary from zero to n.

It will not be necessary to calculate the values of the binomial coefficients — the first multiplier in the equation — since these have been conveniently tabled for small values of n in Table 6 in the Appendix. For larger values of n, the binomial equation becomes unwieldy, and another procedure (to be presented later in this chapter) will be substituted.

17.4 APPLICATION OF THE BINOMIAL PROBABILITY DISTRIBUTION

To make use of the BPD, every object in the population under study must be clearly identifiable as belonging to one of two classes, although the distribution of these objects (the number in each class) may not be known. Second, the researcher must have some basis for arriving at the values of p and q, the probabilities associated with membership in the two classes. The BPD is a *theoretical* probability distribution — a mathematical model for certain kinds of random events. In addition to its use to determine the probability of certain kinds of random events (such as chance scores on objective tests), the BPD may also be used as a model for testing certain kinds of research hypotheses. In the latter application, a sample distribution is compared to the theoretical distribution to see if the sample deviates significantly from the theory. Both applications will be illustrated.

Suppose a teacher constructs a ten-item true-false examination and is interested in determining what kinds of scores students might be expected to get by guessing at the answers. Referring to the general equation of the BPD, we need only substitute the appropriate values to determine the probability of a student getting any particular score, assuming his responses are random. For this example, $n = 10$, p and q are both $1/2$, and r will range from 0 to 10. The probability associated with each score from 0 to 10 is presented in Table 17.1.

Table 17.1 Distribution of chance scores to ten T-F questions

r	$\dbinom{n}{r}$	p^r	q^{n-r}	Probability
10	1	$(1/2)^{10}$	$(1/2)^0$	$1/1024 = .001$
9	10	$(1/2)^9$	$(1/2)^1$	$10/1024 = .010$
8	45	$(1/2)^8$	$(1/2)^2$	$45/1024 = .044$
7	120	$(1/2)^7$	$(1/2)^3$	$120/1024 = .117$
6	210	$(1/2)^6$	$(1/2)^4$	$210/1024 = .205$
5	252	$(1/2)^5$	$(1/2)^5$	$252/1024 = .246$
4	210	$(1/2)^4$	$(1/2)^6$	$210/1024 = .205$
3	120	$(1/2)^3$	$(1/2)^7$	$120/1024 = .117$
2	45	$(1/2)^2$	$(1/2)^8$	$45/1024 = .044$
1	10	$(1/2)^1$	$(1/2)^9$	$10/1024 = .010$
0	1	$(1/2)^0$	$(1/2)^{10}$	$1/1024 = .001$

Sum = 1.000

Let us now suppose that the teacher has decided that a passing grade on this examination is achieved by every student who correctly answers seven or more questions. What is the probability of a student guessing seven or more correct answers? The addition rule of probability applies, and we need only add the probabilities associated with scores of seven or more correct answers to determine that the answer is .172. In other words, the student has about one chance in six of getting seven or more correct answers if his answers are completely random.

Now let us revise the test, using multiple-choice questions with four responses to each question. The theoretical distribution is given in Table 17.2.

Table 17.2   Distribution of chance scores to ten M-C questions

r	$\binom{n}{r}$	p^r	q^{n-r}	Probability
10	1	$(1/4)^{10}$	$(3/4)^0$	$1/1048576 = .000$
9	10	$(1/4)^9$	$(3/4)^1$	$30/1048576 = .000$
8	45	$(1/4)^8$	$(3/4)^2$	$405/1048576 = .000$
7	120	$(1/4)^7$	$(3/4)^3$	$3240/1048576 = .003$
6	210	$(1/4)^6$	$(3/4)^4$	$17010/1048576 = .016$
5	252	$(1/4)^5$	$(3/4)^5$	$61236/1048576 = .058$
4	210	$(1/4)^4$	$(3/4)^6$	$153090/1048576 = .146$
3	120	$(1/4)^3$	$(3/4)^7$	$262440/1048576 = .250$
2	45	$(1/4)^2$	$(3/4)^8$	$295245/1048576 = .281$
1	10	$(1/4)^1$	$(3/4)^9$	$196830/1048576 = .188$
0	1	$(1/4)^0$	$(3/4)^{10}$	$59049/1048576 = .056$

Sum $= .998$

Notice that the probability of seven or more correct answers is now .003. In other words, the student has about one chance in 300 of getting a passing grade if his responses are completely random. Of course, the chance factor in the examination could be reduced even more by using five-response multiple-choice questions or by using more questions.

It was indicated earlier that the BPD may also be used for testing certain kinds of research hypotheses. We shall illustrate the idea with an experiment in extrasensory perception (ESP). Suppose that a certain individual claims to have special powers of this sort, and a psychologist agrees to conduct a brief experiment to test these powers. The supposedly gifted person is blindfolded and placed in a chair with his back to a table. Three objects are placed on the table: a book, a bouquet of flowers, and a reading lamp. The psychologist points to one of these objects and asks the subject to tell which of the three it is. If the subject gave five correct answers in six attempts, how would you assess the outcome of the experiment? Consider the BPD of Table 17.3.

Table 17.3 Chance distribution for ESP experiment

r	$\binom{n}{r}$	p^r	q^{n-r}	Probability
6	1	$(1/3)^6$	$(2/3)^0$	$1/729 = .001$
5	6	$(1/3)^5$	$(2/3)^1$	$12/729 = .016$
4	15	$(1/3)^4$	$(2/3)^2$	$60/729 = .072$
3	20	$(1/3)^3$	$(2/3)^3$	$160/729 = .219$
2	15	$(1/3)^2$	$(2/3)^4$	$240/729 = .329$
1	6	$(1/3)^1$	$(2/3)^5$	$192/729 = .263$
0	1	$(1/3)^0$	$(2/3)^6$	$64/729 = .088$
				Sum $= .998$

The probability that the subject would get five or more correct responses is .017, or about 1 chance in 60, unless he has something more than chance going for him. Of course, much more satisfactory research could have been conducted simply by increasing the number of trials (the sample size) from 6 to 20, for example.

The student's attention is directed to the fact that the use of the binomial will not always require the calculation of the entire distribution as given in the three illustrations. There is no reason, for example, why the analysis of the ESP experiment could not have been completed by calculating only the probabilities for $r = 6$ and $r = 5$.

17.5 THE NORMAL APPROXIMATION OF THE BINOMIAL

The reader's attention is directed to Figures 10.1 and 10.2, which include both the histogram and frequency polygon for the chance distribution of responses to two true-false examinations. Notice the close resemblance of the frequency polygons, especially that for the ten-item examination, to the normal curve. It can be demonstrated mathematically and empirically (that is, by establishing an empirical sampling distribution) that the binomial distribution approaches the normal distribution if the sample size (in this case, the number of test items) is made very large. This suggests the use of the normal distribution as an approximation of the binomial distribution. The probabilities associated with normal distributions are conveniently tabled, and the binomial becomes extremely awkard to work with as sample size increases.

The accuracy of the normal approximation of the binomial is dependent upon the sample size and the value of p. If p is equal to or close to .50, the approximation is quite good for samples as small as 10 in size. However, as p departs from .50, larger samples are required to achieve a good approximation. Research has established that an approximation acceptable for most purposes is achieved when the products np and nq are both equal to or greater than 5. Thus, if p equals .50, n should be 10 or larger; if p or q is as small as .20, then n should be at least 25.

The mean of any binomial distribution is equal to the product np. This is immediately apparent in a symmetrical distribution (where $p = .50$), but is easily demonstrated to be true for other binomial distributions as well. With a little more effort it can be demonstrated that the variance of any binomial distribution is equal to the product npq. Thus, a z-score may be calculated, substituting r for the score, np for the mean, and npq for the variance:

$$z = \frac{r - np}{\sqrt{npq}}$$

If two such z-scores are calculated and entered in a table of areas under the normal curve, the area between the two is the probability associated with all of the scores in that score range. It is imperative that the real limits of scores be used when treating areas as probabilities. For example, to determine the probability of a score of 5, we would determine the area between two z-scores, one corresponding to a score of 4.5 and the other to 5.5. To determine the probability of a score of 5 or larger, we would determine the area to the right of 4.5, and to determine the probability of a score of 5 or smaller we would determine the area to the left of 5.5. The use of real limits in this fashion is essential whenever a continuously distributed model is used to approximate a discrete probability distribution. This is sometimes referred to as a *continuity correction*, or a *Yates correction*, from the man who originated it.

The use of the normal approximation of the binomial will be illustrated with the chance distribution of ten true-false items — a distribution for which the exact probabilities have been calculated and reported in Table 17.1. Suppose the investigator wishes to determine the probability of a score of 5. Then:

$$z = \frac{4.5 - 5}{1.58} = -.316$$

$$z = \frac{5.5 - 5}{1.58} = +.316$$

Rounding the z-scores to two decimal places, and entering these in Table 3 of the Appendix, we find that the area between the two scores is determined to be .251 (that is, .1255 plus .1255). This is a reasonable approximation of the .246 recorded in Table 17.1. As a matter of fact, the error in the approximation is mostly due to rounding of the z-scores.

Now let us use the normal approximation to determine the probability of a score of 7 or more. Only one z-score will be needed in this case, since we are interested in all of the area to the right of the score.

$$z = \frac{6.5 - 5}{1.58} = .95$$

Again referring to Table 3 in the Appendix, we find the area to the right of the score is .171 (from .5000 minus .3289), which is an excellent approximation of the exact probability calculated earlier.

EXERCISES

17.1 Given a true-false examination with six questions:
    (a) What is the probability of a student guessing all six correct answers if his responses are completely random?
    (b) What is the probability of a student giving random responses and scoring none right?
    (c) What is the probability of a student giving random responses and scoring exactly five right?
    (d) What is the probability of a student giving random responses and scoring five or more?

17.2 Repeat Exercise 17.1 for a multiple-choice examination having three responses for each question.

17.3 Refer to the ESP experiment reported in Section 17.4. Suppose the number of trials was increased to ten; what then is the probability of the subject getting eight or more right by chance?

17.4 Refer again to the ESP experiment. Suppose the number of trials was held at six, but the number of objects on the table was increased to four; what then is the probability of the subject getting five or more right by chance?

17.5 Given a 100-item true-false examination:
    (a) What is the probability of a student scoring 70 or higher if all of his responses are completely random?
    (b) What is the probability of a student scoring 60 or higher if all of his responses are completely random?
    (c) What is the probability of a student scoring 50 or higher if all of his responses are completely random?

17.6 Given a 50-item multiple-choice examination in which each question has four responses:
    (a) What is the probability of a student scoring 35 or higher if all of his responses are completely random?

(b)   What is the probability of a student scoring 30 or higher if all of his responses are completely random?

(c)   What is the probability of a student scoring 25 or higher if all of his responses are completely random?

17.7   Which examination has the least chance score factor, the 100-item true-false examination or the 50-item multiple-choice examination?

18

BASIC SAMPLING THEORY

18.1 RANDOM SAMPLING

Typically, scientific research involves intensive study of small groups (called *samples*) in order to draw conclusions about much larger groups (called *populations*). As a general rule, this is a necessary economy in any kind of empirical research, and its validity rests upon the assumption that the samples are genuinely representative of the populations. It is important for the student to understand that the researcher works with samples, but his primary concern is with conclusions that apply to populations.

A population is a collection of objects, events, or individuals having some common characteristic that the researcher is interested in studying. A sample is a smaller group of objects, events, or individuals selected from the population for actual participation in the research. For example, a behavioral scientist might be interested in analyzing the study habits of freshmen students at a large university. It is unlikely that he can collect meaningful data on all of the individuals in the population. He must select a sample to represent the total population and analyze the study habits of this sample.

Statistical inference consists of an accumulation of techniques for drawing inferences or generalizations from samples to populations. Such inferences from samples to populations will always be subject to some error, since each sample from a given population might be expected to have somewhat different characteristics and yield somewhat different results in the experimental setting. Statistical inference uses the mathematics of probability to provide means for estimating this error, that is, for stating quantitatively how much confidence can be placed in the outcome of the experiment.

Invariably, statistical inference will require that the researcher use random samples. A simple random sample is a sample chosen from a population in such a fashion as to insure that every object, event, or individual in the population has equal chance of being drawn for the sample. There are several methods of accomplishing this end, all of which require that a listing be made of every object, event, or individual in the population. For a relatively small population, it is conceivable that the researcher might achieve a random sample by listing all of the objects in the population on separate slips of paper, placing these in a container, then forming the sample by drawing slips from the container. However, most research will be facilitated by using more sophisticated techniques for drawing random samples.

One of the more popular procedures for compiling random samples calls for the use of a table of numbers generated by a random process. Table 1 in the Appendix is such a table. This particular table is unusually compact for a table of random numbers, but it is quite adequate for drawing random samples as large as several hundred from populations consisting of as many as several thousand objects. Its use is outlined as follows:

1. All objects in the population are numbered from 1 to N.
2. The investigator selects a row number and a column number in the table as a point from which to start drawing his sample. The method of selecting row and column numbers should approximate a random process; the particular method to be used is left to the imagination of the reader.
3. Beginning at the point selected above, and using as many columns as are necessary to include numbers as large as N, the investigator systematically draws a list of numbers from the table. He may move to the left or right, upward or downward from his starting point.
4. The list of numbers drawn from the table is then referred to the population listing. Objects in the population whose numbers correspond to the numbers drawn from the table are designated for the sample.

Suppose the population consists of 500 subjects, and the investigator wishes to draw a random sample of size 30. Suppose also that Column 14, Row 9 is selected as the starting point. In order to include numbers as large as 500 in the sample, the investigator will read columns 14 to 16. Thus the first number from the table is 918, but this is too large, so it is cast aside. Proceeding downward from the starting point, the investigator would draw the following sample:

354	297	096	256	293	109	044	411	425	307
175	206	478	482	416	399	213	059	496	485
055	147	446	229	476	215	113	406	325	172

18.2 PARENT AND TARGET POPULATIONS — INFERENCES AND GENERALIZATIONS

Few investigators are in a position where they can draw random samples from the populations which they wish to study and about which they wish to draw conclusions from their research. As a general rule, the population of interest will be much too large and much too inaccessible to draw random samples in the manner described in Section 18.1. The investigator may wish to draw conclusions which are valid for all adolescent females in the world, or for all fourth grade students of arithmetic in the United States, for example. We will use the term *parent population* to refer to that population from which the investigator draws one or more random samples to participate in his research. The parent population will also be referred to as the experimentally accessible population or as the population under study. We will use the term *target population* to refer to that larger group to which the investigator wishes to generalize his findings.

The techniques of statistical inference enable the investigator to draw inferences from samples to parent populations and to do so with mathematically defined precision. It is important to note that the investigator is privileged to draw statistical inferences only to those populations from which he has drawn random samples. Often, he will be able to draw valid generalizations to other individuals — that is, to the target population.

Often in research, an available group of subjects is randomly divided into samples. For example, an available group of 40 subjects might be randomly divided into an experimental group and a control group, each with 20 subjects. The use of inferential statistics with such samples is completely valid. However, the parent population to which statistical inferences may be drawn is that available group of 40 subjects with which the investigator started.

As a general rule, the investigator will not be content to restrict the use of the findings from his research to the parent population. He has a responsibility to get as much mileage out of his research as he can, as long as this does not lead him into serious error. The most valuable behavioral research is invariably that whose findings can be generalized to the greatest number of people. This sort of generalization is not at all uncommon in scientific research. Consider, for example, the important *ex post facto* research, which has led scientists to conclude that there is a relationship between cigarette smoking and lung cancer in the United States. We would anticipate that the findings from this research would generalize to other populations throughout the world. We would expect these various populations to differ on numerous variables from the population that was studied in the original research. Further, some of these variables might have some effect on the dependent variable (the incidence of lung cancer); that is, people in certain parts of the world might be more or less inclined to get lung cancer than people in the United States. However, as long as these other variables do not interact with the independent variable (the use of cigarettes), the findings with respect to the relationship between cigarette smoking and lung cancer may be validly generalized to the other populations.

Let us consider another example. An available group of 40 subjects is randomly divided into two samples of 20 subjects each (an experimental and a control group) to participate in learning research. The investigator finds that the experimental group scores significantly higher on the dependent variable than the control group. The average IQ of the parent population is 114, and we would anticipate that the scores on the dependent variable used in almost any learning research would be influenced by IQ. Can the findings from the research be validly generalized to a population whose average IQ is 97? If there is interaction between IQ and the two methods of learning — that is, if the method used with the experimental group is more or less effective with students of high or low IQ — the results are not generalizable to populations of different average IQ. However, if the differential effectiveness of the two methods of learning is the same for students of high or low IQ, the results of the research are indeed generalizable to populations of different average IQ than the parent population. We would also expect, for example, that learning research with fourth grade students in Colorado would readily generalize to fourth grade students in Wisconsin. There is always danger of error in extending the findings of research in this fashion, just as there is always danger involved in driving an automobile down a crowded highway. The researcher must learn, like the driver, to be on the lookout for road hazards. The problem is one of being both prudent and productive.

A special problem exists for the classroom researcher. He seldom has any authority over the manner in which students are assigned to the various classes, but he often feels that the procedures used to assign students closely approximate random procedures. This is probably an unwarranted assumption, but he feels the need to be productive more than he feels the need to be prudent, so he conducts this research in which he has a target population but no true parent population. He must recognize that he makes use of the techniques of statistical inference at considerable risk of violating the assumption of randomness, and it becomes his responsibility to defend the validity of his samples with whatever logic he can muster. Sometimes the only solution is to suggest the existence of a hypothetical population from which these subjects might be regarded as a random sample.

18.3   THE NOTION OF A SAMPLING DISTRIBUTION

Sampling distributions are frequency distributions constructed by statisticians so that relationships between sample statistics and their corresponding population parameters may be studied. Let us suppose that a statistician is interested in studying the sampling distribution of the mean for samples of size nine drawn from a distribution of percentile ranks. He might proceed in this fashion.

1. He establishes a well-defined population with characteristics suited to his purposes. Although he may use an existing population, more often he will build a population to his own specifications. For example, he might build a population of percentile ranks simply by writing the numbers 1 through 99 on small sheets of paper (one number to each sheet of paper, total of 99 sheets).
2. A very large number of random samples of a given size (in the example, $N = 9$) is drawn from the parent population. In the example, the small sheets of paper could be placed in a cookie jar, stirred thoroughly, and the samples drawn with replacement. The mean for each sample is calculated and recorded. While each sample mean is an estimate of the population mean, we would expect these sample means to vary from one sample to another.
3. A frequency distribution is constructed from the sample means (not the raw scores). Such a frequency distribution is called an *empirical frequency distribution*, because it is constructed from empirically derived data.

With a little effort, the student should be able to demonstrate that the mean (μ) and standard deviation (σ) of the parent population of percentile ranks in the example are 50.0 and 28.6, respectively. The frequency distribution of the sample means will have some very interesting characteristics. (1) The shape of the distribution will be very nearly normal, rather than uniform like the scores from which the means were calculated. (2) The mean of the sampling distribution will be 50.0, the same as the parent population. (3) The standard deviation will be 9.53 or very close to it, which is the standard deviation of the parent population divided by the

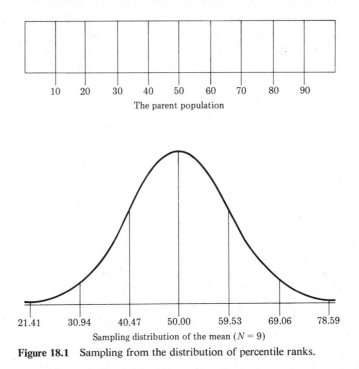

Figure 18.1   Sampling from the distribution of percentile ranks.

square root of the sample size. The situation is illustrated in Figure 18.1. The implications of these properties are far-reaching, and they will be discussed in some detail in the following sections.

Empirical sampling distributions are more likely to be constructed with a computer than with slips of paper in a cookie jar. A computer can be instructed to build a parent population having almost any desired characteristics which the statistician desires. The computer can draw thousands of random samples in a few seconds and calculate the sample statistics as it goes along. Many sampling distributions are derived mathematically, without going through the process of actually drawing the random samples. These are called *theoretical sampling distributions* or simply *sampling distributions*. Some sampling distributions are relatively easy to derive mathematically. For example, Tables 17.1, 17.2, and 17.3 in the chapter on the binomial probability distribution are theoretical sampling distributions. As a general rule, theoretical sampling distributions for continuous variables require a great deal more mathematical sophistication to construct. Many of these are conveniently tabled in the Appendix. Table 3, *Areas under the Normal Curve*, is an example. Sometimes statisticians are confronted with sampling distribution problems which do not readily lend themselves to mathematical solution. Often, the solution lies in drawing thousands of random samples on a computer and building the sampling distribution empirically. Such procedures are called *Monte Carlo* procedures. Monte Carlo studies have been effectively used to study the characteristics of statistical analyses when the mathematical assumptions underlying these analyses have been violated.

18.4   SAMPLING DISTRIBUTIONS AND PROBABILITY

When the theoretical sampling distribution of a statistic is known, the investigator may readily determine the relative frequency with which different sample values of this statistic are expected to occur, provided that these samples are randomly selected from a parent population for which the value of the parameter is specified. Of course, these relative frequencies may be directly equated with the probability of occurrence of different sample values of this statistic. Thus, by drawing a random sample and calculating the value of the statistic, it may be compared to the theoretical sampling distribution, and the probability that it was drawn from a population with a specified value of the parameter may be determined. Similarly, one may draw two or more random samples and determine the probability that they were drawn from the same parent population. This is the essence of statistical hypothesis testing.

The *expected value* of any statistic — the predicted value which would give the least error (see Sections 14.1 and 14.3) for many samples — is the mean of the sampling distribution. The *standard error* of any statistic is the standard deviation of its sampling distribution. The expected value and standard error of theoretical frequency distributions are parameters and will be represented by Greek letters.

Given a collection of normally distributed scores with mean μ and standard deviation σ, the student has learned to transform these into z-scores, which may be referred to Table 3 of the Appendix to determine what proportion of the scores in the collection are above or below any particular score. The equation for this standard score transformation was given as:

$$z = \frac{X - \mu}{\sigma}$$

The student is reminded that the proportion of the area under the normal curve which is to the left of a particular score is the relative frequency of scores in the collection which are lower than this score. Similarly, the proportion of the area which is to the right of a particular score is the relative frequency of scores in the collection which are higher than this score. The student is also reminded that we have defined probability in terms of relative frequency. Let us suppose that the score in the equation above is randomly selected from a population of normally distributed scores with mean μ and standard deviation σ. The probability that the obtained z-score will be equal to or smaller than -1.96 is .025; the probability that the obtained z-score will be equal to or larger than $+1.96$ is also .025; and the probability that the obtained z-score will be equal to or smaller than -1.96 or equal to or larger than $+1.96$ is .05. The student should confirm these by referring to Table 3.

It has been established that the mean of a binomial distribution is np, and the standard deviation is $\sqrt{npq}$. Thus, in the normal approximation of the binomial, a z-score was calculated from:

$$z = \frac{r - np}{\sqrt{npq}}$$

Under the appropriate circumstances (outlined in Section 17.5) a z-score calculated in this manner may be referred to Table 3, and the probability of a particular value of r (or of a higher value or of a lower value) may be determined.

The important idea here is that whenever the normal probability distribution is the appropriate model (and it is for all of the sampling distributions considered in this chapter), and the mean and standard deviation of the sampling distribution are known, a z-score may be calculated from:

$$z = \frac{\text{random variable} - \text{expected value}}{\text{standard error}}$$

These relationships form the basis for the techniques of statistical inference that are described in the following chapters.

18.5   THE SAMPLING DISTRIBUTION OF THE MEAN

The sampling distribution of the mean is of particular interest to the behavioral scientist, since this statistic is typical of most of the samples used in behavioral research and is most often used to represent these samples in statistical analysis.

It can be shown that the expected value of the sample mean is equal to the population mean regardless of the form of distribution of the population and the size of sample drawn. In other words, the sample mean is an unbiased estimate of the population mean and will be used to estimate the population mean when the value of the parameter is unknown.

If the scores of the individuals in the parent population are normally distributed, the sampling distribution of the mean is also normal. Given any population (with any distribution, normal or otherwise) with mean μ and finite variance σ^2, as the sample size increases without limit, the sampling distribution of the mean approaches a normal distribution with mean μ and variance σ^2/N. This is known as the *central limit theorem*, and its importance lies in the fact that it enables the researcher to use the familiar normal probability distribution for testing hypotheses about populations having any form of distribution. The size of sample needed to take advantage of the central limit theorem depends upon the extent to which the population under study deviates from normality. Generally, the sampling distribution of the mean closely approximates the normal for samples of 100 or larger, no matter how radically the parent population deviates from the normal. For most behavioral research, samples of size 30 will be adequate to insure close approximation of the normal by the sampling distribution of the mean. Of course, if the parent population is normal, the distribution of the mean is normal for all sample sizes.

When the variance of the sampling distribution of the mean is to be estimated from sample data, which is usually the case, it will be calculated in the following manner:

$$S_M^2 = \frac{S^2}{N} = \frac{SS}{N(N-1)}$$

The standard error of the sampling distribution of the mean is the square root of that variance:

$$S_M = \sqrt{S_M^2} = \sqrt{\frac{S^2}{N}} = \sqrt{\frac{SS}{N(N-1)}}$$

The reader's attention is directed to the fact that, for a given standard deviation of the scores, the standard error of the mean is a function of sample size. For example, if the standard deviation of the scores is 12 and the sample size 4, the standard error of the mean is 6. Increasing the sample size to 9, with the same standard deviation of the scores, yields a standard error of the mean equal to 4. The student should complete these calculations for himself to verify the relationship. This is an important idea in sampling statistics — that larger samples yield smaller sampling errors.

The utility of this information is best demonstrated with an example. A popular IQ test has been standardized on a population having a mean of 100 and standard deviation of 15. What is the probability that a random sample of size 25 drawn from this population would have a sample mean of 109 or higher? First, we must calculate the standard error of the mean:

$$S_M = \frac{\sigma}{\sqrt{N}} = \frac{15}{5} = 3$$

Then, a z-score is calculated:

$$z = \frac{M - \mu}{S_M} = \frac{109 - 100}{3} = 3$$

The probability associated with a z-score of 3 or higher is determined from Table 3 to be .0013, a highly improbable event.

18.6   THE SAMPLING DISTRIBUTION OF THE DIFFERENCE BETWEEN THE MEANS OF TWO INDEPENDENT SAMPLES

The behavioral scientist is much more likely to be interested in comparing statistics from two independent experimental samples than comparing a single statistic to a hypothesized parameter as described in the previous section. The usual hypothesis is that there is no difference between the two population means; this is a very common *null* hypothesis.

The theoretical sampling distribution of the difference between the means of two independent random samples from normally distributed populations is likewise normal. The mean of this distribution (the expected value) is equal to the difference between the two population means — if the two parameters are equal, the mean of the distribution is zero. The variance of this distribution is equal to the sum of the variances of the two populations:

$$\sigma^2_{M_1 - M_2} = \sigma^2_{M_1} + \sigma^2_{M_2} = \frac{\sigma^2_1}{n_1} + \frac{\sigma^2_2}{n_2}$$

If we assume that the two population variances are equal and substitute σ^2 for σ^2_1 and σ^2_2, the equation becomes:

$$\sigma^2_{M_1 - M_2} = \sigma^2 \left(\frac{1}{n_1} + \frac{1}{n_2} \right)$$

When the variance of the sampling distribution of the difference between the means is to be estimated from sample data (which is usually the case), it will be calculated by:

$$S^2_{M_1 - M_2} = \frac{SS_1 + SS_2}{n_1 + n_2 - 2} \left(\frac{1}{n_1} + \frac{1}{n_2} \right)$$

This equation makes use of information from both samples to estimate the common population variance. The procedure is known as *pooling the variances*.

The standard error of the sampling distribution is, of course, the square root of the variance:

$$S_{M_1 - M_2} = \sqrt{\frac{SS_1 + SS_2}{n_1 + n_2 - 2} \left(\frac{1}{n_1} + \frac{1}{n_2}\right)}$$

18.7 THE SAMPLING DISTRIBUTION OF THE DIFFERENCE BETWEEN THE MEANS OF TWO RELATED SAMPLES

When the behavioral scientist wishes to compare two means from different sets of measures on one group of subjects or from measures on two groups made up of matched pairs of subjects, he must take into account that the two sets of measures are correlated.

If the original measures are normally distributed, the difference between the two sets of measures (let $D_i = X_1 - X_2$ for any pair of measures) and the sampling distribution of the difference between the means ($\bar{D} = M_1 - M_2$) are also normal. The mean of this distribution (the expected value) is equal to the difference between the two population means; if the two parameters are equal, the mean of the distribution is zero. The variance of this distribution is:

$$\sigma_{\bar{D}}^2 = \sigma_{M_1}^2 + \sigma_{M_2}^2 - 2r_{12}\sigma_{M_1}\sigma_{M_2}$$

When this variance must be estimated from sample data, the equation becomes:

$$S_{\bar{D}}^2 = S_{M_1}^2 + S_{M_2}^2 - 2r_{12}S_{M_1}S_{M_2}$$

As a general rule, some computational effort will be saved by approaching the problem in this fashion:

$$S_{\bar{D}}^2 = \frac{SS_D}{N(N-1)} \qquad \text{where } SS_D = \Sigma d^2 = \Sigma D^2 - \frac{(\Sigma D)^2}{N}$$

As before, the standard error of the sampling distribution is the square root of the variance:

$$S_{\bar{D}} = \sqrt{\frac{SS_D}{N(N-1)}}$$

Because the correlation of two related samples will usually be positive, the standard error for two related samples will usually be somewhat smaller than that for two independent samples.

18.8 CONFIDENCE INTERVALS FOR
NORMALLY DISTRIBUTED STATISTICS

Up to this point we have been concerned with statistics which yield a single-valued estimate of the corresponding parameter. Such statistics are called *point estimates*. Whenever a point estimate is reported, it is desirable to report the standard error of the statistic so that the reader has some conception of the possible error involved in using this estimate. For example, if the sample mean is reported as 44, and the standard error of the mean is 4, the reader should be aware that there is approximately a 68 percent probability that the parameter lies within the limits 40 to 48, and approximately a 95 percent probability that the parameter lies within the limits 36 to 52. Other investigators prefer to report *interval estimates* rather than point estimates. An interval estimate includes a range of values with an upper and lower limit such that the value of the parameter is believed to lie somewhere within the interval.

A *confidence interval* is an interval for which it can be asserted with a given probability that the interval will contain the parameter that it is intended to estimate. The probability that the parameter lies within the interval is called the *confidence coefficient* or the *degree of confidence*. It is represented by the lowercase Greek letter gamma (γ) and is often reported as a percentage (100γ percent).

The reader's attention is directed to Figure 18.2, which shows the sampling distribution of the mean. Notice that 95.46 percent of the total area

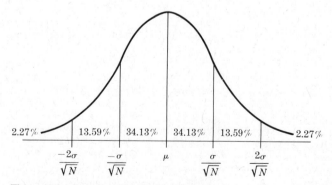

Figure 18.2 Sampling distribution of the mean.

under the curve lies within plus or minus two standard errors of the mean, and that 4.54 percent of the total lies in the extreme areas beyond this interval. Thus, if the point estimate of the mean were 30 and the standard error of the mean 4, one could be 95.46 percent confident that the population mean lies within the range 22 to 38. The more common practice is to establish the 90 percent, 95 percent, or 99 percent confidence interval.

By referring to a table of areas under the normal curve, it may be readily determined that the 95 percent confidence interval (which extends .4750 of the area on either side of the mean) corresponds to plus or minus 1.96 standard errors. Formulas for determining the limits of the confidence interval of the mean are simple modifications of the z-score formula:

$$\text{upper limit} = M + z_c S_M$$
$$\text{lower limit} = M - z_c S_M$$

where z_c = 1.645 for the 90 percent confidence interval
1.960 for the 95 percent confidence interval
2.576 for the 99 percent confidence interval

Now, let us illustrate the use of these formulas with data from a sample of size 25 which yields unbiased estimates of the mean and standard deviation, which are 100 and 15, respectively. The standard error of the mean, then, is 3 (15 divided by the square root of 25). Substituting into the formula, we find that the 90 percent confidence interval is 95.1–104.9; the 95 percent confidence interval is 94.1–105.9; and the 99 percent confidence interval is 92.3–107.7. It is apparent that the size of the interval is a function of the level of confidence required, higher levels of confidence resulting in wider intervals. Of course, a large interval does not provide a very precise estimate. The size of the interval is also a function of sample size. Suppose the sample under discussion were increased to 225, and that the estimates of the mean and standard deviation were the same. The standard error of the mean becomes 1, the 90 percent confidence interval is reduced to 98.4–101.6, the 95 percent confidence interval to 98.0–102.0, and the 99 percent confidence interval to 97.4–102.6. These examples illustrate the very important idea that more precise estimates of population parameters are achieved with larger samples. The use of the normal curve as a model for interval estimation is not recommended for samples smaller than 30 measures. The t-statistic will be substituted for the normal in a later chapter and will permit working with samples of any size.

Confidence intervals may be constructed in this fashion for any statistic that is known to be normally distributed and for which the standard error is known.

18.9   CONFIDENCE INTERVALS FOR THE PEARSON CORRELATION COEFFICIENT

The sampling distribution of the Pearson correlation coefficient (r) is not normal. The form of the sampling distribution is a function of the magnitude and the sign of the coefficient in the population under study. As the magnitude of the coefficient increases, the sampling distribution becomes increasingly skewed — to the left for positive correlations and to the right for negative correlations. The great statistician, Sir Ronald Fisher, demonstrated that a particular mathematical transformation of the correlation coefficient would yield a normally distributed variable with mean of zero:

$$Z_r = \frac{1}{2} \log_e \frac{1 + r}{1 - r}$$

It will not be necessary to complete these transformations algebraically, for Table 11 in the Appendix reports values of Z_r for specified values of r. The standard deviation of Z_r is given by:

$$S_{zr} = \frac{1}{\sqrt{N - 3}}$$

where N is the number of pairs of scores. The limits of the confidence interval for the Pearson correlation coefficient may then be determined from:

$$\text{upper limit} = Z_r + z_c S_{zr}$$
$$\text{lower limit} = Z_r - z_c S_{zr}$$

where z_c is determined as in Section 18.8, and the results are in the scale of the normally distributed variable Z_r. The two values of Z_r thus achieved may be referred back to Table 11 and converted to the corresponding values of r.

For example, with $N = 103$ and $r = 0.600$, we determine that $Z_r = 0.693$ from the table, and that $S_{zr} = 0.100$ from the formula given above. For the 95 percent confidence interval:

$$\text{upper limit} = 0.693 + 1.96(0.100) = 0.889$$
$$\text{lower limit} = 0.693 - 1.96(0.100) = 0.497$$

But these answers are in terms of the normally distributed variable Z_r.

Referring these to the table, the results are now expressed in terms of r:

$$\text{upper limit} = 0.71$$
$$\text{lower limit} = 0.46$$

The student's attention is directed to the fact that although the limits of Z_r are symmetrical about the sample value, the limits of r are not symmetrical about the sample value.

EXERCISES

18.1 A population is known to have mean of 100 and standard deviation of 15.

(a) What is the probability that a random sample of size 25 will have a mean of 106 or higher?

(b) What is the probability that a random sample of size 25 will have a mean of 91 or higher?

(c) What is the probability that a random sample of size 100 will have a mean of 107.5 or higher?

(d) What is the probability that a random sample of size 100 will have a mean of 106 or higher?

(e) What is the probability that a random sample of size 100 will have a mean of 92.5 or higher?

18.2 Given a random sample of 64 subjects with a sample mean of 50 and unbiased estimate of the population variance equal to 100.

(a) Calculate the estimate of the standard error of the mean.

(b) Determine the limits of a 90 percent confidence interval for the mean.

18.3 Given a random sample of 30 subjects with sample mean of 22 and sum of squares equal to 261.

(a) Calculate the estimate of the population standard deviation.

(b) Calculate the estimate of the standard error of the mean.

(c) Determine the limits of a 95 percent confidence interval for the mean.

18.4 Given a random sample of 84 pairs of scores with sample correlation co-efficient of 0.245. Determine the 95 percent confidence interval for the correlation coefficient.

19

STATISTICAL HYPOTHESES

19.1 NULL HYPOTHESES

In order to expand the frontiers of man's knowledge, the scientist must be able to extend with confidence the findings generated by his research far beyond the restricted setting in which most research takes place. The problem is essentially one of drawing conclusions about some specific population from experience which is limited to a relatively small sample chosen to represent that population. Arriving at conclusions of this sort is a necessary hazard with which modern men of·science have learned to live, and it has neither dampened their enthusiasm nor slowed their progress in the expansion of knowledge. With the development of the techniques of statistical inference in the twentieth century, the generalization of research findings from sample to population may be regarded as a calculated risk, with the degree of risk defined in terms of the mathematics of probability.

The most fruitful approach to the testing of statistical hypotheses has been that in which the possible outcomes of an experiment are expressed as a *null hypothesis* and its alternatives. The null hypothesis is a statement to the effect that there is no difference between specified population parameters. Ordinarily, it is a statement to the effect that there is no relationship between the independent and dependent variables in the *population* under study. Of course, this is not usually the anticipated outcome of the experiment; typically, the investigator conducts the experiment because he has reason to believe that manipulation of the independent variable will influence the dependent variable. In other words, it is *rejection* of the null hypothesis which is interpreted as a significant finding. It is important to note that the null hypothesis is a statement about population *parameters;* the hypothesis will be rejected or retained on the basis of sample statistics and a knowledge of the sampling distribution of these statistics.

The logic of hypothesis testing is briefly summarized as follows:

1. A null hypothesis is stated in terms of some parameter of the population under study. The theoretical sampling distribution of that parameter is obtained, usually from the appendix of a statistics textbook.
2. A random sample is drawn, and the sample statistic corresponding to the population parameter is calculated.
3. The sample statistic is referred to the theoretical sampling distribution. If the probability that this statistic was drawn from such a population (that is, one in which the null hypothesis is true) is very small, the null hypothesis is rejected. Just what constitutes a very small probability is determined before the investigator collects his data. The determination is made on the basis of his research experience, the consequences of an erroneous conclusion, and the kind of risk he is prepared to take. This probability is called the *level of significance;* it is represented by the lowercase Greek letter alpha (α), and it is the probability of rejecting a null hypothesis which is in fact true. In other words, it is the probability that the investigator will arrive at a certain kind of erroneous conclusion — the conclusion that a relationship exists between the independent and dependent variables when in fact no such relationship exists in the population under study.

Suppose, for example, that the investigator is interested in studying the relationship between IQ score and academic achievement in American history in the high schools of a very large school district. The scores on both variables for over a million students have been recorded in the central office of the school district. A random sample of 30 pairs of scores is drawn from these central files, and the correlation between the two measures is determined to be .50. The null hypothesis in this case is that the correlation between the two measures is zero in the population under study. To test this hypothesis, the investigator must determine whether a correlation of .50 is sufficiently larger than zero that it is unreasonable to conclude that this correlation is due to sampling error. The sampling distribution of the correlation coefficient is known, and whether this observed coefficient should be attributed to chance may be readily determined, provided that the investigator is prepared to stipulate how big a chance he is prepared to take. Suppose the investigator regards one chance in a hundred as a reasonable risk to take, and from a knowledge of the sampling distribution of the correlation coefficient, he determines that there is only one chance in a hundred that a correlation coefficient of .463

or larger will be produced by a sample of size 30 drawn randomly from a very large population in which the correlation is zero. Because the observed correlation coefficient is larger than .463, the null hypothesis is rejected, and the investigator concludes that there is a relationship between the two variables.

Up to this point, the author has used the term *significant* without specifying precisely what is meant. The term is a technical one that statisticians use in a very precise way. When the null hypothesis is rejected, the finding is said to be statistically significant at the specified level of significance. In the example quoted in the previous paragraph, the finding was significant at the .01 level. Rejection of the null hypothesis implies acceptance of some alternative hypothesis. In the example, the investigator concluded simply that there was a relationship between the parameters — that the correlation between the two parameters was larger than zero. No conclusion was arrived at with respect to the magnitude of the relationship between the parameters. However, the investigator would probably regard the sample statistic as an acceptable estimate of the corresponding population parameter. Of course, he would have greater confidence in the estimate had the statistic been calculated from a larger sample.

When the available evidence does not merit rejection of the null hypothesis, the finding is *nonsignificant*, and the null hypothesis is *retained*. The fact that the investigator did not gather evidence sufficient to reject the null hypothesis does not establish that it is true. At best, when the null hypothesis is retained, it is said to be *tenable*, and the investigator may find it convenient to regard it as true until proved otherwise. As a general rule, the retention of a statistical hypothesis is not a very conclusive finding, and it is very difficult to build a strong argument from a nonsignificant finding.

19.2   THE REGION OF REJECTION

The region of rejection is a proportion of the area in the theoretical sampling distribution which is equal to the level of significance. Generally, this region will be located in one or both tails of the sampling distribution. It represents those sample values that are highly improbable if the null hypothesis is true. The remainder of the area in the theoretical sampling distribution is sometimes referred to as the region of retention. Thus, if a hypothesis is to be tested at the .01 level, 1 percent of the area will be in the region of rejection, and 99 percent will be in the region of retention. Some authors refer to the region of retention as the region of acceptance. This is a poor choice of words, since retention of the null hypothesis should not be equated with acceptance of the null hypothesis.

Table 19.1 is a table of normal areas and z-scores which are presented in a form convenient for hypothesis testing. The information in this table is essentially the same as corresponding parts of Table 3 in the Appendix, except for the manner in which the information is presented. Notice that a distinction is drawn between a *one-tailed* hypothesis test and a *two-tailed* hypothesis test. In the one-tailed test, the area corresponding to the level

Table 19.1   The normal model for hypothesis testing

Level of significance for a one-tailed test	.10	.05	.025	.01	.005	.0005
Level of significance for a two-tailed test	.20	.10	.05	.02	.01	.001
z-score	1.282	1.645	1.960	2.326	2.576	3.291

of significance (the region of rejection) is located at one end of the sampling distribution. If the investigator locates this area in the left tail of the distribution, a z-score equal to or smaller than -1.645 is significant at the .05 level. If this area is located in the right tail, a z-score equal to or greater than $+1.645$ is significant at the .05 level. The student should refer to Table 3 in the Appendix to verify that these z-scores correspond to the 5th and 95th percentiles, respectively. In the two-tailed test, the region of rejection is divided into two equal parts, one at each end of the distribution. In this situation, a z-score equal to or smaller than -1.960 *or* a z-score equal to or larger than $+1.960$ is significant. Again, the student should refer to Table 3 to verify that these z-scores correspond to percentile ranks of 2.5 and 97.5, respectively.

In some research situations, the investigator will have a choice between a one-tailed or a two-tailed test, and his selection will be determined by the manner in which he views his hypothesis and its alternatives. In other situations, the nature of the hypothesis will dictate which type of test is to be used. Generally, when a one-tailed test is the logical choice, it is more likely to produce a significant finding than the two-tailed test. Three examples will be cited to demonstrate how the choice between a one- and two-tailed test is made:

1. When the significance of a correlation coefficient is tested, the null hypothesis states that the correlation between the two variables is zero in the population under study. There are two alternatives to this null hypothesis; the correlation may be either positive or negative. If the investigator is interested in both alternatives, a two-tailed test is appropriate. If, however, the investigator is prepared to specify that only a positive correlation, for example, will be considered significant, a one-tailed test is appropriate. In this situation, the entire region of rejection

will be located at the positive end of the sampling distribution, and any negative correlation coefficient will be regarded as nonsignificant.

2. Suppose the investigator is interested in comparing two different methods of instruction. The usual dependent variable will be an examination given at the end of the experiment, and the typical null hypothesis states that there is no difference in the means of the two populations. Again, there are two alternatives to the null hypothesis: Group 1 may have the higher mean, or Group 2 may have the higher mean. Generally, a two-tailed test will be used in this situation. However, if the investigator is prepared to specify in advance that only one of these two alternatives is of interest and will be regarded as significant, he may specify which alternative and choose a one-tailed test.

3. Consider the situation in which the investigator believes that by showing a certain motion picture to a group of subjects he can cause a positive shift in their attitudes toward members of other races. An attitude scale will be administered before and after the film showing, and the null hypothesis states that there is no difference in the mean scores of the population at the two administrations. Notice that the investigator has stated his interest in a positive shift in attitudes. Only a positive shift in the means will be regarded as significant (a negative shift will be regarded as nonsignificant), and the logical choice is a one-tailed test. Of course, if the investigator had been interested in the possibility of a shift in either direction, a two-tailed test would have been chosen.

When the investigator describes the region of rejection for a statistical test, he will need to indicate the level of significance and whether a one- or two-tailed test is used. The manner in which the region of rejection is described will also be influenced by the type of statistical analysis used. When the normal distribution is used as the theoretical model, the region of rejection will usually be described in terms of z-scores of a certain magnitude. A two-tailed region of rejection at the .02 level would be stated as follows:

$$R: z \leq -2.326, \quad z \geq +2.326$$

The reader's attention is directed to the ESP experiment reported in Section 17.4, and especially to Table 17.3 with the binomial model for this experiment. The null hypothesis for this experiment states that the number of correct responses given by the subject does not differ significantly from the number of correct responses expected when the responses are completely random. The one-tailed test is appropriate, since the investigator is interested only in rejecting the hypothesis if the subject gives an improbable number of correct answers; he is not interested in rejecting the hypothesis if the subject gives an improbable number of *in*correct answers. Suppose the level of significance were set at .05; the region of rejection may be stated in terms of the number of correct answers.

$$R: r \geq 5$$

Notice that the region of rejection in this case is actually about .017; because of the crudeness of the experiment, this is as close to the .05 level as the investigator can get.

19.3 OUTLINE PROCEDURE FOR TESTING STATISTICAL HYPOTHESES

A five-step procedure which is applicable to the testing of a wide variety of statistical hypotheses will be outlined here. Its use will be illustrated with a simple research problem using the normal distribution as a model for the sampling distribution of the means.

A sociologist is interested in studying the relationship between a certain kind of cultural deprivation and adult intelligence test scores. He has identified a population of adult residents of the United States which meets his criterion of cultural deprivation, and he hypothesizes that this population will score below the average on a standardized adult intelligence test. The independent variable is whether or not the individual is a member of the culturally deprived population, and the dependent variable is IQ score as measured by the Wechsler Adult Intelligence Scale (abbreviated WAIS). The WAIS is an individual test of intelligence (it is administered by a professional psychometrist to one individual at a time) which has been standardized on several thousand adults representative of the general populace of the United States. The instrument yields normally distributed scores with mean of 100 and standard deviation of 15 when used with typical adults similar to those in the norm group. The research will be conducted with 100 individuals selected at random from the population of interest.

Step 1 State the null hypothesis and its alternatives in terms of population parameters. It is probably good practice to make both a narrative and a mathematical statement of the null hypothesis. The alternatives may be stated in mathematical form.

In the example, the null hypothesis is that there is no difference in the mean for the culturally deprived population (μ_1) and the mean for the norm group (μ_0). The investigator hypothesizes that the sample will score significantly lower than the norm group, so a one-tailed alternative appears reasonable. The null hypothesis and its alternative may be stated mathematically as follows:

$$H: \mu_1 = \mu_0 \qquad A: \mu_1 < \mu_0$$

Step 2   Choose the appropriate statistic and specify the sample size.

The student is not yet acquainted with the large variety of statistical tests available to fit the many different problems encountered in behavioral research. For this particular example, the normal probability distribution will be used as a model for the sampling distribution of the mean. The sample size has been set at 100. In practical research, the size of the sample is largely determined by economic considerations and the availability of subjects. The fact that the WAIS is an individual test of intelligence is a limiting factor with respect to sample size in this experiment.

Step 3   Specify the level of significance and define the region of rejection.

Suppose, that for whatever reasons he might have, the investigator chooses to test the hypothesis at the .005 level of significance. The level of significance and the region of rejection are reported as:

$$\alpha = .005 \qquad R: z \leq -2.576$$

Step 4   Collect the data and complete the calculations. At this point, the investigator is dealing with sample information.

A mean of 94.3 is calculated from the 100 WAIS scores. The sample mean will now be compared to the sampling distribution of the means. The mean of this distribution is μ_0, which is 100. The standard error of this distribution is $\sigma/\sqrt{N}$, which is 1.5. A z-score may be calculated by taking the deviation of the sample mean from the theoretical mean and dividing by the standard error of the means:

$$z = \frac{M - \mu_0}{\sigma_M} = \frac{94.3 - 100}{1.5} = -3.80$$

Step 5   Complete the test of the hypothesis and interpret the result. If the sample statistic falls within the region of rejection, the null hypothesis is rejected, and the alternative is accepted. If the sample statistic falls within the region of retention, the null hypothesis is retained.

The calculated statistic in this case falls in the region of rejection specified earlier, and the null hypothesis is rejected at the .005 level of significance. Of course, this does not mean that every member of the culturally deprived population has a lower than average IQ, but it does establish, within the limits specified by the level of significance, that the mean for this population is lower than that of the general populace used in norming the instrument.

EXERCISES

19.1 A random sample of 75 subjects yields a mean of 62.70 and an estimate of the standard deviation of 12.99. Test the hypothesis that the sample was drawn from a population with mean of 60.
   (a) Determine the region of rejection for the .05 level.
   (b) Calculate the appropriate test statistic.
   (c) Determine the outcome of the statistical test.

19.2 It is hypothesized that students in a certain school district have greater than average creativity as measured by a standardized test of creativity. A random sample of students from this district yields a mean of 33, and the standard error of the mean is estimated to be 1.2. The national average on this test is 30. Use the normal model to test the null hypothesis.
   (a) Determine the region of rejection for the .01 level.
   (b) Calculate the appropriate test statistic.
   (c) Determine the outcome of the statistical test.

19.3 A school psychologist in a Southwestern community is interested in the academic problems encountered in the public schools by children of Spanish-American heritage. He draws a random sample of 64 children with Spanish surnames from the district files and determines that they averaged 47.75 on an English test which had been standardized with mean of 50 and standard deviation of 10. Use a one-tailed test at the .01 level to test the hypothesis that the population under study is significantly lower than the norm group.
   (a) Determine the region of rejection.
   (b) Calculate the appropriate test statistic.
   (c) Determine the outcome of the statistical test.

19.4 An experiment in ESP is conducted in which a blindfolded subject calls out the color (red or black) of a playing card as it is dealt from a well-shuffled deck. The subject scores 68 right out of 100 trials. Test the hypothesis that he is only guessing. Use the normal approximation of the binomial.
   (a) Determine the region of rejection for the .0005 level.
   (b) Calculate the appropriate test statistic.
   (c) Determine the outcome of the statistical test.

19.5 Refer to the ESP experiment in Section 17.4. Suppose the number of trials were increased to ten, and the subject got eight correct. Test the hypothesis that he is only guessing.

(a) Determine the region of rejection at the .001 level.

(b) Determine the outcome of the statistical test.

SELECTED REFERENCES

BAKAN, DAVID, "The test of significance in psychological research," *Psychological Bulletin*, **6**, 423–437 (1966).

BINDER, ARNOLD, "Further considerations on testing the null hypothesis and the strategy and tactics of investigating theoretical models," *Psychological Review*, **70**, 107–115 (1963).

GRANT, DAVID A., "Testing the null hypothesis and the strategy and tactics of investigating theoretical models," *Psychological Review*, **69**, 54–61 (1962).

NUNNALLY, JUM, "The place of statistics in psychology," *Educational and Psychological Measurement*, **20**, 641–649 (1960).

ROZEBOOM, WILLIAM W., "The fallacy of the null-hypothesis significance test," *Psychological Bulletin*, **57**, 416–428 (1960).

SELVIN, HANAN C., "A critique of tests of significance in survey research," *American Sociological Review*, **22**, 519–527 (October 1957).

TUKEY, JOHN W., "Conclusions vs. decisions," *Technometrics*, **2**, 423–433 (November 1960).

THE PROBLEM OF ERROR IN HYPOTHESIS TESTING

20.1 SAMPLING ERROR

Conclusions drawn about populations from sample information will always be subject to some error as each sample drawn from a given population might be expected to differ somewhat from the parent population. Sampling error is simply the difference between the value of a sample statistic and the corresponding population parameter. While a variety of procedures have been developed for the purpose of minimizing sampling error (matching samples on one or more variables, for example), no procedure available completely eliminates sampling error. The researcher is continually confronted with the possibility that the outcome of his research is attributable to sampling error rather than any real relationship between the dependent and independent variables.

Two kinds of errors are encountered in testing the null hypothesis: (1) rejecting a true hypothesis, and (2) retaining a false hypothesis. These will be treated in considerable detail in the following sections.

20.2 WHEN THE HYPOTHESIS IS REJECTED

Two possibilities exist when the null hypothesis is rejected: (1) the hypothesis is false, or (2) the hypothesis is true.

Consider the situation in which the null hypothesis is in reality false, and the hypothesis is rejected. The estimation of the population parameters by the sample statistics has been sufficiently accurate to produce a valid test of the hypothesis. There is a real relationship between the independent and dependent variables. This is a desirable situation; the investigator has arrived at a true conclusion by rejecting the null hypothesis.

Consider also the situation in which the null hypothesis is in reality true, but the hypothesis is rejected. The estimation of the population parameters by the sample statistics has been sufficiently inaccurate to produce an invalid test of the hypothesis. The investigator concludes that there is a relationship between the independent and dependent variables when no such relationship exists. This is an undesirable situation; the investigator is in error; he has arrived at a false conclusion by rejecting the null hypothesis.

The rejection of a true hypothesis is known as a Type I error. It can only occur when both of these conditions are met: (1) the hypothesis is rejected, and (2) the hypothesis is true. The probability of a Type I error is equal to the level of significance. It is controlled by the investigator; he may set it as high or as low as he wishes. The selection of the .05 level of significance, for example, indicates that there is 1 chance in 20 of a Type I error — 1 chance in 20 that the investigator will reject a true hypothesis.

Typically, when rejection is indicated by a statistical test, it is followed by action on the part of the investigator. Recommendations will be made as a result of the research findings. This is a decision-making situation, and changes in established procedures may be brought about.

The consequences of a Type I error may be great or small, depending upon the kind of action that results from the research findings. If a Type I error leads to further research, for example, it is likely that the subsequent research will eventually culminate in nonsignificant findings. How serious this is depends upon the amount of time, energy, and other resources expended in the subsequent research.

If a Type I error leads to changes in an educational program, for example, it is unlikely that the changes will prove either beneficial or detrimental to the educational program. How serious this is depends upon the amount of time, energy, and other resources expended in making the changes in the educational program.

Obviously, if important and expensive decisions are to be made from significant research findings, then a rigorous level of significance is needed. The probability of a Type I error must be set very small.

20.3 WHEN THE HYPOTHESIS IS RETAINED

Two possibilities exist when the null hypothesis is retained: (1) the hypothesis is false, or (2) the hypothesis is true.

Consider the situation in which the null hypothesis is in reality true, and the hypothesis is retained. The estimation of the population parameters by the sample statistics has been sufficiently accurate to produce a valid test of the hypothesis (assuming the investigator has collected sufficient evidence to test the hypothesis adequately). There is no real relationship between the independent and dependent variables. This is a desirable situation from a statistical point of view; the investigator has arrived at a true conclusion by retaining the null hypothesis.

Consider also the situation in which the null hypothesis is in reality false, but the hypothesis is retained. Either the estimation of the population parameters by the sample statistics has been sufficiently inaccurate to produce an invalid test, or the investigator has not collected enough evidence to test the hypothesis adequately. There is a real relationship between the independent and dependent variables, but the investigator has failed to establish its existence. This is an undesirable situation; the investigator is in error; he has arrived at an erroneous conclusion by retaining the null hypothesis.

The retention of a false hypothesis is known as a Type II error. It can only occur when both of these conditions are met: (1) the hypothesis is retained, and (2) the hypothesis is false. The probability of a Type II error is sometimes represented by the lowercase Greek letter beta (β). The value of α is established by the investigator, but the value of β is a function in part of the unknown value of one or more parameters which determine just how false the hypothesis is. Thus, in the typical research setting, the value of β is unknown. Statisticians, however, have made extensive studies of β under circumstances in which they could specify the values of parameters and the degree of falsity of the null hypothesis. They have established that Type II errors are more likely to occur when sample sizes are too small, and when the value of α is set too rigorously. The more the investigator guards against Type I error the greater the likelihood of Type II error if the hypothesis is false. Although there is a sense in which the value α should be chosen to minimize the probability of Type II error as well as Type I error, the most important factor in guarding against Type II error is adequate sample size.

Typically, when retention is indicated by a statistical test, it is followed by inaction on the part of the investigator. No recommendations will be made as a result of the research findings; this is not a decision-making situation, and no changes in established procedures are likely to be brought about.

The consequences of a Type II error may be great or small, depending upon what truth is lost as a result of the erroneous nonsignificant finding. If a Type II error results in the loss of an important idea — an idea which might eventually be of benefit to mankind — the loss could be very great. The investigator who is working on the frontiers of human knowledge, working with new ideas having only vague referents in practical reality, may be more concerned about Type II errors than Type I errors. He may choose to test his hypotheses at the .10 or even the .20 level of significance, for he knows that subsequent research will eventually reveal Type I errors made in the primitive stages of research. No such safeguard exists with respect to Type II errors, however, since these generally discourage further research.

Consider the effects of a Type II error upon research which, for example, is intended to have immediate impact upon educational practices. The investigator may be reasonably sure that the loss is not great, for it is unlikely that research of great practical significance will lead to a statistically nonsignificant finding. This last statement refers only to the situation in which the hypothesis is false; a nonsignificant finding when the hypothesis is true may be of great practical significance.

20.4   CHOOSING THE LEVEL OF SIGNIFICANCE

The level of significance can be intelligently chosen only by weighing the consequences of the two kinds of error. The probability of Type I error is readily controlled by the investigator and is determined by his selection of the level of significance. However, the probability of Type II error is unknown, and, when all other things are equal, the smaller the probability of a Type I error, the greater the probability of a Type II error.

Most behavioral research is conducted at the .01 and .05 levels of significance. However, in exploratory research, the .10 and .20 levels may be more appropriate. When the research leads to the implementation of programs of great cost, it is necessary to guard more rigorously against the Type I error, and the .001 level may be chosen.

Consider the situation in which the investigator may be testing many hypotheses. If 100 hypotheses were tested at the .05 level, the rejection of five true hypotheses should be anticipated. For this reason, when many hypotheses are tested in a given piece of research, it is probably desirable to choose a more rigorous level of significance.

The investigator may reduce the probability of a Type II error by increasing the size of his sample or by using a more powerful statistical test. Sample size and the power of a statistical test are considered in the following sections.

20.5   SAMPLE SIZE

Sampling error has been defined as the difference between the value of a sample statistic and the corresponding population parameter. It should be immediately apparent that sampling error is a function of sample size, and that the error tends to be smaller for larger samples. This relationship was illustrated in Section 18.5. The formulas used for determining the limits of confidence intervals may be modified slightly to provide a formula for sample size, provided that the investigator is prepared to specify how much error is acceptable and how much confidence is required. Let e_m be the maximum acceptable error and z_c the z-score corresponding to the required level of confidence. Then, for the sampling distribution of the means:

$$e_m = z_c \sigma_M \qquad \text{where } \sigma_M = \frac{\sigma}{\sqrt{N}}$$

With a little algebra, one may demonstrate that:

$$N = \left(\frac{z_c \sigma}{e_m}\right)^2$$

For example, suppose the investigator desires to obtain an estimate of the mean IQ (where $\sigma = 15$) for a population of interest, and that he wishes to be 95 percent confident that the population mean is within three points of the sample mean. The required sample size is:

$$N = \left(\frac{1.96 \times 15}{3}\right)^2 = 96$$

This calculation is, of course, dependent upon a knowledge of the standard deviation of the population under study. This information is not usually available and must be estimated from sample information. Typically, the investigator is not prepared to estimate the standard deviation from sample data until after the research is completed. This is no problem if he is willing to state the maximum acceptable error as a fraction of the standard deviation. Suppose, for instance, the investigator specifies that e_m is to be $\sigma/10$ at the 95 percent level of confidence; the calculation becomes:

$$N = (1.96 \times 10)^2 = 384$$

These calculations are based upon the assumption that the investigator is sampling from a population of infinite size — a common assumption in statistical inference.

The author prefers to approach the problem of sample size with some rules of thumb which he believes are appropriate for determining sample size in most behavioral research:

1. Although many of the examples and exercises in this text utilize extremely small samples, this is done for pedagogical reasons. The use of statistical analyses with samples smaller than 10 in size is not recommended.
2. In simple experimental research with tight experimental controls (matched pairs, for example), successful research may be conducted with samples as small as 10 to 20 in size.
3. In *ex post facto* research, in most experimental research, and in all research where the dependent variable is of low reliability, the use of samples 30 or larger in size is recommended.
4. Whenever samples are to be broken into subsamples (males and females, for example), and generalizations are to be drawn from these subsamples, recommendations with respect to minimum sample size apply to the subsamples.
5. In multivariable research (multiple regression, for example), the sample size should be several times (preferably 10 or more times) as large as the number of variables.
6. There are few occasions in behavioral research where samples smaller than 30 or larger than 500 in size can be justified. The use of samples of size 30 or larger usually insures for the investigator the benefits of the central limit theorem (see Section 18.5 for a discussion of this theorem). A sample of size 500 assures that the sampling error will not exceed $\sigma/10$ about 98 percent of the time. Within these limits — from about 30 to 500 — the use of a sample about one-tenth as large as the parent population is recommended.
7. Generally, the choice of sample size is as much a function of budgetary considerations as it is of statistical considerations. Larger samples, when these can be afforded, will ordinarily be preferred over smaller ones; however, well-chosen small samples will ordinarily be preferred over poorly chosen large ones.

20.6 THE POWER OF A STATISTICAL TEST

The *power* of a statistical test of some hypothesis is defined as the probability that it will reject the hypothesis when the hypothesis is false. The power of a test plus β (the probability of a Type II error) is equal to one. Like the value of β, the power of a statistical test in a given practical research setting is ordinarily in part a function of the unknown value of one or more parameters. While the power cannot be determined for parameters whose values are unknown, it may be determined if these and certain other conditions are specified. There are two situations where power determinations may be made: (1) the contrived situation set up by the statistician to study the characteristics of a given statistical test, and (2) the practical research situation in which the investigator is prepared to specify the value of one or more parameters as conditions for rejecting the null hypothesis. Suppose, for example, that the null hypothesis is $\mu = 100$, and the investigator is only interested in rejecting the hypothesis if $\mu = 105$ or larger. For a given sample size, the power is readily determined. For a given power, the required sample size is readily determined.

Generally, the power of a given statistical test is a function of the sample size, of the level of significance used, and of just how false the hypothesis is. These ideas are illustrated for the test of a hypothesis about the mean of a normally distributed population in Figure 20.1. The graph consists of a family of power curves, each curve representing a different sample size, for the null hypothesis $\mu = \mu_0$. The level of significance has been set at .05, and a two-tailed test is used. Notice that the probability of rejecting the null hypothesis when the hypothesis is true (the probability of a Type I error) is represented by the convergence of the curves in the center of the graph. Suppose, however, that the population mean is equal to $\mu_0 + \sigma/2$. With a sample of size 10, the probability of rejecting the hypothesis (which is false) is .37, and the probability of retaining the hypothesis (the probability of a Type II error) is .63. When the sample size is increased to 20, the probability of rejecting the hypothesis is .62, and the probability of retaining the hypothesis is .38. Of course, when it is appropriate to use, a one-tailed test is more powerful than a two-tailed test in the same situation.

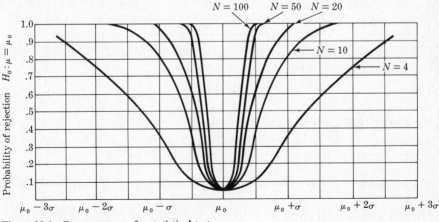

Figure 20.1 Power curves of a statistical test.

Some statistical tests are more powerful than others. The t-tests and the F-tests are usually more powerful than other statistical tests of the more common null hypotheses. The *relative power* (or *power-efficiency*, as it is sometimes called) is a comparison of the ability of a statistical test to detect a false hypothesis to the ability of a standard test (such as the t-test or the F-test) to detect the same false hypothesis. The usual procedure for expressing relative power is to form a ratio of the sample size needed to reject the hypothesis with the standard test to the sample size needed to reject the hypothesis with the test under consideration. Thus, if it should be determined that a given false hypothesis is rejected by the t-test with a sample of size 90, but it requires a sample of size 100 to reject the same hypothesis with the Mann-Whitney U-test, the Mann-Whitney is said to be 90 percent as powerful as the t-test.

Summarizing, the following factors influence the power of a given statistical test:

1. *Choice of statistical procedure.* Some statistical tests are more powerful than others. Where they are appropriate to use, the parametric tests will usually be more powerful than the nonparametric alternatives.
2. *Sample size.* Larger sample sizes give greater power.
3. *Value of α.* The more rigorous (the smaller the size of) the value of α, the less power the statistical test will have.
4. *Location of the region of rejection.* A one-tailed test, where it is appropriate, will be more powerful than a two-tailed test.
5. *Size of the parameters.* The more false the null hypothesis the more power the statistical test will have.

Textbooks in experimental design ordinarily report power curves for the common parametric tests. These may be used to determine the sample size needed to detect a specified false hypothesis, and this approach to the determination of sample size has much to commend it. However, a discussion of these topics is beyond the scope of this text, and it is reserved for a course of study in experimental design.

SELECTED REFERENCES

Cohen, Jacob, *Statistical Power Analysis for the Behavioral Sciences.* New York: Academic Press, 1969.

Kaiser, Henry F., "Directional statistical decisions," *Psychological Review*, **67**, 160–167 (1960).

Kimmel, Herbert D., "Three criteria for the use of one-tailed tests," *Psychological Bulletin*, **54**, 351–353 (1957).

21

INTRODUCTION TO EXPERIMENTAL DESIGN

21.1 FUNDAMENTAL PRINCIPLES OF EXPERIMENTAL DESIGN

The author has defined scientific research as the systematic and empirical study of relationships among variables. In our consideration of experimental design, this definition will be expanded and explored in depth. First, we will need to review the notion that experimental design is concerned with three kinds of variables: (1) independent variables whose effects are under investigation; (2) independent variables whose effects are to be controlled; and (3) dependent variables which are observed in order to determine the effects of variation in the independent variables.

Invariably, scientific research will take place within a narrow context with the intent to draw conclusions about a much larger context. The techniques of statistical inference enable the investigator to draw inferences from small groups, called samples, to larger groups, called populations — from individuals who are direct participants in the research to individuals who do not participate. Inherent in this procedure is the random assignment of individuals to samples and the use of the mathematics of probability to define the degree of confidence that the investigator has in drawing inferences from the samples to the populations.

For purposes of statistical analysis, the research problem or question will usually be expressed as a null hypothesis — a statement to the effect that there is no significant relationship between the independent and dependent variables. The usual intent in this case is to disprove the null hypothesis. The null hypothesis usually takes one of three forms: (1) there is no correlation between the independent and dependent variables; (2) with respect to the dependent variable, there is no difference in the distribution of the populations of interest; and (3) with respect to the dependent variable, there is no difference in the means of the populations of interest. Null hypotheses are concerned with population parameters and will be expressed algebraically with Greek letters; the decision as to whether to retain or reject the null hypothesis will be determined on the basis of sample statistics, which are represented by Latin letters.

Suppose, for example, that the investigator is exploring the possibility of using an admissions test of some sort for purposes of screening candidates for admission to some educational program. The critical issue is usually this: Can the admissions test predict performance in the educational program? If performance is measured by an achievement test, the investigator is interested in determining whether there is a significant correlation between the two tests. The Greek letter rho (ρ) will replace the Latin letter r representing the correlation coefficient, and the null hypothesis will be expressed as $\rho = 0$. Suppose, however, that the investigator is interested in determining which of two educational experiences is more effective. Typically, two samples are chosen. They are subjected to two different learning experiences (the independent variable), and they are administered an achievement test (the dependent variable) at the end of the learning experiences. The critical issue is usually this: Is there any difference in the average achievement of the two groups? The null hypothesis will be expressed as $\mu_1 - \mu_2 = 0$, or as $\mu_1 = \mu_2$.

In summary, the fundamental principles of experimental design are the following:

1. The major purpose of an experiment is to describe the effect of the independent variable on the dependent variable.
2. The term "independent variable" refers to any induced or selected variation in the experimental methods or materials or subjects whose effect is to be observed and evaluated.
3. Evaluation of the effects of the independent variable necessitates a criterion — the dependent variable. The outcome of the experiment will be described in terms of variation in the dependent variable. This will most often be in the form of a prediction equation, a comparison of means or medians of the various experimental groups, or a comparison of the frequency of occurrence of certain events.
4. Provision is made for controlling the effects of other variables whose presence might influence the variation of the dependent variable and thus confound the results.
5. Invariably, the investigator will wish to generalize the results of his research to individuals who did not participate in the research. The techniques of statistical inference (utilizing the mathematics of probability) give scientific objectivity to this process of generalization.

21.2   EXPERIMENTAL VERSUS *EX POST FACTO* RESEARCH

The distinction between experimental research and *ex post facto* research needs to be observed. In experimental research, the independent variable whose effects are to be studied is manipulated by the experimenter; other independent variables are carefully controlled (often by holding them constant); and the dependent variable is the criterion by which he determines whether or not anything of consequence happens when he manipulates the independent variable. Experimental research is generally regarded as the only genuinely satisfactory way to establish cause-and-effect relationships. There is an important class of behavioral research in which the investigator does not manipulate the independent variable, he merely selects the independent variable. Research in which the investigator selects rather than manipulates the independent variable is called *ex post facto* research.

While experimental research is regarded as the most highly developed and formalized of all scientific endeavor, it has certain disadvantages in the study of human behavior. First, some of the most important variables of interest to the behavioral scientist do not lend themselves to experimental manipulation. Suppose, for example, that the investigator wishes to study the relationship between the extent of a certain kind of brain damage (the independent variable) and verbal learning skills (the dependent variable). He is hardly in a position to systematically vary the amount of brain damage in his subjects! Instead, he would probably seek to select subjects who were already brain damaged and to classify them according to the extent of the brain damage. Second, the determinants of human behavior are ordinarily extremely complex, involving a great deal of interaction of a multitude of variables, many of which are unobserved. This real life environment cannot be duplicated in the laboratory; consequently, behavioral research conducted in the laboratory with rigid experimental controls is sometimes a sterile endeavor. Whether a child has been reared in riches or in poverty or has lived in a democracy or under a dictatorship for the first ten years of his life is likely to be a more important determinant of his behavior in any situation than a treatment administered by a researcher. Most classroom research, for example, demonstrates that the achievement of the student is determined much more by the experiences he had before he enrolled in a given course of study than the experiences he entered into while in the classroom. Some of the most valuable research in the behavioral sciences takes place without the intervention of the investigator, who merely observes and analyzes relationships among variables in their natural settings.

There is always a danger in *ex post facto* research that the effects observed are caused by some variable other than the independent variable whose effects the investigator wishes to study. For example, a number of *ex post facto* studies have been made of the relationship between the amount of time freshman college students spend in study (the independent variable) and the grades they receive in college (the dependent variable). Typically, students involved in research of this sort keep a diary of the hours spent in study. As a general rule, it can be predicted that the more the student studies, the lower his grades will be — there is a negative correlation between amount of time spent in study and grades achieved in freshman college courses! It would be ridiculous to conclude that there is a simple cause-and-effect relationship between the two variables and thus conclude that students should study less to learn more. There is a third variable which needs to be taken into consideration in research of this sort — the intelligence of the student. It is well-established that bright students often study less than not-so-bright students yet get better grades in school. Experimental research into the relationship between time spent in study and college grades would require that the investigator manipulate the amount of time which each student spends in study while controlling in some fashion the effects of intelligence.

Experimental research differs from *ex post facto* research in three important respects:

1. In experimental research, the investigator manipulates the independent variable whose effects are to be studied. In *ex post facto* research, the investigator merely selects the independent variable whose effects are to be studied.
2. As a general rule, in experimental research, the investigator has a wider range of options with respect to the manner in which he controls the effects of other independent variables.
3. Finally, experimental research ordinarily yields more definitive conclusions, especially where it is the investigator's intention to explore cause-and-effect relationships.

There is an important dimension in which experimental and *ex post facto* research do not differ, and it is important that the student keep this fact in mind as he studies the various experimental designs. As a general rule, the data from *ex post facto* research will be subjected to precisely the same kind of statistical analysis as the data from similar experimental research. For this reason, much of the discussion that comes under the rubric of experimental design is equally applicable to *ex post facto* research. The model for research given in Figure 21.1 is equally applicable to both experimental and *ex post facto* research.

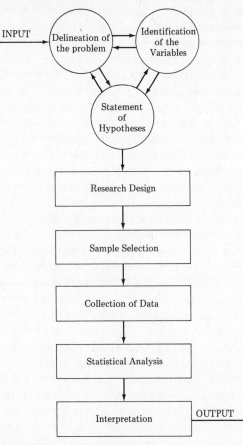

Figure 21.1 Model for research.

21.3 METHODS OF CONTROLLING INDEPENDENT VARIABLES

In the author's classification of variables that enter into scientific research, he has proposed two categories of independent variables: (1) independent variables whose effects are to be studied, and (2) independent variables whose effects are to be controlled. As a general rule, the distinction between these two kinds of independent variables will be clearly evident in a given research, and we will proceed on the assumption that the investigator has classified his research variables in this fashion. The reader should be aware, however, that the distinction may be a little blurred in some research, especially that involving a large number of variables and utilizing multiple regression analysis.

Other authors use a variety of names to define or describe those independent variables whose effects are to be controlled. Sometimes they are called extraneous, intervening, irrelevant, nuisance, or organismic variables. These terms have different meanings, however, and they are indeed used in different ways by the different authors. None of these terms conveys precisely what this author is trying to say, and they will be avoided.

Obviously, those variables whose effects are to be controlled play an important part in sound experimental design, and we shall be deeply concerned with the various methods used to control their effects. These may be classified under four headings: (1) holding the variable constant, (2) randomizing the effects of the variable, (3) matching of subjects with respect to the variable, and (4) controlling the variable statistically. A brief discussion of each of these methods follows.

Controlling a variable by holding it constant is a time-honored technique in laboratory research; however, it should be noted that all research uses this technique to some degree simply by the choice of the problem. In the classic laboratory research procedure, the ideal was to vary one independent variable (the one whose effects are to be studied) while holding all others constant. This is really quite impractical when dealing with human subjects, as they vary in so many different ways. Nevertheless, the investigator may be able to identify a number of important variables which can be controlled in this fashion, and all investigators should be on the alert for opportunities to use this method of experimental control.

Suppose, for example, that the investigator is interested in controlling the effects of human intelligence in a given learning research. He might choose (1) to work with subjects of average intelligence, thus holding the variable relatively constant, or (2) to divide his subjects into groups of high, average, or low intelligence, then analyze the results for each group separately. Similarly, an investigator might control the effects of sex (1) by working only with male or with female subjects, or (2) by analyzing the results for males and females separately. Another investigator might choose to control the effects of age by working with subjects of a given age or age range.

All behavioral research should use the method of randomization. When subjects are randomly chosen or randomly assigned to experimental treatments, the effects of all independent variables other than the treatment variable (even variables of whose existence the investigator is not aware) are controlled. The techniques of statistical inference (using the mathematics of probability) take into consideration the effects of all randomized variables. It is important to note, however, that randomization is most effective when used with large samples and/or other methods of control discussed here. Suppose, for example, that a basketball coach wished to do a little experimental study of the relative effectiveness of zone versus man-to-man defense (the independent variable whose effects are to be studied). Suppose also that he has ten basketball players, who are randomly assigned to two teams of 5 men each. If one of the 10 players is so much better than the other players that his team always wins, the random assignment of this player does little to control the effects of his superior skill (the independent variable whose effects are to be controlled). This is an ever-present danger in small sample research and one to which the investigator must be alert. If the research involved 100 basketball players, the effects of this one player would not be so great.

A number of procedures are available that involve the matching of subjects with respect to some variable. Suppose that an investigator has 20 subjects, 10 males and 10 females, available for research which requires two samples. Instead of randomly assigning the 20 subjects to the two samples, he might first separate them into two groups with respect to sex, then randomly assign 5 males and 5 females to each sample. Thus, if the sex of the subject has some effect on the dependent variable, it will have equal effect for each sample. Samples drawn in this fashion are called *stratified samples*. A similar approach might be used to insure that two samples have equal numbers of students of high, average, and low intelligence. With some effort, an investigator may succeed in stratifying his samples with respect to two or more variables. A closely related procedure makes use of matched pairs. Suppose the investigator has access to 20 mentally retarded youngsters of about the same age, and that he wishes to do two-sample learning research with these subjects. It is unlikely that he will be able to increase the size of the available group, and the dangers of small sample research are present — two or three youngsters of unusually high or low ability randomly assigned to one sample could determine the outcome of the research. The subjects should first be separated into pairs as closely matched with respect to intelligence (and other variables if practical) as possible. Then one subject from each pair should be randomly assigned to each sample. The result is two samples closely matched with respect to one or more variables.

There is a third technique that can be used to achieve greater control, but it is applicable to a narrow class of experiments. Suppose a psychiatrist is investigating the effects of two tranquilizer drugs, and that the effects of the drugs are known to wear off in a reasonable length of time. Half his patients could be given one drug, the other half the other drug. When the effects of the drugs have worn off, the experiment could be repeated, but with the drugs switched. Thus, each patient participates in each phase of the experiment; the two samples are identical except for the effects of the brief passage of time.

The methods of experimental control discussed in the preceding paragraph are often difficult to introduce, especially in classroom research.

Many variables, however, can be controlled statistically without these difficulties. Whenever two variables are correlated, one can be used to predict performance on the other. Thus, IQ scores can be used to predict performance on any educational achievement test, including those which are used as dependent variables in classroom learning research. To the extent that IQ scores account for a portion of the variance of the dependent variable, the effects of IQ can be systematically controlled by eliminating this portion of the variance from the final analysis.

The use of randomization is required if the techniques of statistical inference are to be used. Invariably a good research design will also make provision for holding one or more variables constant. Often one or more additional methods of controlling variables will be used in the same design. The chapters which follow will discuss methods of controlling variables in somewhat greater detail.

21.4   SELECTION OF A STATISTICAL ANALYSIS

There are literally dozens of different techniques for the analysis of research data presented in the following chapters of this book. Still other techniques are presented in books of a more specialized nature, such as those on non-parametric statistics, regression analysis, variance analysis, and multi-variate analysis. The question arises now, given all of these options, just how does one go about choosing the appropriate statistical analysis? The choice will be based upon the following considerations:

1. *The level of measure of the dependent variable.* All statistical procedures are classified according to the level of measurement of the dependent variable. There are nonparametric statistics suitable for use with nominal and ordinal data, and there are parametric statistics suitable for use with interval and ratio data. If the dependent variable is measured on a nominal scale, the choice of a statistical procedure is greatly restricted. If the dependent variable is measured on an ordinal scale, there is some debate as to whether the investigator is restricted to use of those nonparametric statistics intended for use with ordinal data. Some statisticians feel this is a critical distinction which needs to be observed. However, the position taken here is that the distinction between ordinal and interval data may be ignored under most circumstances. The distinction between interval and ratio data is not a factor in the selection of a statistical procedure — the parametric statistics work equally well with both.

2. *The distribution of the dependent variable.* Most statistical procedures assume that the measures are continuously distributed and are invalidated by gross violation of this assumption. Of course, the student

is now aware that all measures are discrete but most are "sufficiently continuous" that they may be analyzed as continuous variables. In the event that the measures are so discrete that this distinction cannot be ignored, they will ordinarily be analyzed as nominal data whether they are nominal or not.

The parametric statistics also require that the sampling distribution of the statistic closely approximate a normal distribution. The distribution of the statistic is in part a function of the distribution of the variable. If the dependent variable is normally distributed, there need be no concern about the assumption of normality. Most hypotheses tested with the parametric statistics are hypotheses about means, and the student is already aware that the sampling distribution of the mean tends to be normal even when the measures from which the mean is calculated are not normal. Nevertheless, there are circumstances when violation of the assumption of normality will require that the investigator shift to a nonparametric (usually ordinal) statistic.

There is yet another restriction commonly placed upon the parametric statistics. As a general rule, parametric tests involving more than one sample will assume that the samples are drawn from populations having equal variances. There are numerous circumstances in which the violation of this assumption will favor the choice of a nonparametric statistic (again, usually the choice will go to one designed for use with ordinal data).

3. *The number of samples.* The choice of a statistical procedure will also be determined in part by the number of samples. Some statistical analyses are designed for use with one sample, some are designed for use with two samples, and some for use with two or more samples.

4. *The methods of experimental or statistical control.* All of the techniques of statistical inference will assume that the subjects have been drawn at random from the parent populations, or have at least been assigned at random to the various samples. We can also anticipate that the research will have some variables controlled by holding them constant. The use of matching techniques of any kind will, however, influence the choice of a statistical analysis, as only certain statistical procedures make provision for analyzing data organized in this fashion. The same is true with the use of statistical controls; only certain statistical analyses make provision for this sort of control.

The nonparametric statistics are ordinarily restricted to the simpler statistical analyses. Whenever the more sophisticated techniques of experimental or statistical control are required, the parametric statistics will be preferred.

The simplest research is that in which only two variables are identified and studied by the investigator: (1) the independent variable whose effects are to be studied, and (2) the dependent variable, which is observed to determine the effects of variation in the independent variable. As a general rule, behavioral research will involve additional variables: (1) sometimes the effects of two or more independent variables are studied simultaneously, and/or (2) often the effects of one or more independent variables are to be controlled. We will refer to research of this sort as *multivariable* research. All of the research and analysis techniques discussed here assume that there is a single dependent variable — such research is referred to as *univariate* research. When several dependent variables are analyzed simultaneously, *multivariate* analysis is required, but this is beyond the scope of this text.

AN OVERVIEW OF
EXPERIMENTAL DESIGN

22.1 THE SIMPLE RANDOMIZED DESIGN

The simplest type of experimental design is called the *simple randomized* design. Only the independent variable, whose effects are to be studied, and the dependent variable are identified by the investigator and enter into the analysis of the data. Control of the effects of other variables is accomplished through the process of randomization and possibly by holding some of the variables constant.

Figure 22.1 illustrates the simple randomized design as used in *ex post facto* research. In this situation the investigator is interested in studying the relationship between membership in two or more populations (the independent variable) and scores on some criterion (the dependent variable). Notice that for each population under study, there is a mean parameter value (μ_j) on the dependent variable. We will assume that the populations are very large and that it would be impractical to collect all of the data necessary to determine the values of the parameters. The null hypothesis is that $\mu_1 = \mu_2 = \mu_3$ in the three-sample case. Samples are randomly drawn, one from each population, and the corresponding statistic (M_j) for each sample is calculated. The techniques of statistical inference (using the mathematics of probability) enable the investigator to infer from the sample statistics whether the null hypothesis about the population parameters should be retained or rejected.

Figure 22.2 illustrates the simple randomized design as used in true experimental research. A single parent population exists, from which two or more samples are randomly drawn. (The parent population might

197

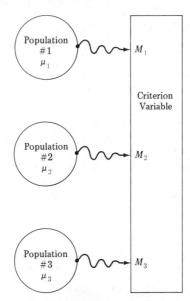

Figure 22.1 Simple randomized design — *ex post facto* research.

consist of an available colony of white rats or an available group of school-children — their random assignment to the various samples insures the validity of the experiment but may restrict the inferences drawn to a relatively narrow population.) The samples are then subjected to various experimental treatments (the independent variable whose effects are to be studied), one sample to each treatment. In some research, one of the samples may be assigned as a control group receiving no treatment at all, but this does not affect the analysis in any way. The treatments may be, for example, two or more different kinds of learning experiences. At the conclusion of the experiment, a single criterion variable is administered to all subjects, and the various sample means (M_j's) are calculated. The student should note that the results would be completely meaningless if different criteria were administered to the various samples. As before, the null hypothesis is $\mu_1 = \mu_2 = \mu_3$ for the three-sample case.

Now, a problem arises: Just precisely what are these parameters, and what populations do they represent? The samples, having undergone the treatments administered, are no longer simple random samples of the parent population. It will be useful to think of each sample as representing a hypothetical population, a population just like the parent population had it undergone the same treatment as the sample. Thus the null hypothesis

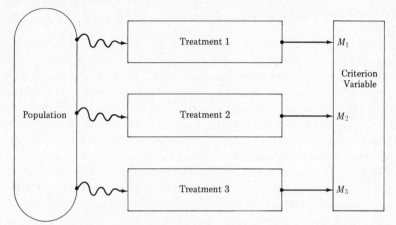

Figure 22.2   Simple randomized design — experimental research.

is concerned with hypothetical parameters from hypothetical populations. Again, the techniques of statistical inference enable the investigator to infer from the sample statistics whether the null hypothesis about the population parameters should be retained or rejected.

Now, let us turn our attention to the problem of selecting a statistical procedure for analyzing the data yielded by research with the simple randomized design:

1. *Dependent variable is nominal.* As a general rule, this will call for the chi-square test for contingency table. While there are a few alternatives to the chi-square test, these need not be considered here.
2. *Dependent variable is ordinal.* In the two-sample situation, the Mann-Whitney *U*-test will usually be preferred, although there are a few situations where the Kolmogorov-Smirnov two-sample test may be more appropriate. The Kruskal-Wallis test is suitable for use with two or more samples.
3. *Dependent variable is interval.* In the two-sample situation, the *t*-test for two independent samples will usually be preferred. The analysis of variance is the traditional alternative to the *t*-test when a test for two or more samples is needed; however, the analysis of regression will do just as well.

The simple randomized design is often referred to as the completely randomized design; however, this is a more general term which also applies to the factorial designs which follow.

22.2 FACTORIAL DESIGNS

Often the behavioral scientist will wish to conduct research in which the effects of two independent variables are studied simultaneously. In such research, the independent variables are called *factors*, and the design is said to be a *factorial* design. Two benefits accrue from using factorial designs. (1) The investigator is conducting two experiments simultaneously with the same subjects; the savings in time, expense, and other resources may be appreciable. (2) The factorial design permits the investigator to study the possible interaction between the effects of the two independent variables. If the two variables are interdependent in the effects which they produce, this would not be apparent if two separate simple randomized designs were used. While the use of factorial designs may be extended to three or even more independent variables, there are computational and conceptual problems associated with these designs that are beyond the scope of this book; thus our discussion will be limited to two-factor designs. Like the simple randomized design, the two-factor design is a completely randomized design; there is no provision for controlling the effects of other independent variables other than the use of randomization and the possible holding of one or more variables constant.

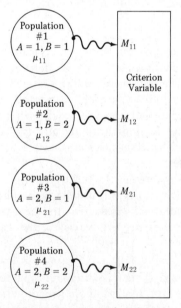

Figure 22.3 Factorial design — *ex post facto* research.

Figure 22.3 illustrates the use of a factorial design in *ex post facto* research. We will refer to the two independent variables as the A-factor and the B-factor. Each of these factors exists at two levels in the example, so that we have a two-by-two factorial design. Suppose that the investigator were a sociologist interested in studying the effects of race and sex (the two independent variables whose effects are to be studied) upon attitudes toward the so-called women's liberation movement (the dependent variable). Let the A-factor represent the race of the subject (A_1 for white, A_2 for black) and the B-factor the sex of the subject (B_1 for male, B_2 for female). Thus four populations are defined: population one (A_1B_1) is white male, population two (A_1B_2) is white female, population three (A_2B_1) is black male, and population four (A_2B_2) is black female. Four samples are randomly drawn, one from each population, and the attitude scale is administered. The null hypothesis for no difference between whites and blacks is

$$\mu_{11} + \mu_{12} = \mu_{21} + \mu_{22}.$$

The null hypothesis for no difference between the males and females is

$$\mu_{11} + \mu_{21} = \mu_{12} + \mu_{22}.$$

With some effort, it may be demonstrated that the hypothesis of no interaction is

$$\mu_{11} - \mu_{12} = \mu_{21} - \mu_{22},$$

which is equivalent to saying the difference between the white males and females is the same as the difference between the black males and females. With a little more effort the student may demonstrate that this is precisely the same as saying that the difference between the two groups of males is the same as the difference between the two groups of females. As before, the techniques of statistical inference use sample statistics to draw inferences about the population parameters and to retain or reject the various hypotheses.

Figure 22.4 illustrates the use of a factorial design in experimental research. Suppose that the investigator were a psychiatrist who wished to study simultaneously the effects of a certain drug and electroshock therapy. Let the A-factor be the electroshock variable (A_1 for shock, A_2 for no shock) and the B-factor be the drug variable (B_1 for drug, B_2 for no drug). Four samples are randomly drawn: the first sample (A_1B_1) receives both the drug and the electroshock therapy, the second sample (A_1B_2) only the electroshock therapy, the third sample (A_2B_1) only the drug, and the fourth sample (A_2B_2) receives neither the drug nor the shock treatment. The

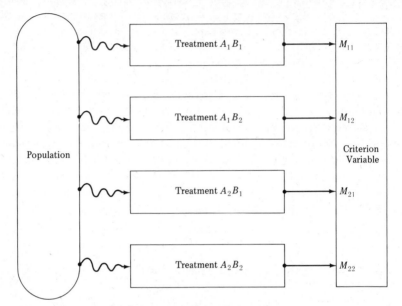

Figure 22.4   Factorial design — experimental research.

hypotheses are the same as for the *ex post facto* research, but the populations and parameters are hypothetical as with other experimental research. The null hypotheses are retained or rejected on the basis of inferential analyses of the sample data.

The data for factorial designs are often organized into a double-entry table, with one of the independent variables cast into columns and the other into rows. Each sample is represented by a cell in the table. For the example:

	Drug	*No drug*
Electroshock	A_1B_1 (μ_{11})	A_1B_2 (μ_{12})
No shock	A_2B_1 (μ_{21})	A_2B_2 (μ_{22})

A cursory examination of the table should reveal the economies achieved in the factorial design. All of the subjects participate in an experiment comparing electroshock to no shock therapy, and all of the subjects participate in an experiment comparing the drug to no drug.

Traditionally, factorial designs have been analyzed with the analysis of variance; however, they may also be conveniently analyzed with regression analysis.

22.3   RANDOMIZED BLOCKS DESIGNS

The simple randomized design can be greatly improved through the use of samples matched with respect to some independent variable whose effects the investigator wishes to control. The design may call for sampling in such a fashion as to ensure that each sample has the same number of male and female subjects, or that each sample has the same number of subjects of high, medium, and low ability on some trait that the investigator believes is related to performance on the dependent variable. The procedure calls for stratifying the population with respect to the variable whose effects are to be controlled, then drawing subjects from the various subsets of the population so as to ensure that the distribution of the trait in each sample grossly matches the distribution of the trait in the population. Designs of this sort are most often called *randomized blocks* designs (although some prefer to call them treatment-by-levels designs), and the variable whose effects are to be controlled is most often called the *blocking* variable.

Suppose, for example, that the investigator wishes to do research into the relative effectiveness of two instructional strategies (B_1 and B_2), and that he wishes to control the effects of intelligence as measured by an IQ test of some sort. Suppose also that he has 60 schoolchildren available for participation in the research. Using their IQ scores, the 60 children could be divided into three groups of 20 each, one group of high ability students (A_1), one of medium ability (A_2), and one of low ability (A_3). The students of high ability are randomly divided into two groups of 10 each and assigned to the two instructional strategies (producing A_1B_1 and A_1B_2). The process is repeated for students of medium ability (producing A_2B_1 and A_2B_2) and for the students of low ability (producing A_3B_1 and A_3B_2). The data may be arranged in rows and columsn as follows:

	B_1	B_2
A_1	A_1B_1 (μ_{11})	A_1B_2 (μ_{12})
A_2	A_2B_1 (μ_{21})	A_2B_2 (μ_{22})
A_3	A_3B_1 (μ_{31})	A_3B_2 (μ_{32})

Now, it is the primary intent of the investigator to determine the relative effectiveness of the two learning strategies (B_1 and B_2), so the major null hypothesis is:

$$\mu_{11} + \mu_{21} + \mu_{31} = \mu_{12} + \mu_{22} + \mu_{32}.$$

The student should learn to think of this as the hypothesis of equal column means. Whether or not the hypotheses of equal row means or of interaction are of interest will depend upon the particular research.

The procedure for the randomized blocks design is illustrated for a two-by-two design in Figure 22.5. The student may find it profitable to regard the blocking variable as an *ex post facto* variable and the treatment variable as an experimental variable. Although the analysis will require that groups A_1B_1 and A_2B_1 be kept separate (but only for purposes of analyzing the data), in the conduct of the experiment it may be highly desirable to regard these as a single sample (combining the two to form a single group receiving treatment B_1). The same applies to the research described in the preceding paragraph — the students may be divided into only two classes while the experiment is carried out.

A randomized blocks design of the sort described has three distinct advantages over a simple randomized design. (1) Some control over the blocking variable is achieved by ensuring that the samples are at least grossly matched on this variable. (2) Whenever samples are matched in this fashion, the analysis is more sensitive to possible relations between the independent variable whose effects are to be studied and the dependent variable. (3) The data for the various levels of the blocking variable may be analyzed separately. In the example given above, the investigator can

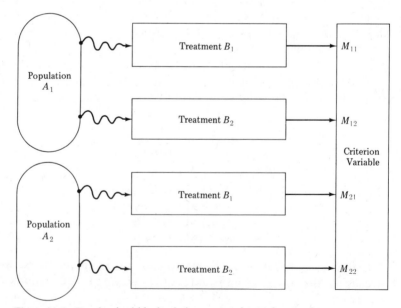

Figure 22.5 Randomized blocks design — experimental research.

draw separate conclusions about the relative effectiveness of the two instructional strategies with respect to students of high ability, to those of medium ability, and to those of low ability.

The statistical analysis for randomized blocks designs of the sort described above is precisely the same as for factorial designs. Either the analysis of variance or the analysis of regression will be used.

The matched-pairs design is a special case of the randomized blocks design. It is commonly used when the subjects are paired in nature (as in research with twins) or when the number of subjects available for research is few and there is a need for tight experimental control of one or more variables. Suppose, for example, that the investigator has just 20 subjects available for research, and he wishes to do two-sample research with tight control of human intelligence. Instead of stratifying this tiny population at just three levels (high, medium, and low, as before), he should stratify them at ten levels with just one subject in each cell when the data are arranged in a ten-by-two table. The subjects are first separated into pairs as closely matched with respect to intelligence as possible; then one subject from each pair is randomly assigned to each sample. The result is two samples closely matched with respect to intelligence. After the samples are constituted, the experiment is carried on just as though it were a simple randomized design. In the final data analysis, however, the subjects are re-paired, and the analysis is performed on the differences in the pairs. Matched-pairs research is not ordinarily performed with a nominal criterion; when an ordinal criterion is used, the Wilcoxon matched-pairs signed-ranks test may be used. As a general rule, the parametric t-test for two related samples will be preferred. If the research is done with matched trios or higher order grouping, the analysis of variance is the likely statistical procedure. Regression analysis with data of this sort (where there is one entry per cell) is very clumsy and will be avoided.

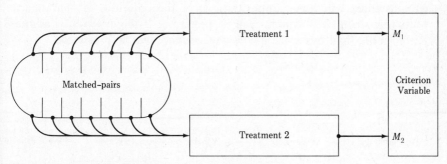

Figure 22.6   Matched-pairs design — experimental research.

22.4 REPEATED MEASUREMENTS DESIGNS

There is a special class of experiments in which it is possible to administer the various experimental treatments to each subject sequentially. The criterion variable is administered to each subject following each experimental treatment. Such a design is called a *repeated measurements* design. Its use is limited to those situations in which the effects of all treatments (with the possible exception of the last one) are known to be temporary and do not influence the effects of the following treatment. A similar limitation is made with the repeated administration of the criterion; if the effects of one measurement will influence scores on subsequent measurements, the design is not valid. The design has also obvious limitations with respect to use when learning is taking place. In view of the fact that only one sample is used throughout, the degree of control of a wide range of variables is unique; when applicable, the repeated measurements design is unmatched.

Figure 22.7 illustrates the repeated measurements design with two treatments and two administrations of the criterion. A single random sample is drawn, it is subjected to the first treatment, this is followed by the first administration of the criterion, then the sample is subjected to the second treatment, and this is followed by the second administration of the same criterion previously used. Occasionally, the repeated measurements design may be encountered in which the first treatment is omitted; that is, the sample acts as its own control, and the first administration of the criterion is to the sample as the control group, the second is to the sample as the experimental group.

The statistical analysis of the data from a repeated measurements design is precisely the same as that used with a matched-pairs design. If the criterion is administered only twice, and if the data are ordinal, the Wilcoxon matched-pairs signed-ranks test may be used. However, under most circumstances with two administrations of the criterion, the parametric *t*-test for two related samples will be preferred. If the criterion is administered more than twice, the analysis of variance will be preferred.

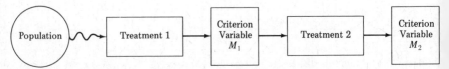

Figure 22.7 Repeated measurements design — experimental research.

22.5   EXPERIMENTAL DESIGNS WITH NESTED VARIABLES

Consider the situation in which classroom learning research is conducted with two different learning strategies (B_1 and B_2). Four classroom teachers are to participate (A_1, A_2, A_3, and A_4). Ideally, each teacher should teach two classes, one with each learning strategy. Thus, a four-by-two factorial or randomized blocks design would be used. Whether the design is factorial or randomized blocks in a situation like this is largely a matter of how the investigator sees the role of the teacher. Let us suppose that (for whatever reasons might exist) each of the four teachers can only teach one class, and each teacher must be assigned to only one of the two learning strategies. The situation is diagrammed in Figure 22.8 and by the following four-celled diagram:

	B_1	B_2
A_1	A_1B_1 (μ_{11})	
A_2	A_2B_1 (μ_{21})	
A_3		A_3B_2 (μ_{32})
A_4		A_4B_2 (μ_{42})

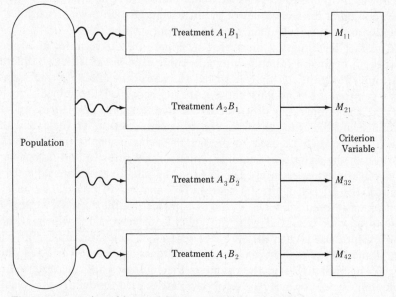

Figure 22.8   Design with nested factor — experimental research.

Notice that the columns are formed as usual in either factorial or randomized blocks designs; however, there is only one cell in each row — rows, as such, do not exist and are not subject to statistical analysis in the design with nested variable as outlined here. The A-factor is said to be nested within the B-factor. Educational researchers have been known to analyze data from such research as though they had a simple randomized design, ignoring the existence of the teacher variable. The danger in this is that the teacher variable may appreciably affect the outcome of the experiment. A valid analysis can be achieved only if the nested factor is taken into consideration in the analysis. Either the analysis of variance or the analysis of regression may be conveniently used.

22.6 DESIGNS UTILIZING THE ANALYSIS OF COVARIANCE

The analysis of variance involves the analysis (or dividing into component parts) of the variance of the criterion variable. The student first encountered this idea when he or she learned that r^2 is that proportion of the variance of one variable that is accounted for by another — in prediction, it is the proportion of dependent variable that is accounted for by the independent variable. In the analysis of variance with a factorial design, it is possible to determine that the A-factor accounts for a given proportion of the criterion variance, that the B-factor accounts for so much more, and that the interaction of the two factors accounts for so much more. As with the use of r^2, there will always be some variance unaccounted for. The analysis of covariance combines the techniques of prediction and the analysis of variance to identify and control the effects of one or more variables.

Consider the situation illustrated by Figure 22.9. Again, let us suppose that the investigator is conducting a learning experiment with two different learning strategies, and that he wishes to control the effects of intelligence. Two samples are randomly chosen, the IQ scores of the subjects are recorded, the samples are subjected to the two learning experiences, and are administered the criterion at the end of the experiment. To the extent that the IQ scores can be used to predict the achievement scores, there is a portion of the criterion variability which existed prior to and independent of the learning experiences. The data in the final analysis are adjusted to control the effects of the IQ scores. The analysis of covariance may be conveniently used with two or more samples. Mathematically equivalent results may be achieved with the analysis of regression, which is more conveniently used when the effects of several variables are to be controlled simultaneously. Properly, the term covariates is applied to variables which vary concomitantly — in the example, the IQ scores and the criterion. In common usage, however, the term covariate is used to identify the variable whose effects are being controlled.

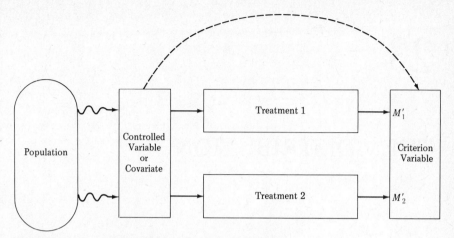

Figure 22.9 Experimental design with analysis of covariance.

SELECTED REFERENCES

CAMPBELL, DONALD T., and JULIAN C. STANLEY, *Experimental and Quasi-Experimental Designs for Research*. (Reprinted from *Handbook of Research on Teaching*.) Chicago: Rand McNally, 1966.

DAYTON, C. MITCHELL, *The Design of Educational Experiments*. New York: McGraw-Hill, 1970.

EDWARDS, ALLEN L., *Experimental Design in Psychological Research*, 3rd ed. New York: Holt, Rinehart and Winston, 1968.

KELLY, FRANCIS J., et al., *Research Design in the Behavioral Sciences — Multiple Regression Approach*. Carbondale: Southern Illinois University Press, 1969.

KERLINGER, FRED N., *Foundations of Behavioral Research*. New York: Holt, Rinehart and Winston, 1965.

KIRK, ROGER E., *Experimental Design Procedures for the Behavioral Sciences*. Belmont: Brooks/Cole, 1968.

LINDQUIST, E. F., *Design and Analysis of Experiments in Psychology and Education*. Boston: Houghton Mifflin, 1953.

MYERS, JEROME L., *Fundamentals of Experimental Design*. Boston: Allyn and Bacon, 1966.

WINER, B. J., *Statistical Principles in Experimental Design*, 2nd ed. New York: McGraw-Hill, 1971.

23

THE *t*-DISTRIBUTION

23.1 THE DEVELOPMENT OF SMALL-SAMPLE STATISTICS

The early history of statistical inference is that of large-sample statistics using the normal curve as a model for the sampling distribution. Although the normal curve provides a probability distribution that is fairly well approximated when drawing large samples, it has proved inappropriate for small samples, especially those smaller than 30 in size. The problem arises from the fact that there are usually two distinct sources of sampling error in the typical test of a practical research hypothesis. The first of these, estimation of the parameters identified in the statement of the null hypothesis, is taken into consideration in the normal model of the sampling distribution. The second source of sampling error, estimation of the standard error of the sampling distribution, is not. When the standard error is estimated from information provided by a small sample, the error may be of considerable consequence, especially in the tails of the sampling distribution where hypotheses are tested.

Shortly after the turn of the century, a young chemist named William S. Gossett was employed by a brewery in Dublin, Ireland. Gossett was concerned with quality control, and he approached the problem by drawing samples and analyzing these with the traditional statistical procedures. He soon recognized the inadequacy of the normal curve as a probability model for small samples, but lacking the mathematical sophistication needed to derive new models, he determined to establish the necessary sampling distributions empirically.

Gossett secured the records of Bertillon measurements for some 3000 British criminals. The Bertillon system was the first scientific method for criminal identification and was based largely on classification of body measurements. He recorded the measurements on cards, building an approximately normal population from which he could conveniently draw random samples of various sizes. By drawing large numbers of samples of a given size, he was able to establish empirically the sampling distribution of the means for that size sample. He established a different sampling distribution for each sample size of interest to him, and he discovered that the distributions for small samples differed markedly from the normal. Later, with the assistance of a university professor, Gossett was able to derive a general mathematical expression for these sampling distributions, and published the results in 1908 under the pen name, "Student." His findings received little attention until they were included in the first textbook of modern statistics, written almost 20 years later by Sir Ronald A. Fisher. The development of Student's t-distribution was the major breakthrough from classical to modern statistical inference.

23.2 THE STUDENT'S t-DISTRIBUTION

A t may be calculated which closely resembles the z used in testing the hypothesis about the mean in Chapter 19. The numerator of the t will be the deviation of the sample mean from the hypothesized mean, and the denominator will be the estimate of the standard error of the mean:

$$t = \frac{M - \mu}{S_M} \quad \text{where } S_M = \sqrt{\frac{SS}{N(N-1)}}$$

The equation of the t-distribution is much too complex to be treated here. There is a family of t-curves, each determined by a single parameter, its degrees of freedom, a concept closely related to sample size. Figure 23.1 is a graphic illustration of the t-distribution for one, three, five, and infinite degrees of freedom. The distribution for infinite degrees of freedom is identical to the normal. Table 7 in the Appendix gives the distribution of t for many different degrees of freedom and for those levels of significance commonly used in testing hypotheses.

Like the standard normal curve, the various t-distributions are symmetrical and bell-shaped with mean of zero and unit standard deviation. When the number of degrees of freedom (abbreviated df) is small, the proportion of the area in the tails is much greater than that of the normal distribution. For example, .046 of the area of the normal curve lies beyond plus or minus two standard deviations from the mean, but the corre-

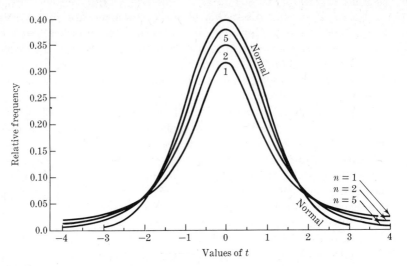

Figure 23.1 The t-distribution for various degrees of freedom.

sponding points on the t-curve for three degrees of freedom mark off .140 of the area under the curve. This suggests that there is greater probability of error when dealing with small samples, and rejection of the null hypothesis is not easily achieved.

23.3 DEGREES OF FREEDOM

The number of degrees of freedom of a statistic depends upon the number of observations (N) and the nature of the problem. The term "degrees of freedom' is a descriptive one which suggests its general nature, but a full appreciation of the term is not likely to be realized without some very serious study of theoretical statistics. This will not be undertaken in this text. The approach used here shall be:

1. The term will be defined, even though it is anticipated that the definition will not be particularly meaningful to many students.
2. For each type of statistical calculation encountered in this text, the number of degrees of freedom will be clearly specified.
3. The manner in which the definition was used to arrive at the number of degrees of freedom will be explained.

First, the definition: The number of degrees of freedom is equal to the number of observations minus the number of algebraically independent linear restrictions placed upon them. Generally, these restrictions will be identified as statistics which must be calculated from sample data before it is possible to calculate the statistic used in testing the hypothesis. Consider the estimation of the standard error of the mean:

$$S_M = \sqrt{\frac{SS}{N(N-1)}} \quad \text{where } SS = \Sigma (X_i - M)^2$$

Notice that it is necessary to estimate the mean from sample data before the standard error of the mean can be estimated. The price for this procedure is one degree of freedom, and the degrees of freedom for the standard error of the mean is $N - 1$.

Another way of looking at degrees of freedom is to think in terms of the number of values of the variable that are free to vary. Suppose, for example, that a sample of size four is drawn, the scores are 2, 3, 5, 6, and the mean is 4. The deviation scores are -2, -1, $+1$, $+2$, and the algebraic sum of the deviation scores is, of course, zero. As a consequence, if any three of the deviation scores are calculated, the value of the fourth is not free to vary; its value must be such that the algebraic sum of the deviation scores is zero. This is equivalent to saying that if the mean is specified, the value of the last score drawn is not free to vary. The degrees of freedom in the example are three, as only three scores are free to vary.

When the sum of the squared deviations about the mean was calculated, only one parameter was estimated from sample data, and the degrees of freedom for the standard error of the mean were determined to be $N - 1$. When the sum of the squared deviations about the regression line was calculated, however, both the slope of the line and its intercept with the Y-axis were estimated from sample data, and the degrees of freedom for the standard error of estimate were determined to be $N - 2$.

It is customary to arrange the values of t in tabular form for all degrees of freedom from one to 30, for a few larger values, and for infinity. The t-distribution for infinite degrees of freedom is identical to the normal and will be substituted for the normal in hypothesis testing. Where the table is incomplete and does not include the exact degrees of freedom specified for a given problem, the appropriate thing to do is to use the table value closest to the desired value, but on the low side. Discarding degrees of freedom in this fashion gives a conservative approach to hypothesis testing, and rejection will be slightly less likely. This recommendation is not limited to the t-distribution, but will be used with other statistics as well.

23.4 THE TEST OF A HYPOTHESIS ABOUT THE MEAN

The t-distribution may be used as a model for testing a hypothesis about the mean of a normally distributed population, just as the normal distribution was used in Chapter 19, but with the additional advantage that the *t*-test is valid for samples of any size. In order to use this test, the investigator must be able to specify some hypothesized mean to which he wishes to compare the mean of the population under study. Typically, the procedure will be used to determine whether the population of interest differs significantly from the norm group used in standardizing the measuring instrument.

The test assumes that the sampling distribution of the mean is normal. This is true for samples of any size drawn from normal populations, and it is satisfactorily approximated by samples of adequate size from non-normal populations. If the departure is moderate, as in the case of the typical examination prepared by a classroom teacher, a sample of 30 measures is probably adequate. If the departure from normality is more than moderate, a larger sample is required. A sample of size 100 is adequate for almost any research situation.

The procedure for conducting the investigation is outlined as follows:

1. A random sample of measures is drawn from the population under study, and the mean of this sample (M) is calculated.
2. With μ representing the hypothesized mean and S_M representing an estimate of the standard error of the mean, the t is calculated by:

$$t = \frac{M - \mu}{S_M} \quad \text{where } S_M = \sqrt{\frac{SS}{N(N - 1)}}$$

3. At some predetermined level of significance, the calculated t is compared to the tabled value with the appropriate number of degrees of freedom ($df = N - 1$). A one- or two-tailed test may be used. If the calculated value equals or exceeds the tabled value, the finding is significant, and the hypothesis is rejected; the difference between the two means is significant at the specified level. If the calculated value is smaller than the tabled value, the hypothesis is retained; there is no significant difference between the means (see Table 7).

The procedure outlined here is sometimes referred to as the "one-sample *t*-test." The situation is illustrated by Figure 23.2, for which the null hypothesis is:

$$\mu_1 = \mu_0.$$

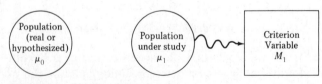

Figure 23.2   One-sample *t*-test.

23.5   EXAMPLE

The one-sample t-test will be illustrated with the same example used in Section 19.3 to illustrate the use of the normal model. In the example, a sociologist is exploring the relationship between a certain kind of cultural deprivation and adult intelligence test scores. He hypothesizes that the mean score on the Wechsler Adult Intelligence Scale for the population under study is lower than that of the population on which the test was standardized.

1. H: $\mu_1 = \mu_0$ A: $\mu_1 < \mu_0$
2. One-sample t-test, $N = 100$.
3. .005 level, one-tailed, $df = 98$.
   R: $t \leq -2.660$ (the tabled value for $df = 60$ is used).
4. Let us assume that the sample mean (M) is 94.3 as before, and that the estimate of the standard deviation (S) is 15. In the example using the normal model, the standard deviation of the norm group ($\sigma = 15$) was used.

$$S_M = \frac{S}{\sqrt{N}} = \frac{15}{\sqrt{100}} = 1.5$$

$$t = \frac{M - \mu_0}{S_M} = \frac{94.3 - 100}{1.5} = -3.80$$

5. Reject the null hypothesis. The population under study does have a lower mean IQ than the population on which the test was standardized.

The use of the t-test rather than the normal model for situations of this sort is recommended.

23.6   INTERVAL ESTIMATION

The t-distribution is to be preferred over the normal for constructing confidence intervals for small samples. While confidence intervals may be constructed for all of the common statistics, the confidence interval for the mean is that which is most often desired. The formulas for the upper and lower limits are the same as those used when calculating confidence limits with the normal model, except that the appropriate t is substituted for the z.

As always, two pieces of information—the level of significance and the degrees of freedom — will be needed to enter the t-table. The level of significance, in this case, will be equal to one minus the level of confidence, and the two-tailed t will be used. The degrees of freedom will be the same as that used in calculating the unbiased estimate of the standard error of the mean, and this will usually be $N - 1$.

The formulas for the confidence limits are as follows:

$$\text{upper limit} = M + t_c S_M$$
$$\text{lower limit} = M - t_c S_M$$

The data which served to illustrate the use of the normal model for interval estimation will also be used to illustrate the use of the t for this purpose. A sample of size 25 was drawn, the sample mean was 100, and the standard error of the mean was determined to be 3. For the 90 percent confidence interval with 24 degrees of freedom, the t is 1.711, and the confidence limits of the mean are 94.9–105.1 for the example. This interval is slightly larger than that yielded by the normal model; it would have been much larger had a smaller sample been used.

EXERCISES

23.1 Suppose an investigator wishes to test the hypothesis that the mean of a certain population is 30. A random sample of ten measures is drawn:

$$5, 15, 20, 20, 25, 25, 30, 35, 35, 40$$

Using the five-step procedure for hypothesis testing, test the hypothesis that there is no difference between the population mean and the hypothesized mean. Use the .05 level and a two-tailed test.

23.2 Calculate the 95 percent confidence limits for the mean of Exercise 23.1.

23.3 A junior high school counselor hypothesizes that entering seventh grade students in his school score above the national norm (grade equivalent 7.00) on a certain section of the Iowa Tests of Basic Skills. A random sample of 25 recent test scores is drawn from student files, and it yields a mean of 7.80 and sum of squares of 150.
(a) Determine the region of rejection for the 0.05 level.
(b) Calculate the appropriate error term for the test of the null hypothesis.
(c) Calculate the t-statistic from the sample data.
(d) Determine the outcome of the statistical test.
(e) Establish the 95 percent confidence limits for the sample mean.

SELECTED REFERENCE

WALKER, HELEN M., "Degrees of freedom," *Journal of Educational Psychology*, **31**, 253–269 (April 1940).

24

THE *t*-TEST FOR TWO INDEPENDENT SAMPLES

24.1 FUNCTION

The situation in which the researcher assigns one group of subjects to an experimental treatment and uses another group of similar subjects as a control group is one of the most popular experimental designs. This is equivalent to using two experimental treatments, and our discussion will not distinguish between the use of a control group or of a second experimental group. For example, in a learning experiment, the experimental group might be subjected to a new instructional technique, although the control group receives the traditional instruction. The traditional instruction is simply a second experimental treatment. In a medical experiment, one group of subjects might receive a new medicine, while the second group receives a placebo (a medicine with no active ingredients). Whether the investigator chooses to regard the subjects receiving the placebo as a control group or as a second experimental group is of no consequence. The experimental treatment or other variable which distinguishes the two groups from each other is the independent variable.

At the conclusion of the experiment, all subjects in both groups are measured with the same criterion — a psychological test or other instrument intended to determine whether anything of consequence took place as a result of the experiment. The criterion is also called the dependent variable. The *t*-test for two independent samples is used to determine whether the criterion means for the two groups differ significantly.

It is also permissible to use the *t*-test for two independent samples to examine existing differences between random samples from two different populations. No experimental treatments are administered by the investigator in this situation; he is seeking to determine whether the two populations differ significantly on some criterion variable. This is equivalent to testing the null hypothesis that, with respect to the criterion variable, the two populations are in fact a single population. Thus an investigator might be interested in determining whether a collection of children should be regarded as a single population, or as two populations, one male and the other female. Similarly, with respect to a given criterion, one might be interested in determining whether a given group of psychology students is simply a group of psychology students, or two groups, one consisting of psychology majors and the other consisting of nonpsychology majors.

The examples cited in the first paragraph of this section represent experimental research. The situation is illustrated by Figure 24.1. The null hypothesis is $\mu_1 = \mu_2$; however, the student is reminded that these are parameters of hypothetical populations. The examples cited in the third paragraph of this section represent *ex post facto* research. The situation is illustrated by Figure 24.2. As before, the null hypothesis is $\mu_1 = \mu_2$; these, however,

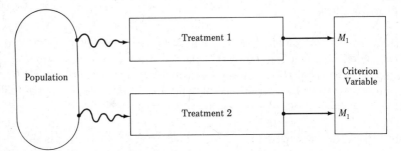

Figure 24.1   Simple randomized design — experimental research.

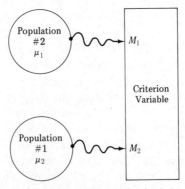

Figure 24.2   Simple randomized design — *ex post facto* research.

are real parameters from actual populations. Both the experimental and the *ex post facto* applications may be classed as simple (or completely) randomized designs.

24.2 RATIONALE

In order to test a hypothesis about the difference between the means of two independent samples, one must first be able to estimate the standard error of the difference between the means of two independent samples. This statistic was identified in the chapter on sampling theory, and it may be estimated from:

$$S_{M1-M2} = \sqrt{\frac{SS_1 + SS_2}{n_1 + n_2 - 2}\left(\frac{1}{n_1} + \frac{1}{n_2}\right)}$$

The reader should note that there are two important assumptions underlying the use of this estimate: (1) the distribution of the measures in both samples is normal, and (2) the variances of the two populations are equal. Moderate departures from these assumptions have proved to be of no practical consequence. When the two samples are of equal or near equal size, the statistical test is quite insensitive to violations of these assumptions.

Dividing the difference between the two means by the standard error of the difference yields a statistic that is distributed as the *t*-distribution if the two population means are equal. To the extent that the two means are not equal, we would expect the value of the calculated *t* to be inflated, and the probability of rejecting the null hypothesis becomes greater than the level of significance.

24.3 PROCEDURE

1. Two random samples are independently drawn from the population under study, and these are subjected to two different experimental treatments.
2. At the conclusion of the experiment, a single criterion measure is administered to all subjects, and the mean of each group is calculated. The standard error of the difference between the means is calculated, using the formula given in the previous section.
3. The *t* is calculated by:

$$t = \frac{M_1 - M_2}{S_{M_1 - M_2}}$$

4. The calculated t is compared to the tabled value at the desired level of significance and with degrees of freedom equal to $n_1 + n_2 - 2$. A one- or two-tailed test may be used, although the two-tailed test is more common with this design. If the calculated statistic equals or exceeds the tabled value, the hypothesis is rejected, and the observed difference between the two means is a significant one. The two experimental treatments are determined to produce different results at the specified level of significance, and the treatment which produced the higher mean is usually regarded as superior, although the exact interpretation is dependent upon the nature of the dependent variable. If the calculated value is smaller than the tabled value, the hypothesis is retained, and no significant difference between the means has been established.

24.4 EXAMPLE

A researcher is studying the effects of two different diets on weight gain of white rats. Two random samples of size ten each are chosen from an available colony of adolescent male white rats. One sample is fed Diet A and the other Diet B during the experimental period. The rats are weighed before the experiment and again following the experiment; the criterion variable is gain in weight. The test of the null hypothesis is recorded below, complete with all sample data:

1. H: $\mu_A = \mu_B$ A: $\mu_A \neq \mu_B$
2. t-test for independent samples, $n_A = n_B = 10$.
3. .05 level, two-tailed, $df = 18$.
    R: $t \leq -2.101$ or $t \geq +2.101$.
4. Sample A: 1, 2, 3, 4, 4, 5, 5, 8, 9, 9

$$n_A = 10, \quad M_A = 5, \quad SS_A = 72$$

Sample B: 4, 6, 7, 7, 8, 8, 9, 10, 10, 11

$$n_B = 10, \quad M_B = 8, \quad SS_B = 40$$

$$S_{M_A - M_B} = \sqrt{\frac{72 + 40}{10 + 10 - 2} \left(\frac{1}{10} + \frac{1}{10}\right)} = 1.12$$

$$t = \frac{5 - 8}{1.12} = -2.67$$

5. Reject the null hypothesis. There is a significant difference in the means of the two samples. The data *suggest* that Diet B secures greater weight gains than Diet A.

24.5 DISCUSSION

The t-test for two independent samples is one of the most popular statistical tests. It is a very powerful test that lends itself to a wide variety of research problems.

The matter of interpreting the results from the statistical test deserves special consideration. If the finding is nonsignificant, the investigator is not privileged to conclude that μ_1 is indeed equal to μ_2. While this might actually be the case, it is also likely that the investigator simply did not collect enough evidence to detect the existing difference between the two parameters. Larger samples (perhaps much larger) might reveal that there is indeed some difference between the two. If the finding is *statistically significant*, one usually has to decide whether the observed difference is *practically significant*. Assuming the investigator has some notion as to what constitutes practical significance, an examination of the sample means (which are unbiased estimates of the corresponding parameters) may enable him to arrive at some reasonable conclusion.

The t-test does not directly address itself to the degree of relationship between the independent and dependent variables. This is often a matter of no small importance in interpreting the outcome of the research. The following solution is suggested:

1. Let Y_i equal the criterion score for each subject.
2. Assign to each subject in the first sample a score of $X_i = 1$, and to each subject in the second sample a score of $X_i = 0$. Thus, each subject has a pair of scores.
3. Calculate either the Pearson correlation coefficient or the mathematically equivalent point-biserial coefficient from the paired scores.
4. The square of this correlation coefficient (r^2) is that proportion of the criterion variance that is accounted for by the independent variable.

The following mathematically equivalent procedure may prove computationally less demanding:

$$r^2 = \frac{t^2}{N - 2 + t^2}$$

where t is the test statistic calculated from the sample data and $N = n_1 + n_2$.

There are certain assumptions underlying the use of the t-test for two independent samples, and it is important to note that these assumptions apply to the two populations specified in the statement of the null hypothesis. In the experimental setting where the investigator actually manipulates the independent variable, the two populations do not exist in reality — they are hypothetical populations which have been subjected to experimentation, and from which the two samples (after experimentation) could be regarded as random samples. After experimentation, the two samples are no longer random samples from the original parent population. The use of the t-test for two independent samples assumes that the dependent variable is measured on at least an interval scale (at least some authors feel it does), that it is normally distributed in the two populations, and that the two populations have equal variances. A final assumption that the means of the two populations are equal is necessary for the distribution of the calculated t to match that of the tabled t. When the hypothesis is rejected, we prefer to assume that it is this final assumption which has been violated. Fortunately, this t-test is much more sensitive to violation of the assumption of equal means than to violation of the assumptions of normality or of homogeneous variances. Tests for normality and for equal variances are discussed in other parts of this text (see Sections 29.5, 32.5, and 35.2). Unfortunately, tests for normality are not very effective with small samples, and the benefits of the central limit theorem eliminate the need for such tests with large samples. The tests for equal variances are very sensitive to departures from normality and are of limited value for testing the assumption of homogeneous variances. Generally, research with respect to the t-distribution suggests that when the investigator is working with samples of the same size or nearly the same size, he may ignore these assumptions unless he has reason to believe that his measures deviate greatly from them. The one exception appears to be the situation in which the assumptions of normality and homogeneous variances are both violated, in which case the probability of a Type I error is likely to be somewhat larger (perhaps twice as large) than the level of significance.

The Mann-Whitney U-test is a nonparametric alternative to the t-test which may be used with ordinal measures or with data that deviate from the normal or in the situation where heterogeneous variances are encountered. It is almost as powerful as the t, but it becomes less powerful if many tie scores are encountered. The median test, which uses the chi-square distribution, is another nonparametric alternative, but it is not as powerful as the Mann-Whitney.

EXERCISES

24.1 Air Force psychologists are conducting research into the relative effectiveness of two methods of training navigators. The first method makes use of computer-simulated flight; the second uses traditional classroom instruction. Two random samples of size ten each are selected for participation in the experiment; however, two of the men in the first group get chicken pox and are dropped from the experiment. The criterion score is a performance test which has been scaled to yield a stanine score. The criterion scores for the two samples are as follows:

Group I (simulated flight):   2, 5, 5, 6, 6, 7, 8, 9

Group II (traditional):   1, 2, 2, 3, 4, 4, 5, 5, 6, 8

Use the five-step procedure for testing the null hypothesis at the .05 level.

24.2 Refer to Exercise 12.3. Use the *t*-test for two independent samples to test the hypothesis of no difference between boys and girls at the .05 level. Use a one-tailed alternative that girls will score higher than boys.

24.3 Given the following data:

$$n_1 = 30, \quad M_1 = 48.2, \quad SS_1 = 1740$$
$$n_2 = 29, \quad M_2 = 52.4, \quad SS_2 = 1622$$

test the hypothesis of equal means using a two-tailed test at the .05 level.

24.4 An experiment was conducted to determine the relative effectiveness of two different methods of instruction. An available group of students was randomly divided into two groups of equal size, one receiving instruction by Method A and the other by Method B. A test was given at the end of the experiment; the mean for Method A was 53 and the sum of squares was 1011, the mean for Method B was 59 and the sum of squares was 879. The total number of students involved was 30.

(a) Determine the region of rejection for the .01 level.
(b) Calculate the appropriate error term for the test of the hypothesis of equal means.
(c) Calculate the *t*-statistic for the sample data.
(d) Determine the outcome of the statistical test.

SELECTED REFERENCE

BONEAU, C. ALAN, "The effects of violations of assumptions underlying the *t* test," *Psychological Bulletin*, **57**, 49–64 (1960).

25

THE t-TEST FOR TWO RELATED SAMPLES

25.1 FUNCTION

The t-test for two independent samples is used with simple randomized designs — designs in which randomization is the only provision for controlling the contribution of irrelevant variables. When some additional provision is made for the control of these variables, the size of the error term is reduced, and a more precise experimental design is achieved.

Under certain circumstances, it is possible to increase greatly the precision of an experiment by switching from two independent samples to two related samples. This calls for the t-test for two related samples and fits a number of experimental designs.

One way to increase the precision of an experiment is to draw a single random sample and administer both experimental treatments (at different times) to the same group of subjects. The criterion measure must be administered following each treatment. Of course, this procedure is valid only when there is no carry-over from one treatment to the next and from one measurement to the next. This design is not likely to be practical in research where learning takes place. However, it is quite practical in some psychophysical and medical experiments. Whenever this procedure is valid, an extremely precise research design is possible due to the unique control of irrelevant variables. Designs in which the criterion is administered more than once to the same group of subjects are called *repeated measurements* designs (Figure 25.1).

Occasionally, the experimenter will be able to use an experimental group as its own control. In this case, the criterion measure is administered to

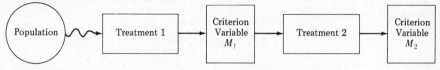

Figure 25.1 Repeated measurements design — experimental research.

the sample before and after the experimental treatment. The test of significance is used to determine whether any change has taken place. Of course, the experimenter must be on the lookout for effects other than the experimental treatment which might cause a significant shift in the criterion variable.

Finally, this type of analysis may be used with an experiment in which two random samples have been matched on one or more variables. Typically, the design calls for identifying a variable which is known to correlate with the criterion (IQ or a pretest on the criterion are common in learning research). The population is then split into pairs on the basis of the pretest measure, each subject being matched as closely as possible with another subject (Figure 25.2). A random sample of pairs of subjects is drawn, and one member of each pair is randomly assigned to each of the two treatments. This design is especially valuable to the investigator who must work with a very small population — handicapped children, for example. It is not likely to be as precise as a repeated measurements design, but it is applicable to learning research.

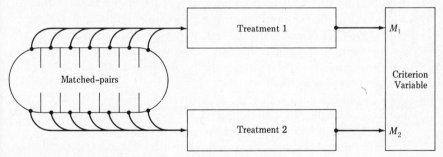

Figure 25.2 Matched-pairs design — experimental research.

25.2 RATIONALE

In order to test a hypothesis about the difference between the means of two related samples, one must first be able to estimate the standard error of the difference between the means of two related samples. This statistic was identified in the chapter on sampling theory, and it may be estimated from:

$$S_{\bar{D}} = \sqrt{\frac{\Sigma \, d^2}{N(N-1)}}$$

where

$$\Sigma \, d^2 = \Sigma \, (D - \bar{D})^2 = \Sigma \, D^2 - \frac{(\Sigma \, D)^2}{N}$$

The symbol D represents the difference between a pair of scores, $\bar{D}$ the mean of these differences or $M_2 - M_1$, and N the number of *pairs* of scores. The degrees of freedom for this error term is $N - 1$, since it is calculated from N differences, and $\bar{D}$ is estimated from sample data.

To use this as the error term in a *t*-test, one must assume that the distribution of the difference scores is normal — a reasonable assumption in practically any behavioral research using this statistic.

25.3   PROCEDURE

1. The procedure for conducting the experiment is largely determined by the selection of one of the three designs discussed in Section 25.1.

   Where two experimental treatments are administered to the same group of subjects, a single random sample is drawn, the two treatments administered (at different times) to all subjects, and a single criterion measure is administered twice, once after each treatment. It is usually desirable to randomize the order of the two treatments, with half of the subjects getting one treatment first and the other half getting the other treatment first. However, this cannot be expected to compensate completely for any carryover from treatment to treatment or measure to measure, and the design should not be used if carryover is suspected.

   In the situation in which the experimental group is used as its own control, a single random sample is drawn, and the criterion is administered both before and after the experimental treatment.

   When matched pairs are used with two different experimental treatments, paired random samples are drawn, as described in Section 25.1. The conduct of the experiment will be much like that with two independent samples; however, the criterion measures will be paired for purposes of analysis.

2. At the conclusion of the experiment, the difference between each pair of scores and the difference between the two means are calculated. When a group is used as its own control, the first score should probably be subtracted from the second (and the first mean from the second), since this yields a positive difference for a gain on the criterion.

The standard error is calculated using the formula given in Section 25.2.
3. The t is calculated by:

$$t = \frac{\bar{D}}{S_{\bar{D}}} \quad \text{where } \bar{D} = M_2 - M_1 \quad \text{and} \quad D = X_2 - X_1$$

4. The calculated t is compared to the tabled value at the desired level of significance and with degrees of freedom equal to $N - 1$ (where N is the number of pairs of scores). A one- or two-tailed test may be used. If the calculated statistic equals or exceeds the tabled value, the hypothesis is rejected, and it is concluded that the observed difference between the two means is a significant one. The exact interpretation of a significant finding is dependent upon the experimental design used and the nature of the criterion variable. If the calculated t is smaller than the tabled value, the hypothesis is retained, and no significant difference between the two means has been established.

25.4 EXAMPLE

A researcher is studying the relative merits of two different methods of teaching mentally retarded children to solve simple problems. There are 20 such children available to participate in the research, and these are grouped in ten pairs, with the two children in each pair matched as closely as possible on such variables as age, IQ, social development, and so on. The children are randomly assigned to two training classes, with one member of each pair going into each class. One class is taught by Method A and the other by Method B. At the end of the training period, each child is administered an examination consisting of 20 problems, and his score is recorded as the number of problems solved correctly. The data yielded by the experiment are recorded in Table 25.1.

Table 25.1 Data from the problem-solving experiment

Pair	Method A	Method B	D	D²
1	12	16	4	16
2	11	9	-2	4
3	12	15	3	9
4	10	10	0	0
5	10	9	-1	1
6	9	12	3	9
7	7	9	2	4
8	6	10	4	16
9	4	9	5	25
10	4	3	-1	1
			$\Sigma D = 17$	$\Sigma D^2 = 85$

The test of the null hypothesis is outlined below:

1. $\mu_A = \mu_B$ $A: \mu_A \neq \mu_B$
2. t-test for related samples, $N = 10$.
3. .05 level, two-tailed test, $df = 9$.
   R: $t \leq -2.262$ or $t \geq +2.262$.
4. $\bar{D} = 17/10 = 1.70$.

$$\Sigma d^2 = 85 - \frac{17^2}{10} = 56.1$$

$$S_{\bar{D}} = \sqrt{\frac{56.1}{10(10 - 1)}} = .79$$

$$t = \frac{1.70}{.79} = 2.15$$

5. Retain the null hypothesis. The difference between the sample means is nonsignificant.

25.5   DISCUSSION

The t-test for related measures is a very powerful statistical test that may be used with certain kinds of experimental designs. Its great advantage over the t-test for independent samples is the increase in precision accomplished by the reduction in sampling error. Its use with two experimental treatments in the repeated measurements design is restricted to the situation in which the treatment effects are known to be temporary. There must be no carryover from treatment to treatment or from measure to measure. Obviously, this design is of limited value in learning research.

The use of this statistical test assumes that the distribution of the differences between the two sets of criterion measures is normal in the populations specified by the null hypothesis. This is a reasonable assumption in practically any behavioral research and should be of little concern to the investigator unless he has reason to believe otherwise. The use of the test is believed by some authors to require measurement on at least an interval scale.

Of the available nonparametric alternatives to the t-test for related samples, the Wilcoxon matched-pairs signed-ranks test is probably the best. It is intended for use with ordinal data.

EXERCISES

25.1 An investigator has developed an individualized program of instruction in problem solving which he believes will raise the IQ of an individual. A random sample of eight children is chosen from a junior high school physical education class to participate in the instructional program. Their IQ scores are assessed before and after the instructional program, using equivalent forms of the same instrument. The scores are recorded below:

Before: 81   84   90   97   108   111   118   124

After: 89   88   94   96   118   111   121   121

Use the five-step procedure for testing the null hypothesis at the .05 level.

25.2   A sociologist is studying the effects of a certain motion picture film upon the attitudes of black students toward white students. He hypothesizes that viewing the film will cause the scores of the students on a certain attitude scale to shift downward. The scores of 12 participating students are recorded below:

Before: 10   13   18   12   9   8   14   12   17   20   11   7

After: 5   9   13   17   4   5   11   14   13   18   10   7

Use the five-step procedure for testing the null hypothesis at the .05 level.

26

THE MANN-WHITNEY *U*-TEST

26.1 FUNCTION

The Mann-Whitney *U*-test is a nonparametric alternative to the *t*-test for two independent samples. It is commonly used where the experimenter draws two random samples from the same parent population, subjects each to a different experimental treatment, and compares the two on a single criterion. Of course, the experimenter may designate one of his samples as a control group, in which case the experimental treatment is no treatment. The *U*-test may also be used in the situation in which independent random samples are drawn from two different parent populations and compared on a single criterion to determine whether the two populations differ.

The Mann-Whitney *U*-test requires data on at least an ordinal scale, and these data are assumed to be continuously distributed. It does not require normality of distribution nor homogeneity of variance for the groups under study. This is one of the most useful of the nonparametric tests; it is almost as powerful as the *t*-test under common research conditions, and the computations are quite simple. It is especially useful with small samples, since violations of the assumptions underlying the *t*-test are most likely to be detrimental and most likely to go undetected with small samples.

26.2 RATIONALE

Suppose a measure on some specified continuous variable is available for every member of a population of interest. Let us draw two independent random samples from this population, and let n_a be the size of Sample A, n_b the size of Sample B, and N the sum of n_a plus n_b.

Now let us combine the two sets of scores and rank them from 1 to N, using the subscripts a and b to identify each rank as belonging to a particular sample. If the original scores were truly continuously distributed (no tie scores), the number of possible combinations of ranks is equal to

$$\binom{N}{n_a} \quad \text{or} \quad \binom{N}{n_b}$$

these being mathematically equivalent. If the two samples were indeed independent random samples from the same parent population, each of these combinations is equally probable. If, however, the samples came from two populations which differ on the criterion variable, there will be a tendency for the ranks of Sample A to cluster at one end of the distribution and those of Sample B to cluster at the other end.

Let U_a be the number of times a rank from Sample A is preceded by a rank from Sample B, and U_b the number of times a rank from Sample B is preceded by a rank from Sample A. For example, consider this set of data.

$$1_a, 2_a, 3_b, 4_a, 5_b, 6_b$$
$$U_a = 0 + 0 + 1 = 1$$
$$U_b = 2 + 3 + 3 = 8$$

Let us illustrate the calculations by concentrating on U_b: the first b-score (3_b) is preceded by two a-scores, the second b-score (5_b) by three a-scores, and the third b-score (6_b) by three a-scores. Thus the value of U_b is 8. With a little algebra, one may demonstrate that:

$$U_a = n_a n_b - U_b \quad \text{and} \quad U_b = n_a n_b - U_a$$

We will concentrate our attention on the smaller of these two statistics, and designate it as U.

The value of U will be zero if all of the ranks from Sample A are clustered at one end of the distribution and all of the ranks for Sample B at the other; that is, U equal to zero represents the greatest difference possible between the two samples. The value of U will increase as the ranks are intermixed, that is, as the samples become more alike.

Refer to the data given earlier; the number of combinations for N equal to six and n_a equal to three is 20. The probability that all of the ranks of Sample B would be larger than the ranks of Sample A is 1/20. The value of U in this case would be zero. Thus, U equal to zero would be the region of rejection for a one-tailed test at the .05 level or for a two-tailed test at the .10 level. In similar fashion, one could establish the probabilities associated with different values of U (values less extreme than zero) and for different size samples. The values of U needed to reject the null hypothesis at various levels of significance and for various sample sizes will be found in Table 8 of the Appendix.

It will, as a general rule, be more convenient to determine the value of U from these formulas:

$$U_a = n_a n_b + \frac{n_b(n_b + 1)}{2} - \Sigma R_b$$

$$U_b = n_a n_b + \frac{n_a(n_a + 1)}{2} - \Sigma R_a$$

where ΣR_a is the sum of the ranks for Sample A and ΣR_b is the sum of ranks for Sample B.

Referring to the example given earlier, we see that

$$U_a = 9 + \frac{12}{2} - 14 = 1$$

$$U_b = 9 + \frac{12}{2} - 7 = 8$$

Thus, the value of U is 1, as before.

26.3 PROCEDURE

1. Two random samples are independently drawn from the population under study, and these are subjected to two different experimental treatments. Let n_a be the size of one sample, n_b the size of the other, and N the sum of n_a plus n_b.
2. At the conclusion of the experiment, a single criterion measure is administered to all subjects. The measures from the two samples are combined in a single-ordered series, and ranked from 1 to N, using the subscripts a and b to identify each rank as belonging to a particular sample.
3. Determine the values of U_a and U_b, as described in Section 26.2. Let U equal the smaller of these.
4. The calculated U is compared to the tabled value at the desired level of significance and for the appropriate sample sizes. A one- or two-tailed test may be used. If the calculated value is *equal* to or *smaller* than the tabled value, the hypothesis is rejected; the observed differences in the ranks are significant. The two experimental treatments are determined to produce different results at the specified level of significance, and the treatment which produced the higher ranks is usually regarded as superior, although the exact interpretation is

dependent upon the nature of the criterion variable. If the calculated value is *larger* than the tabled value, the hypothesis is retained, and no significant difference in the ranks has been established.

Table 8 in the Appendix uses the symbol m to represent the size of the larger sample and the symbol n to represent the size of the smaller sample. If the samples are the same size, the investigator may use either of these for n_a and the other for n_b. For example, if m equals 30 and n equals 20, the critical value of U for a one-tailed test at the .05 level is 216. Any value of U equal to or smaller than 216 is significant for the one-tailed test at the .05 level with samples of these sizes.

26.4   EXAMPLE

The experiment which served to illustrate the use of the t-test for two independent samples (see Section 24.4) will also be used to illustrate the use of the Mann-Whitney U-test.

1. *H*:   No difference in the ranks for the two diets.
   A: Higher ranks are associated with one of the diets.
2. Mann-Whitney U-test, $n_a = n_b = 10$.
3. .05 level, two-tailed test.
   R: $U \leq 23$.
4.

All scores	All ranks	A-ranks	B-ranks
1a	1a	1	
2a	2a	2	
3a	3a	3	
4a	5a	5	
4a	5a	5	
4b	5b		5
5a	7.5a	7.5	
5a	7.5a	7.5	
6b	9b		9
7b	10.5b		10.5
7b	10.5b		10.5
8a	13a	13	
8b	13b		13
8b	13b		13
9a	16a	16	
9a	16a	16	
9b	16b		16
10b	18.5b		18.5
10b	18.5b		18.5
11b	20b		20

$$\Sigma R_a = 76 \qquad \Sigma R_b = 134$$

Notice that in the case of tie scores, the ranks are evenly divided between them. For example, the 18th and 19th ranks are evenly divided for the two scores of 10.

$$U_a = 100 + \frac{110}{2} - 134 = 21$$

$$U_b = 100 + \frac{110}{2} - 76 = 79$$

Checking the calculations:

$$U_a + U_b = n_a n_b$$
$$21 + 79 = 100$$
$$U = 21$$

5. Reject the null hypothesis. There is a significant difference in the ranks of the two samples. The data suggest that Diet B secures greater weight gains than Diet A.

26.5 NORMAL APPROXIMATION WITH TIE CORRECTION

Whenever there are substantial numbers of tie scores or when the samples are too large for the use of Table 8 in the Appendix, the normal approximation with tie correction will be preferred to the statistical test outlined earlier in this chapter.

Whenever large numbers of scores are to be converted to ranks, the following strategy is recommended: (1) Arrange the data from both samples into a frequency distribution with the values of cf and f for each score interval conveniently available, as in Table 3.7. (2) The rank for each score interval may be calculated from:

$$\text{rank} = cf - \frac{f}{2} + 0.5$$

The value of U may be calculated as in Section 26.4, and the normal approximation is calculated from:

$$z = \frac{U - (n_a n_b / 2)}{\sqrt{\dfrac{n_a n_b (n_a + n_b + 1)}{12}}}$$

The student should recognize that the formula has a random variable, minus its expected value, divided by its standard error. When tie scores are

encountered, the effect is to inflate the error term and reduce the size of the
z-score. A tie correction factor for the error term may be calculated as
follows: (1) For each score level where ties are encountered, the quantity
$T = (f^3 - f)/12$ should be calculated, where f is the frequency of the score
interval (the number of tied scores). (2) The quantity ΣT should be calculated by summing all the values of T. The normal approximation with tie
correction may be calculated from:

$$z = \frac{U - (n_1 n_2 / 2)}{\sqrt{\left(\dfrac{n_1 n_2}{N^2 - N}\right)\left(\dfrac{N^3 - N}{12} - \Sigma T\right)}}$$

where $N = n_1 + n_2$. The value of z may be referred to Table 3 in the
Appendix to determine its probability of occurrence, or it may be referred
to Table 21.1 for hypothesis testing. In the latter case, the hypothesis is
rejected if the calculated z equals or exceeds the tabled value.

The procedure is illustrated for the data of Sections 24.4 and 26.4:

X	f	T	cf	Rank	A-ranks	B-ranks
11	1	0.0	20	20.0		20.0
10	2	0.5	19	18.5		37.0
9	3	2.0	17	16.0	32.0	16.0
8	3	2.0	14	13.0	13.0	26.0
7	2	0.5	11	10.5		21.0
6	1	0.0	9	9.0		9.0
5	2	0.5	8	7.5	15.0	
4	3	2.0	6	5.0	10.0	5.0
3	1	0.0	3	3.0	3.0	
2	1	0.0	2	2.0	2.0	
1	1	0.0	1	1.0	1.0	
	20	7.5			76.0	134.0

The calculation of the statistic without tie correction is:

$$z = \frac{21 - 50}{\sqrt{175}} = \frac{-29}{13.23} = -2.19$$

The calculation of the statistic with tie correction is:

$$z = \frac{21 - 50}{\sqrt{173.05}} = \frac{-29}{13.15} = -2.20$$

The finding is significant as before.

26.6   DISCUSSION

The Mann-Whitney U-test is one of the more popular nonparametric tests. It is a suitable alternative to the t-test for two independent samples when the assumptions underlying the t-test cannot be met. The U-test is almost as powerful as the t-test (about 95 percent relative power with typical research samples) and does not require homogeneity of variance nor normality of distribution. It does, however, have the assumption of continuity of distribution, and when large numbers of tie scores are encountered the exact test is much less powerful. It is especially appropriate for use with small samples when there is the greatest danger from violating the assumptions underlying the equivalent t-test.

The normal approximation with tie correction is recommended when the samples are large or when large numbers of tie scores are encountered. It has been demonstrated that the approximation is excellent for samples as small as 10 in size, and the tie correction factor is effective even when large numbers of ties are encountered. Although the tie correction factor does not appear to alter the value of a given statistic appreciably, it has been demonstrated that its long-range effects are profound, and its use is highly recommended.

EXERCISES

26.1   Refer to Exercise 24.1. Substitute the Mann-Whitney U-test for the t-test for two independent samples.

26.2   Refer to Exercise 24.2. Substitute the Mann-Whitney U-test for the t-test for two independent samples.

26.3   Given the following independent random samples which have been subjected to experimentation, and in which the individuals have been rated according to the scores recorded:

Sample I: 38, 39, 44, 47, 50, 51, 52, 59, 60, 61, 73, 74, 78, 84, 90
Sample II: 42, 43, 54, 62, 67, 69, 70, 75, 80, 81, 86, 89, 91, 97, 98

Use the Mann-Whitney U-test to determine whether there is a significant difference in the ranks of the two samples. Use a two-tailed test at the .05 level.

SELECTED REFERENCES

BRADLEY, J. V., *Distribution-Free Statistical Tests*. Englewood Cliffs, N. J.: Prentice-Hall, 1968.

CHANDA, K. C., "On the efficiency of the two-sample Mann-Whitney test for discrete populations," *The Annals of Mathematical Statistics*, **34**, 612–617 (1963).

CONOVER, W. J., *Practical Nonparametric Statistics*. New York: John Wiley, 1971.

MANN, H. B., and D. R. WHITNEY, "On a test of whether one of two random variables is stochastically larger than the other," *The Annals of Mathematical Statistics*, **18**, 50–60 (1947).

McNEIL, D. R., "Efficiency loss due to grouping in distribution-free tests," *Journal of the American Statistical Association*, **62**, 954–965 (1967).

PUTTER, J., "The treatment of ties in some nonparametric tests," *The Annals of Mathematical Statistics*, **26**, 368–386 (1955).

SIEGEL, S., *Nonparametric Statistics for the Behavioral Sciences*. New York: Mc-Graw-Hill, 1956.

WILCOXON, F., "Individual comparisons by ranking methods," *Biometrics*, **1**, 80–83 (1945).

WOODS, DONALDSON G., "A comparison of the relative power and robustness of Mann-Whitney, Kruskal-Wallis, and analysis of variance tests employing discrete score point scales common to educational research," unpublished Ph.D. dissertation, Kansas State University, 1972.

27

THE WILCOXON
MATCHED-PAIRS
SIGNED-RANKS TEST

27.1 FUNCTION

The Wilcoxon matched-pairs signed-ranks test is a nonparametric alternative to the t-test for two related samples. It may be used in either repeated measurements or matched-pairs types of designs. For a discussion of these designs, see Section 25.1. The Wilcoxon test requires data on at least an ordinal scale, and these data are assumed to be continuously distributed. The test does not require normality of distribution.

27.2 RATIONALE

Suppose that a random sample of paired measures is available from some population of interest. Let d_i be the difference between any pair of measures. Rank these difference scores from one to N (where N is the number of pairs), with respect to magnitude but without respect to sign (for example, 0, $+1$, -2, -3, $+4$, and so on). After ranking the difference scores in this fashion, separate the ranks into two groups, those corresponding to the positive difference scores and those corresponding to the negative difference scores. Let T_a be the sum of the ranks for the positive differences, T_b the sum of the ranks for the negative differences, and T be equal to the smaller of these two.

There are 2^N unique sets of signed ranks in the situation described. If the relationship between the scores in each pair is a completely random one, each of these 2^N sets is equally probable. If N is six, for example, there are 64 sets, and the probability that T_a will be zero is 1/64; the probability that it will be one is also 1/64, and the probability that it will be one or zero is 2/64. The probability that T_b will be zero is 1/64, and the probability that either T_a or T_b will be zero is 2/64. Following this pattern, the sampling distribution of T could be established for any sample size, and a table could be constructed for testing both one- and two-tailed hypotheses at any desired levels of significance. Table 9 in the Appendix is such a table.

If the relationship between the scores in each pair is a completely random one, the expected values of T_a and T_b would be the same, and the value of T would be maximum under these circumstances. If, however, there is a systematic tendency for the positive differences to be greater than or less than the negative differences, T will tend to be smaller, with T equal to zero representing two maximally different samples.

27.3 PROCEDURE

1. The procedure for conducting the experiment is largely determined by the experimental design (see Section 25.1). In any event, a collection of paired measures is produced by the experiment.
2. Let d_i be the difference between any pair of scores. Rank these differences from 1 to N with respect to their absolute magnitudes (without respect to sign). For example, consider this collection where N is eight:

Difference:	-1	-2	$+4$	-6	$+7$	$+8$	$+10$	-12
Ranks:	1	2	3	4	5	6	7	8

3. Separate the ranks into two groups, those corresponding to the positive differences and those corresponding to the negative differences. Take the sum of each of these two groups of ranks, and let T equal the smaller sum. In the example above, the sum of the ranks for the positive differences is 21; for the negative differences it is 15; T is 15.
4. The value of T is compared to the tabled values at the desired level of significance and for the appropriate value of N. A one- or two-tailed test may be used. If the calculated value is equal to or smaller than the tabled value, the hypothesis is rejected, and the observed differences in the ranks are significant. The exact interpretation of a significant finding is dependent upon the experimental design used and the nature of the dependent variable. If the calculated value is larger than the tabled value the hypothesis is retained, and no significant difference in the ranks has been established.

27.4 EXAMPLE

The experiment that served to illustrate the use of the t-test for two related samples (see Section 25.4) will also be used to illustrate the Wilcoxon matched-pairs signed-ranks test.

1. H: No difference in the ranks for the two methods.
    A: Higher ranks are associated with one of the methods.
2. Wilcoxon matched-pairs signed-ranks test, $N = 10$.
3. .05 level, two-tailed test.
    R: $T \leq 8$.
4.

A-scores	B-scores	d_i	Rank	Positive	Negative
12	16	4	7.5	7.5	
11	9	−2	3.5		3.5
12	15	3	5.5	5.5	
10	10	0	0.0		
10	9	−1	1.5		1.5
9	12	3	5.5	5.5	
7	9	2	3.5	3.5	
6	10	4	7.5	7.5	
4	9	5	9.0	9.0	
4	3	−1	1.5		1.5
				$T_a = 38.5$	$T_b = 6.5$

$T = 6.5$, but notice that one of the difference scores is zero and not ranked. Because it is neither positive nor negative, it is discarded from the analysis, and N is reduced to 9. A new region of rejection must be determined from Table 9. R: $T \leq 6$

5. Retain. No significant difference in the ranks was found. This is consistent with the earlier finding using the t-test.

27.5 DISCUSSION

An important assumption underlying the Wilcoxon test and many other nonparametric statistics is that of continuity of distribution. The presence of tie scores suggests that this assumption has been violated. Whether or not this is of any consequence depends upon the manner of calculating the test statistic. Two kinds of ties are possible with the Wilcoxon, and both are illustrated in the example quoted: (1) tie scores for a given pair resulting in a zero difference, and (2) tie ranks resulting from pairs having the same differences. The first type of tie is of great importance. The usual recommendation is to discard these from the analysis and reduce N accordingly. This procedure cannot be justified, since the presence of these ties is evidence for retention of the null hypothesis. If there is an even number of these ties, they may be evenly divided among the positives and negatives; if there is an odd number, one may be discarded and the remainder divided. A better procedure might call for randomly assigning them to the positives and negatives. An even more acceptable procedure would be to assign them to the algebraic sign most conducive to retention of the null hypothesis. Each of these suggestions is based upon the assumption that the data are truly continuous but crudely measured. None of these is a very satisfying solution, and the presence of this type of tie is a strong argument against the use of this statistical test. The second type of tie is less of a problem, since the corresponding ranks may be evenly divided among the tied differences.

The Wilcoxon matched-pairs signed-ranks test has been suggested as a nonparametric alternative to the t-test for two related samples when the underlying assumptions for the latter test cannot be met. The relative power of the Wilcoxon is on the order of 95 percent with small samples and somewhat less with larger ones. The Wilcoxon, however, is not without assumptions of its own — the assumption of continuity of distribution giving rise to the problems noted in the previous paragraph.

The student's attention is directed to the fact that the assumptions underlying the t-test for related measures are not nearly so likely to be a problem as those underlying the t-test for independent samples. This fact, coupled with the problems encountered with tie scores in the Wilcoxon test, suggests that the Wilcoxon will not be nearly so popular as the Mann-Whitney in behavioral research.

EXERCISES

27.1   Refer to Exercise 25.1. Substitute the Wilcoxon matched-pairs signed-ranks test for the t-test for related samples.

27.2   Refer to Exercise 25.2. Substitute the Wilcoxon matched-pairs signed-ranks test for the t-test for related samples.

28

THE CHI-SQUARE DISTRIBUTION

28.1 THE CONTINUOUS DISTRIBUTION OF CHI-SQUARE

Consider the situation in which a very large number of random samples of the same size are drawn from a normally distributed population. If each measure is expressed as a relative deviation from the mean (a z-score), and if the sum of the squares of these relative deviations is obtained for each sample, the sampling distribution of these sums of squares, first derived by Karl Pearson, is known as the *chi-square distribution*. These ideas may be expressed symbolically by:

$$\chi^2 = \sum \left(\frac{X_i - \mu}{\sigma} \right)^2 \quad \text{with } df = N$$

Under typical sampling conditions, this becomes:

$$\chi^2 = \sum \left(\frac{X_i - M}{\sigma} \right)^2 \quad \text{with } df = N - 1$$

Like the t-distribution, the chi-square distribution is a function of sample size; there is a family of chi-square curves, each determined by a single parameter, its degrees of freedom. The term "degrees of freedom" refers to the number of independent squares in the sum. Unlike the t-distribution, the chi-square distribution is sharply skewed to the right for small samples; however, as N increases, the distribution tends toward symmetry and becomes nearly bell-shaped for large samples. All chi-square values are

242

positive, and they range continuously from zero to infinity. The mean of any chi-square distribution is equal to its degrees of freedom, and the mode of each distribution is at $df - 2$, except for the curve with one degree of freedom. The distribution of chi-square for representative values of degrees of freedom is graphically illustrated in Figure 28.1.

Because chi-square distributions are skewed, and the mean is different for each distribution, the method of presenting the distributions in tabular form differs from that used with the t-distributions. The student's attention is directed to Table 10 in the Appendix. The probabilities recorded across the top of the table are those associated with random chi-squares equal to or larger than the tabled values. Most applications of the chi-square statistic will call for a one-tailed test, and only those values recorded on the right-hand side of Table 10 will be used. Any probability recorded at the top of the table may be chosen for the level of significance. In the event that a two-tailed test is desired, the determination of the region of rejection requires a little more effort. Suppose a two-tailed test at the .02 level is desired, and the number of degrees of freedom is 30. The region of rejection will include all chi-squares equal to or smaller than 14.953 ($P = .99$) plus all chi-squares equal to or greater than 50.892 ($P = .01$).

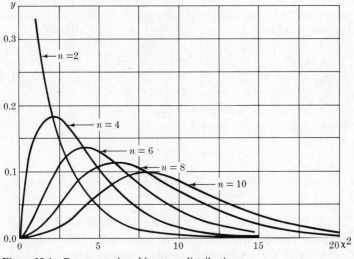

Figure 28.1 Representative chi-square distributions.

28.2 TEST OF A HYPOTHESIS ABOUT THE VARIANCE

Consider the situation in which a random sample is drawn from a normally distributed population, and the investigator wishes to test the hypothesis that the population variance is equal to some specified value, σ^2. The unbiased estimate of the parameter may be obtained from:

$$S^2 = \frac{SS}{N - 1}$$

Notice that $SS = (N - 1)S^2$.

We have previously defined the chi-square statistic as:

$$\chi^2 = \sum \left(\frac{X_i - M}{\sigma}\right)^2 \quad \text{with } df = N - 1$$

Then, with a little simple algebra, one may demonstrate that:

$$\chi^2 = \frac{(N - 1)S^2}{\sigma^2}$$

This value may be compared to the tabled value with the appropriate number of degrees of freedom and the desired level of significance to determine whether the sample value departs significantly from the hypothesized value. This is probably the only situation in which the reader will encounter a two-tailed test with the chi-square statistic.

Suppose, for example, that an investigator wishes to determine whether the IQs of the students in a certain school district are more or less variable than those of the norm group ($\sigma^2 = 225$). A sample of size 30 is to be drawn. If a two-tailed test at the .02 level is desired, a chi-square equal to or smaller than 14.256 or a chi-square equal to or greater than 49.588 will be significant. Let us suppose that the variance estimate from the sample data is 360:

$$\chi^2 = \frac{(30 - 1)360}{225} = 46.4$$

The verdict is to retain the null hypothesis; the available evidence does not establish that the variance of the population under study differs significantly from that of the norm group.

The test of a hypothesis about the variance may be either a one- or two-tailed test. Unfortunately, the statistical test is extremely sensitive to departures from normality. Unless the investigator has compelling evidence that the population from which the sample is drawn is normally distributed, he has no way of knowing whether rejection of the null hypothesis results from unequal variances or non-normality.

28.3 CHI-SQUARE APPROXIMATION OF THE MULTINOMIAL DISTRIBUTION

The general equation of the binomial probability distribution has been previously identified as:

$$P(r \text{ in } n) = \binom{n}{r} p^r q^{n-r}$$

Suppose we effect a change in notation such that n_1 is substituted for r, n_2 for $n - r$, N for n, P_1 for p, and P_2 for q. Thus the equation becomes:

$$P(n_1 \text{ in } N) = \binom{N}{n_1} (P_1)^{n_1}(P_2)^{n_2} \quad \text{where} \quad \binom{N}{n_1} = \frac{N!}{n_1!n_2!}$$

This notation is convenient for extending the expression from binomial populations to multinomial populations. Consider, for example, the expression extended to a trinomial population:

$$\frac{N!}{n_1!n_2!n_3!} (P_1)^{n_1}(P_2)^{n_2}(P_3)^{n_3}$$

This expression may be extended to as many categories as are needed. The student should readily see that the ability to test hypotheses about nominal variables with no restriction on the number of categories (such as we encountered with the binomial) would be a very valuable asset indeed. Unfortunately, the calculation of multinomial probabilities, even with very small samples, is laborious and often prohibitive. Furthermore, these calculations yield point probabilities, and the determination of the region of rejection for testing a given hypothesis is likely to require many such calculations.

Consider, for example, the very simple situation in which three categories are used, the three probabilities are equal ($P_1 = P_2 = P_3 = 1/3$), and the sample size is 3. The theoretical sampling distribution of the three objects is given in Table 28.1.

As the number of categories and the sample size are increased, not only do the calculations expand unendingly, but the identification of a useful region of rejection becomes increasingly impractical. If the sample size is adequate, both of these difficulties may be overcome through the use of the chi-square approximation of the multinomial, which is equivalent to the normal approximation of the binomial. The next two chapters are devoted to this application of the chi-square statistic.

Table 28.1 Multinomial distribution with $N = 3$

Distribution	Probability
0 0 3	1 $(1/3)^0$ $(1/3)^0$ $(1/3)^3 = 1/27$
0 1 2	3 $(1/3)^0$ $(1/3)^1$ $(1/3)^2 = 3/27$
1 0 2	3 $(1/3)^1$ $(1/3)^0$ $(1/3)^2 = 3/27$
0 2 1	3 $(1/3)^0$ $(1/3)^2$ $(1/3)^1 = 3/27$
1 1 1	6 $(1/3)^1$ $(1/3)^1$ $(1/3)^1 = 6/27$
0 3 0	1 $(1/3)^0$ $(1/3)^3$ $(1/3)^0 = 1/27$
1 2 0	3 $(1/3)^1$ $(1/3)^2$ $(1/3)^0 = 3/27$
2 0 1	3 $(1/3)^2$ $(1/3)^0$ $(1/3)^1 = 3/27$
2 1 0	3 $(1/3)^2$ $(1/3)^1$ $(1/3)^0 = 3/27$
3 0 0	1 $(1/3)^3$ $(1/3)^0$ $(1/3)^0 = 1/27$

$$27/27$$

EXERCISES

28.1 Given a random sample of 20 subjects from a normally distributed population. The unbiased estimate of the population variance is 257.5. Test the hypothesis that the variance is equal to 225. Use a one-tailed alternative at the .05 level.

28.2 Given a random sample of 30 subjects from a normally distributed population. The unbiased estimate of the population variance is 187.4. Test the hypothesis that the population variance is 225. Use a two-tailed test at the .04 level.

28.3 Given a random sample of 40 subjects from a normally distributed population. The unbiased estimate of the population variance is 252.9. Test the hypothesis that the variance is equal to 225. Use a two-tailed test at the .04 level.

SELECTED REFERENCES

BRADLEY, J. V., *Distribution-Free Statistical Tests*. Englewood Cliffs, N. J.: Prentice-Hall, 1968.

CONOVER, W. J., *Practical Nonparametric Statistics*. New York: John Wiley, 1971.

LANCASTER, H. O., *The Chi-squared Distribution*. New York: John Wiley, 1969.

SIEGEL, S., *Nonparametric Statistics for the Behavioral Sciences*. New York: McGraw-Hill, 1956.

29

CHI-SQUARE TESTS OF GOODNESS OF FIT

29.1 FUNCTION

The behavioral scientist often encounters research problems in which the variables are expressed as the number of individuals, objects, or events falling in each of several categories. We have been referring to these as *nominal* variables. When there are only two categories, the binomial probability distribution provides a useful model for testing some hypotheses involving nominal variables; however, when there are more than two categories, the binomial model is no longer applicable. The chi-square statistic provides a generalized procedure for testing hypotheses about the distributions of nominal and higher order data.

Chi-square tests of goodness of fit, sometimes called *one-sample chi-square tests*, are used to determine whether an observed frequency distribution departs significantly from a hypothesized frequency distribution. These tests perform essentially the same function as the binomial probability distribution in hypothesis testing, but the chi-square tests of goodness of fit are not restricted to two categories.

29.2 RATIONALE

The binomial probability distribution is a special case of the multinomial probability distribution, the binomial being limited to two categories and the multinomial having no such limitation. While the multinomial probability distribution is a useful model for certain statistical analyses, its

247

calculation tends to be laborious and often prohibitive, and it yields point probabilities which are not readily combined in a fashion useful to the behavioral scientist. For these reasons the multinomial probability distribution is seldom used by the behavioral scientist. The chi-square statistic, however, provides a relatively simple and extremely useful approximation of the multinomial which is equivalent to the normal approximation of the binomial.

Typically, the number of degrees of freedom in a chi-square goodness-of-fit test will be equal to the number of categories (cells) less one. The loss of one degree of freedom is due to the restriction that the sum of the expected frequencies must equal the sum of the observed frequencies. This is equivalent to specifying the frequency in the last cell; it is not free to vary.

29.3 PROCEDURE

1. The application of a chi-square goodness-of-fit test greatly resembles the use of the binomial probability distribution in hypothesis testing. A theoretical distribution is either hypothesized or is known to exist in the completely random situation. That is, the investigator must establish a hypothetical frequency distribution based upon whatever information he may have. Three examples will be considered: (a) the investigator might hypothesize that the population is uniformly distributed on the trait under study; (b) he might hypothesize that the population is normally distributed on the trait under study; or (c) he might hypothesize a distribution structured around scientific speculation, such as a hypothesis about the distribution of a hereditary trait. The categories may be defined on a nominal, ordinal, or interval scale, but the categories must be mutually exclusive and collectively exhaustive.

2. The common practice is to represent each category as a cell in a table such as this one containing four cells:

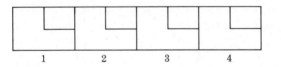

1   2   3   4

The theoretical (or expected) frequencies will be placed in the small boxes in the upper right corners of the cells; the observed frequencies will be placed in the main parts of the cells.

3. Expected frequencies are calculated as follows: Let p_j be the probability (theoretical relative frequency) of an object falling in the jth

cell. Then the expected frequency for the jth cell (E_j) is equal to $p_j N$, where N is the sample size. Of course, the sum of the p_j's must equal one, and the sum of the expected frequencies must equal N.

4. A random sample of objects is drawn from the population being studied, and the frequency distribution of the sample is determined. The observed frequencies are recorded in their respective cells.

5. The chi-square is calculated as follows:

$$\chi^2 = \sum_{j=1}^{k} \frac{(O_j - E_j)^2}{E_j}$$

where O_j = the observed frequency for the jth cell
E_j = the expected frequency for the jth cell
k = the number of cells
$df = k - 1$

6. If the number of degrees of freedom is one, a continuity correction will be used in the calculation of chi-square:

$$\chi^2 = \sum_{j=1}^{k} \frac{(|O_j - E_j| - \frac{1}{2})^2}{E_j}$$

7. At some predetermined level of significance, the calculated chi-square is compared with the tabled value having $k - 1$ degrees of freedom. Only a one-tailed test is appropriate. If the calculated statistic equals or exceeds the tabled value, the finding is significant, and the hypothesis of no difference between the population distribution and the hypothesized distribution is rejected. The population distribution is determined to differ from the hypothesized distribution at the specified level of significance. If the calculated value is smaller than the tabled value, the hypothesis is retained, and no significant difference in the two distributions has been determined to exist.

29.4 EXAMPLE

A high school algebra teacher suspects that his students tend to be either especially bright or especially stupid. A standardized algebra test is administered to 60 students, and their scores are recorded as percentile ranks based on national norms. The percentile ranks were, of course, uniformly distributed in the norm group. The teacher hypothesizes that his students are not distributed in the same fashion as the norm group. The test of the hypothesis is outlined below:

1. H: The students are uniformly distributed.
       A: The students are not uniformly distributed.
2.   Chi-square goodness of fit, $N = 60$.
3.   .05 level, $df = 9$ (10 cells will be used).
       R: $\chi^2 \geq 16.919$.
4.   The expected frequencies in a uniform distribution are all the same and may be calculated from: $E_j = N/k$. With a sample of size 60 evenly divided among 10 cells, each expected frequency will be 6. The 10 cells will be partitioned at the 10th, 20th, 30th, 40th, 50th, 60th, 70th, 80th, and 90th centiles. A frequency distribution is constructed with the data grouped into 10 classes with the class limits just described. Both the expected and observed frequencies are recorded in the table below:

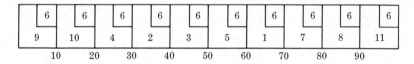

The chi-square is calculated from:

$$\chi^2 = \sum \frac{(O_i - E_i)^2}{E_i} = \frac{9}{6} + \frac{16}{6} + \frac{4}{6} + \frac{16}{6} + \frac{9}{6} + \frac{1}{6}$$

$$+ \frac{25}{6} + \frac{1}{6} + \frac{4}{6} + \frac{25}{6} = \frac{110}{6} = 18.3$$

5.   Reject the null hypothesis. The sample deviates significantly from the hypothesized uniform distribution. The data appear to bear out the teacher's suspicions. However, rejection of the null hypothesis does not necessarily establish that the students are concentrated in the extreme cells; it establishes only that they are not distributed in the same fashion as the norm group.

29.5   CHI-SQUARE TEST FOR GOODNESS OF FIT TO NORMAL

The procedure used to test for uniformity of distribution requires only slight modification to test for normality of distribution. In this situation, the normal deviates (z-scores) corresponding to the percentile ranks are determined from Table 3 in the Appendix, and these are used for the class

limits. If the sample mean and standard deviation are used for the calculation of standard scores for the sample, two additional degrees of freedom
are lost ($df = k - 3$). If the instrument is scaled to yield standard scores
based on national norms, and these are used to determine the class limits,
the degrees of freedom are as before ($df = k - 1$), but we are now testing
a hypothesis of normality with specified mean and standard deviation.
Conceivably, the population under study might be normally distributed,
and yet the hypothesis might be rejected due to difference in means or
standard deviations.

The traditional statistical procedure for testing goodness of fit to normal
has used a model in which the cell limits are defined by dividing a standard
score scale into equal parts (a linear score scale model). For example, an
eight-celled model might have the cell limits defined as -3.00, -2.00,
-1.00, 0.00, $+1.00$, $+2.00$, $+3.00$. Models of this sort have been criticized
because the expected frequencies in the tails of the distribution tend to be
very small. An alternative model in which cell limits are defined by dividing
the area under the curve into equal parts (an equal areas model) has proven
to be quite superior. In addition to overcoming the problem of small
expected frequencies in the tails, this procedure focuses on the added power
characteristic of the chi-square approximation with uniform expected
frequencies. The test appears to be optimal with approximately 20 cells,
and the author suggests that chi-square tests of goodness of fit to normal
be standardized with the use of the equal areas model having 20 cells. The
cell limits for this model are -1.645, -1.282, -1.036, -0.842, -0.675,
-0.524, -0.385, -0.253, -0.126, 0.00, $+0.126$, $+0.253$, $+0.385$, $+0.524$,
$+0.675$, $+0.842$, $+1.036$, $+1.282$, $+1.645$. An acceptable approximation
is achieved using this model with samples when $N = 20$ or more. The equal
areas model has the added computational convenience of a common
denominator ($E_j = N/k$) for all cells.

29.6 DISCUSSION

Chi-square goodness-of-fit tests are useful procedures for testing whether a
distribution of measures in a sample deviates significantly from a hypothesized distribution. The data may be recorded on nominal or higher
order scales.

The use of the chi-square statistic in goodness-of-fit tests is an approximation of the multinomial probability distribution and carries with it some
restrictions with respect to sample size if a good approximation is to be
achieved. Traditionally, statisticians have recommended that expected
frequencies should be equal to or greater than five in order to achieve an
acceptable approximation. Recent research, however, indicates that this is

not at all necessary when the number of degrees of freedom is two or more. The goodness-of-fit test with one degree of freedom tends to be a bit erratic in behavior and generally inferior to tests with two or more degrees of freedom. The binomial probability distribution should probably be substituted for chi-square tests of goodness of fit with one degree of freedom. The chi-square test functions best with equal cell frequencies, and expected frequencies as small as one provide a good approximation under these circumstances.

Whenever the investigator is free to establish the cell limits at his discretion (this is usually the case with all except nominal data), the equal expected frequencies model is recommended. With moderate departures from equal cell frequencies (the sort of thing commonly encountered in research), the average expected frequency may be as low as one for tests at the .05 level, but should be held to two or more for tests at the .01 level. If the departures from the equal expected frequencies are extreme, these recommendations should be doubled. There is a distinct tendency for the test to become liberal (to reject too many true hypotheses) if expected frequencies are permitted to fall below these recommendations.

Sometimes it is suggested that adjacent cells be combined to increase the expected frequencies. This practice introduces an arbitrary component into the analysis and its use is not recommended. The binomial probability distribution has no restrictions with respect to minimum expected frequencies and may be substituted for the chi-square in goodness-of-fit situations with only two cells. If the data are ordinal and continuously distributed, the Kolmogorov statistic may be the appropriate alternative. However, its use is not recommended if the data are grouped or if there are many tie scores. For most behavioral science applications, the chi-square test is preferred.

EXERCISES

29.1 Given the following collection of scores which were randomly drawn from a much larger collection:

$$
\begin{array}{cccc}
-1.90 & +0.05 & +0.60 & +1.00 \\
-1.00 & +0.05 & +0.74 & +1.09 \\
-0.38 & +0.19 & +0.76 & +1.12 \\
-0.08 & +0.40 & +0.85 & +1.27 \\
-0.04 & +0.56 & +0.98 & +1.28
\end{array}
$$

Test the hypothesis that the parent population is normally distributed with mean of zero and standard deviation of one. Use the .05 level of significance.

29.2 Given the following collection of scores which were randomly drawn from a much larger collection:

81	73	67	61	54	51	44	39	33	21
80	73	65	60	54	50	43	37	32	20
78	72	64	58	54	48	43	36	32	20
76	71	64	57	54	48	43	34	31	20
75	69	62	57	53	47	41	33	28	19
74	69	61	55	52	47	40	33	25	19
73	68	61	55	52	46	40	33	23	18

Test the hypothesis that the parent population is normally distributed with mean of 50 and standard deviation of 10. Use the .05 level of significance.

29.3 A high school genetics student hypothesizes that the offspring of a certain pair of guinea pigs will be distributed in the ratio 9:3:3:1 with respect to color and texture of coat. He predicts that 6.25 percent of the offspring will have smooth white coats, 18.75 percent smooth black coats, 18.75 percent rough white coats, and 56.25 percent rough black coats. A total of 96 progeny are raised from this pair, and they include 4 with smooth white coats, 14 smooth black, 23 rough white, and 55 rough black. Test this hypothesis at the .05 level.

SELECTED REFERENCES

COCHRAN, WILLIAM G., "The chi-square test of goodness of fit," *Annals of Mathematical Statistics*, **23**, 315–345 (September 1952).

COCHRAN, WILLIAM G., "Some methods for strengthening the common χ^2 tests," *Biometrics*, **10**, 417–451 (December 1954).

KITTLESON, HOWARD M., and JOHN T. ROSCOE, "An empirical comparison of four chi-square and Kolmogorov models for testing goodness of fit to normal," Paper presented at the annual convention of the American Educational Research Association, Chicago, April 1972.

ROSCOE, JOHN T., and JACKSON A. BYARS, "An investigation of the restraints with respect to sample size commonly imposed on the use of the chi-square statistic," *Journal of the American Statistical Association*, **66**, 755–759 (1971).

SLAKTER, MALCOM J., "A comparison of the Pearson chi-square and Kolmogorov goodness of fit tests with respect to validity," *Journal of the American Statistical Association*, **60**, 854–858 (1965).

WATSON, G. S., "The chi-square goodness of fit test for normal distributions," *Biometrika*, **44**, 336–348 (1957).

30

CHI-SQUARE TESTS FOR CONTINGENCY TABLES

30.1 FUNCTION

Chi-square tests for contingency tables are extremely useful statistical procedures for determining whether two nominal (or higher level) measures are related. If one of the variables is group membership and the other a dependent variable, the test may be used to analyze data from a simple randomized design, and the research may be either experimental or *ex post facto*.

The data are organized into a contingency table, and the statistical test is made to determine whether classification on the row variable is independent of classification on the column variable. For example, suppose that the row variable were used to classify the subjects with respect to religious affiliation and the column variable with respect to political affiliation. The chi-square test is used to determine whether there is a relationship between the two nominal variables. The research question may be seen as: "With respect to religious affiliation, do Democrats and Republicans represent two different populations or a single population?" Or: "With respect to political affiliation, do Catholics and Protestants represent two different populations or a single population?"

There is no restriction with respect to the number of categories in either the row or column variable when the chi-square statistic is used to analyze data in a contingency table. There are, however, restrictions with respect to sample size similar to those encountered in the chi-square tests of goodness of fit.

30.2 RATIONALE

Chi-square tests for contingency tables are an extension of the chi-square approximation of the multinomial to the situation in which a double classification system is used and the expected frequencies are derived from sample data.

Let us suppose, for example, that each of the persons in a certain population can be classified as either Republican or Democrat and that each can also be classified as either Catholic or Protestant. A random sample of 100 persons is drawn, which is found to be evenly divided with respect to political affiliation, there being 50 Democrats and 50 Republicans. The sample is also determined to have 40 Catholics and 60 Protestants. If the two classification systems are independent of each other, we would expect the 40 Catholics to be evenly divided with respect to political affiliation, and we would also expect the 60 Protestants to be evenly divided with respect to political affiliation. The expected frequencies are 20 Catholic Republicans, 20 Catholic Democrats, 30 Protestant Republicans, and 30 Protestant Democrats. The situation is illustrated in the table below:

	Republicans	Democrats	
Catholics	20	20	40
Protestants	30	30	60
	50	50	100

Notice that no sample information has been recorded other than the marginal frequencies (the marginal frequencies are the row and column sums). The expected frequencies are derived from the marginal frequencies. These expected frequencies may be calculated from the formula:

$$E_{ij} = \frac{R_i C_j}{N}$$

where E_{ij} = the expected frequency for cell in row i, column j
R_i = the sum of the frequencies in row i
C_j = the sum of the frequencies in column j
N = the sum of the frequencies for all cells

A quick check on the calculations will verify that the sums of the expected frequencies in each row and column are equal to the marginal frequencies. These restrictions determine the number of degrees of freedom to be

$(r - 1)(c - 1)$, where r is the number of rows and c is the number of columns. With a little effort, the student should be able to see that there is one cell in each row and column which is not free to vary. In the example above, the number of degrees of freedom is one. —

The chi-square statistic is calculated by summing over all cells:

$$\chi^2 = \sum \sum \frac{(O_{ij} - E_{ij})^2}{E_{ij}}$$

Statisticians classify contingency table analyses on the basis of whether the marginal frequencies are fixed or random. Three categories exist:

1. *Random marginal frequencies for both rows and columns.* Suppose, in the example cited earlier, that the subjects were drawn at random from a single parent population, and that the investigator did not know in advance whether a given subject would be a Democrat or a Republican, a Catholic or a Protestant. The marginal frequencies in such a situation would not be determined until the sampling is completed. Such a contingency table is called a *bivariate frequency table* (because there are two random variables), and the statistical test is called a *test of independence.*

2. *Random marginal frequencies for either the rows or the columns with fixed frequencies for the other.* Suppose, in the example cited earlier, that two samples were drawn, one from a parent population of Democrats and the other from a parent population of Republicans. The investigator would determine the size of the two samples, and the column marginal frequencies would be fixed by his decision. The row marginal frequencies would be determined by the religious preference of those people sampled; they would be random rather than fixed. The statistical test in this situation is usually referred to as a *test for homogeneity.* Often the row and column variables will be thought of as independent and dependent variables (the variable which is used to define the populations is usually regarded as the independent variable). The strategy applies to both experimental and *ex post facto* research.

3. *Fixed marginal frequencies for both rows and columns.* The median test, to be discussed in detail later, is an example of a contingency table analysis in which both the row and column marginal frequencies are fixed.

Although this classification scheme has important implications in statistical theory and would influence the manner in which exact probabilities are calculated, the distinction will not influence the manner in which the chi-square approximation of the multinomial is calculated.

30.3 PROCEDURE

1. Two methods of sampling are commonly used in chi-square tests for contingency tables. (a) A single random sample may be drawn from a population that can be classified on two different variables. The two variables become the row and column variables in a bivariate frequency table. (b) A random sample may be drawn from each of two or more populations, all of which can be classified on the same variable. In this situation, the various populations will be used as the second variable in constructing the bivariate frequency table. The analysis is the same in either case, and often the interpretation is the same.

2. A bivariate frequency table is constructed, and the marginal frequencies are determined. The expected frequencies are calculated in the fashion described in Section 30.2 using the formula:

$$E_{ij} = \frac{R_i C_j}{N}$$

3. The observed frequencies are recorded in the table, and the chi-square statistic is calculated from the formula:

$$\chi^2 = \sum \sum \frac{(O_{ij} - E_{ij})^2}{E_{ij}}$$

4. In the case of a two-by-two table (the case where $df = 1$), the use of a continuity correction factor is sometimes suggested. The available evidence indicates that the correction factor does not improve the approximation and its use is not recommended. Some computational effort may be saved, however, if the data are organized into a two-by-two table with cell frequencies represented by the letters A, B, C, and D:

A	B
C	D

The following formula is mathematically equivalent to that given earlier:

$$\chi^2 = \frac{N(AD - BC)^2}{(A + B)(C + D)(A + C)(B + D)}$$

5. At some predetermined level of significance, the calculated chi-square is compared with the tabled value having $(r - 1)(c - 1)$ degrees of freedom. Only a one-tailed test is appropriate. If the calculated statistic equals or exceeds the tabled value, the finding is significant and the hypothesis of independence is rejected; the row and column variables are determined to be dependent at the specified level of significance. The exact interpretation will be dependent upon the nature of the row and column variables. If the calculated statistic is smaller than the tabled value, no significant relationship between the row and column variables has been determined to exist.

30.4 EXAMPLE

A group of college students is asked to answer the following multiple-choice question in a values inventory:

What do you believe is the true nature of God?

1. I believe in a personal God who has revealed Himself in the Bible.
2. There is a God, Father of all men, who is common to all religious faiths. It is not particularly important whether a man is a Christian, Jew, Moslem, Hindu, and so on.
3. I believe in a Supreme Being or First Cause, but I cannot believe in a personal God.
4. The nature of God is not (or cannot) be known by man.
5. There is no God.

The respondents were also identified, by sex, and the following bivariate frequency table was constructed from the data:

Responses:	1	2	3	4	5	
Males	18	40	7	4	1	70
	12	33	17	8	0	
Females	36	80	14	8	2	140
	42	87	4	4	3	
	54	120	21	12	3	210

The test of the hypothesis is outlined below. The investigator would probably choose to think of sex as the independent variable and response as the dependent variable. He is seeking to determine whether the sex of the respondent influences the response to the question.

1. H: Sex of respondent and response are independent.
   A: Response is dependent upon the sex of the respondent.
2. Chi-square test of independence, $N = 210$, 2×5 table.
3. .01 level, $df = 4$.
   R: $\chi^2 \geq 13.277$.
4. The calculation of the expected frequencies will be illustrated with that for $cell_{11}$:

$$E_{11} = \frac{R_1 C_1}{N} = \frac{(70)(54)}{210} = 18$$

The calculation of the chi-square is as follows:

$$\chi^2 = \frac{36}{18} + \frac{49}{40} + \frac{100}{7} + \frac{16}{4} + \frac{1}{1} + \frac{36}{36} + \frac{49}{80} + \frac{100}{14} + \frac{16}{8} + \frac{1}{2} = 33.77$$

5. Reject the null hypothesis. There is a significant relationship between sex of the respondent and response to the question. While the test of independence does not specifically identify the difference between the responses of the two groups, an examination of the available responses and of the bivariate frequency table suggests that the females are more inclined to believe in a personal God than are the males.

30.5 THE MEDIAN TEST

The median test has been suggested as a nonparametric alternative to the t-test for two independent samples when the assumptions underlying both the t-test and the Mann-Whitney U-test cannot be met. It is simply an application of the chi-square test of independence to the situation in which the investigator has ordinal data and wishes to determine whether there is a significant difference in the scores of two or more samples. The scores for all samples are arranged in a single ordered series and their common median determined (see Section 7.4 for instructions for calculating the value of the median). Then a contingency table is constructed with one column for each sample and with two rows; the top row is for individuals who scored above the median, the bottom for individuals who scored below the median. The number of persons in each sample who scored above and below the common median is entered in the appropriate cells in the table, and the expected frequencies and chi-square statistic are calculated as in the previous section. Consider the following table:

Group:	I		II		III		
Above		8		10		9	27
	12		6		9		
Below		8		10		9	27
	4		14		9		
	16		20		18		54

The chi-square for these data is 7.2 with two degrees of freedom, which is significant at the .05 level. Although one may safely conclude that Group I placed significantly more individuals above the common median than did Group II, the test of the hypothesis establishes nothing with respect to the relationship between Group I and Group III, nor with respect to the relationship between Group II and Group III. This is a problem common to the testing of hypotheses about more than two samples. Consider also the possibility that the individuals in Group I who scored above the median were just barely above the median, and that those in Group II who scored above the median were the highest scorers of all. Similarly, those in Group I who scored below the median could have been the lowest scorers of all, although those in Group II could have been just below the median. The possibility exists that the mean for Group II is actually higher than the mean for Group I, even though Group I placed more subjects above the common median. The investigator who uses the median test must carefully consider the hypothesis which he wishes to test and whether or not the median test actually tests that hypothesis.

Generally, the median test will not prove nearly as satisfactory as the Mann-Whitney U-test; however, it may be preferred when the measurement is crude and it has the additional advantage of not being restricted to the two-sample situation.

30.6   THE CONTINGENCY COEFFICIENT

The contingency coefficient is an index of relationship between two variables for the situation in which the data are organized into a bivariate frequency table. It is the only correlation coefficient appropriate for use with nominal data, with the possible exception of the phi coefficient, which is limited to two-by-two tables.

The magnitude of a chi-square statistic calculated from a bivariate frequency table is a function of the relationship between the row and column variables. This fact is utilized in the calculation of the contingency coefficient from the formula:

$$C = \sqrt{\frac{\chi^2}{N + \chi^2}}$$

Whether the relationship between the two variables is a significant one may be tested in the same fashion as in the test of independence, referring the chi-square to Table 10.

Unfortunately, the size of the contingency coefficient is a function of the number of cells in the table, and under no circumstances is it possible for the coefficient to be unity, even though a perfect relationship may exist. The maximum value of the contingency coefficient for a two-by-two table is .707, for a three-by-three table .816, for a four-by-four table .866, and for a five-by-five table .894. The maximum value for a ten-by-ten table is .949. In addition to the fact that it is not directly comparable with other correlation coefficients, the contingency coefficient requires the large sample sizes characteristic of the chi-square statistic, and for these reasons it has not proved very popular.

The author suggests a slight modification of the formula for the contingency coefficient so that a perfect relationship will produce a coefficient of unity:

$$C_r = C \sqrt{\frac{k}{k-1}} = \sqrt{\left(\frac{k}{k-1}\right)\left(\frac{x^2}{N+x^2}\right)}$$

The constant k is the number of cells in each column or row (a symmetrical frequency table is assumed). Thus k equals four for a four-by-four table.

Cramér has also suggested an alternative to the contingency coefficient to produce a coefficient of unity for a perfect relationship. This is known as Cramér's statistic:

$$C = \sqrt{\frac{x^2}{N(L-1)}}$$

The constant L is the number of cells in each column or row, whichever is smaller. Thus L equals three for a three-by-five table.

It should be noted that these two procedures do not yield the same coefficient if the relationship is imperfect. Both are regarded as superior to the traditional contingency coefficient; however, neither has had wide acceptance up to this time.

30.7 DISCUSSION

Chi-square tests for contingency tables are by far the most valuable of the nonparametric procedures available to the behavioral scientist. They are used to test the significance of relationship between the row and column variables. The data may be recorded on nominal or higher order scales.

The use of the chi-square statistic with contingency tables is an approximation of the multinomial probability distribution and carries with it some restrictions with respect to sample size. These are summarized as follows:

1. In two-by-two tables (the situation where $df = 1$), a good approximation is achieved if the average expected frequency is 7.5 or higher. For smaller expected frequencies, the investigator should shift to the Fisher exact test.
2. If expected frequencies are approximately equal throughout the contingency table, an average expected frequency of two is adequate for tests at the .05 level. This should be increased to four for tests at the .01 level.
3. For moderate departures from the ideal of equal expected frequencies, an average expected frequency of four is adequate for tests at the .05 level, and six is adequate for tests at the .01 level. If the departures are extreme, an average expected frequency of six is needed to ensure a good approximation at the .05 level, while ten is needed at the .01 level. Smaller values than those recommended here will ordinarily result in a conservative test.

The use of the continuity correction factor for the two-by-two table is not recommended. Collapsing of cells (the combining of adjacent cells) to increase expected frequencies is regarded as an arbitrary practice and extremely questionable.

If the data are ordinal and continuously distributed, the Kolmogorov-Smirnov statistic may be an appropriate alternative in the two-sample situation. Its use is limited to two samples, however, and it is not recommended for use with grouped data or in cases where there are many tie scores.

EXERCISES

30.1 A multiple-choice type of question with respect to the desirability of teacher tenure is given to several groups of interested persons. Three responses to the question were available: (a) agree, (b) no opinion, and (c) disagree. A group of teachers split on the question, with 75 choosing agree, 10 no opinion, and 5 disagree. A group of school administrators was divided on the issue, with 20 choosing agree, 10 choosing disagree, and none choosing no opinion. A group of businessmen was evenly divided on the issue, with 10 choosing each response. Construct a bivariate frequency table from the data, and test the hypothesis of independence at the .05 level.

30.2 Random samples of students are chosen from the public high school and the parochial high school of a certain community. These are then classified into five socioeconomic classes according to parent's occupation. The 30 students from the parochial school include 2 whose fathers were classified professional or managerial, 0 semiprofessional, 12 skilled workers, 14 semiskilled, and 2 unskilled. The 60 students from the public school were classified 4 professional or managerial, 9 semiprofessional, 18 skilled workers, 22 semiskilled, and 7 unskilled. Construct a bivariate frequency table from the data, and test the hypothesis of independence at the .05 level.

30.3 Refer to Exercise 26.3. Use the median test at the .05 level to determine whether there is a significant difference in the two groups.

30.4 A certain projective instrument is intended for use in classifying personality disorders, and scores yielded by the instrument are used to rate the subject as having Type A, B, C, or D personality. Two psychologists rate each of a group of 100 individuals who are undergoing therapy. The rating of one psychologist is used as the row variable and that of the other psychologist as the column variable in the following contingency table. Calculate the contingency coefficient for these data, and test the significance of relationship. Use the .01 level.

	A	B	C	D
A′	12	4	8	0
B′	7	18	2	3
C′	2	2	15	6
D′	0	1	3	17

SELECTED REFERENCES

BRADLEY, J. V., *Distribution-Free Statistical Tests*. Englewood Cliffs, N. J.: Prentice-Hall, 1968.

CONOVER, W. J., *Practical Nonparametric Statistics*. New York: John Wiley, 1971.

LANCASTER, H. O., *The Chi-Squared Distribution*. New York: John Wiley, 1969.

LEWONTIN, R. C., and J. FELSENSTEIN, "The robustness of homogeneity tests in $2 \times N$ tables," *Biometrics*, **21**, 19-33 (March 1965).

ROSCOE, JOHN T., and JACKSON A. BYARS, "An investigation of the restraints with respect to sample size commonly imposed on the use of the chi-square statistic," *Journal of the American Statistical Association*, **66**, 755-759 (1971).

31

TESTS OF SIGNIFICANCE
OF CORRELATION

31.1 INTRODUCTION

Occasionally, one reads or hears that a correlation coefficient on the order of .80 or .90 signifies a high degree of relationship, while a coefficient of .20 or .30 signifies a low relationship. In statistical inference, one cannot attach meaning to a correlation coefficient without taking into consideration the size of the sample and the sampling distribution of the coefficient. Consider, for example, the ten pairs of scores reported in Figure 12.5:

$$X\text{-score:} \quad 0 \quad 2 \quad 3 \quad 4 \quad 4 \quad 5 \quad 5 \quad 8 \quad 9 \quad 10$$
$$Y\text{-score:} \quad 3 \quad 8 \quad 5 \quad 6 \quad 5 \quad 4 \quad 4 \quad 7 \quad 1 \quad 7$$

A Pearson product moment correlation coefficient calculated from these ten pairs of scores is equal to zero. Suppose, however, that a sample of three pairs of scores was drawn at random from this collection, and that the three pairs happened to be:

$$X\text{-score:} \quad 0 \quad 4 \quad 8$$
$$Y\text{-score:} \quad 3 \quad 5 \quad 7$$

These three scores happen to lie on a straight line, and a Pearson coefficient calculated from them is equal to plus one. This suggests that a test of significance of the relationship between the variables is needed whenever the investigator draws conclusions about population correlations from sample data. As a general rule, the investigator will test the null hypothesis that the population correlation coefficient is zero.

31.2 THE PEARSON AND POINT BISERIAL CORRELATION COEFFICIENTS

The form of the sampling distribution of the Pearson correlation coefficient is a function of the magnitude and the sign of the coefficient in the population under study. As the magnitude of the coefficient increases, the sampling distribution becomes increasingly skewed — to the left for positive correlations and to the right for negative correlations. However, when the population value is zero, the sampling distribution is symmetrical and approximately normal, with the approximation improving with increased sample size. For this reason, the t-distribution is an appropriate model for testing the null hypothesis that the population correlation coefficient is zero, but not for testing the hypothesis of no difference between two correlation coefficients. The t-statistic may be calculated from:

$$t = r \sqrt{\frac{N - 2}{1 - r^2}} \quad \text{with } df = N - 2$$

The calculated value may be compared to the tabled t with $N - 2$ degrees of freedom, where N is the number of pairs of measures. A significant relationship exists if the observed value equals or exceeds the tabled value at the desired level of significance. A one- or two-tailed test may be used. The one-tailed test is often appropriate in learning research, since the investigator will generally be prepared to specify in advance that only a positive coefficient is of interest. However, if the investigator is interested in either a positive or negative coefficient, the two-tailed test must be used.

The formula given above may also be used to transform critical values of t to critical values of r, so that the region of rejection may be stated in terms of r instead of t. The formula for calculating critical values of r is:

$$r = \sqrt{\frac{t^2}{N - 2 + t^2}} \quad \text{with } df = N - 2$$

This formula has been used to construct Table 12 in the Appendix. A significant r is one equal to or larger than the tabled value with $N - 2$ degrees of freedom, where N is the number of pairs of scores. The table is complete enough for most behavioral research, but additional values may be calculated using the formula if they are needed. The significance of the point biserial coefficient is tested in precisely the same way as the Pearson.

The Pearson product moment correlation coefficient as a population parameter is usually represented by the lowercase Greek letter rho (ρ). The null hypothesis is $\rho = 0$. The hypothesis tested is one of linear relationship. It is possible for a strong curvilinear relationship to exist and retain the null hypothesis. The discussion of tests for curvilinear relationship is reserved for the chapter on regression analysis.

The preceding paragraphs suggest the interesting possibility of substituting the point biserial or the Pearson coefficient for the t-statistic in experimental analysis. For example, a binary variable may be substituted for group membership in the t-test for two independent samples. Let us assign a zero to each member of the control group and a one to each member of the experimental group. These group scores are then paired with the criterion scores (giving n_1 plus n_2, or N pairs of scores), and a point biserial or Pearson correlation coefficient is calculated from the paired measures. When this correlation coefficient is tested for statistical significance, the result is precisely the same as the t-test for two independent samples. In addition to the test of significance, the investigator has an index of relationship between the two variables and may meaningfully speak of the proportion of the variance of the criterion accounted for by membership in the two groups.

31.3   THE DIFFERENCE BETWEEN TWO PEARSON COEFFICIENTS FROM INDEPENDENT SAMPLES

Consider the situation in which the correlation of two variables is obtained for two independent samples, and the investigator wishes to determine whether there is a significant difference in the two correlation coefficients. The null hypothesis is $\rho_1 = \rho_2$.

The procedure for testing the null hypothesis is as follows:

1. Using Table 11, convert the values of r_1 and r_2 to normal deviates in the form of Z_{r1} and Z_{r2}.
2. Calculate the error term from:

$$S_{zr1-zr2} = \sqrt{\frac{1}{N_1 - 3} + \frac{1}{N_2 - 3}}$$

3. Calculate a normal deviate with mean of zero and unit standard deviation from:

$$z = \frac{Z_{r1} - Z_{r2}}{S_{zr1-zr2}}$$

4. The probability of z may be determined from Table 3 in the Appendix, or it may be referred to Table 19.1 for the test of the null hypothesis. Either a one- or two-tailed test may be used.

Suppose, for example, that for a sample of 28 adolescent males, the correlation of the subjects' IQ scores with those of their fathers is determined to be .80, and for a similar but unrelated sample of 28 adolescent females, the correlation of the subjects' IQ scores with those of their fathers is determined to be .60. Then

$$z = \frac{1.093 - 0.693}{0.1414} = 2.87$$

Given a two-tailed test at the .05 level, the finding is significant, suggesting that there is a higher degree of relationship between the IQ scores of fathers and sons than between fathers and daughters.

31.4 THE DIFFERENCE BETWEEN TWO PEARSON COEFFICIENTS FROM RELATED SAMPLES

Consider the situation in which the intercorrelations of three variables (r_{12}, r_{13}, and r_{23}) are calculated for a given sample, and the investigator wishes to determine whether there is a significant difference in two of the correlation coefficients. Suppose the null hypothesis were $\rho_{12} = \rho_{13}$. This situation differs considerably from that in Section 31.3, and the hypothesis is tested by:

$$z = \frac{\sqrt{N-1}(r_{12} - r_{13})}{\sqrt{(1 - r_{12}^2)^2 + (1 - r_{13}^2)^2 - 2r_{23}^3 - (2r_{23} - r_{12}r_{13})(1 - r_{12}^2 - r_{13}^2 - r_{23}^2)}}$$

The probability of z may be determined from Table 3 in the Appendix, or it may be referred to Table 19.1 for the test of the null hypothesis. Either a one- or two-tailed test may be used.

Suppose, for example, that for a sample of 28 adolescent males, the correlation of the subjects' IQ scores with those of their fathers is determined to be .80, and the correlation of the same subjects' IQ scores with those of their mothers is determined to be .60, and the correlation of the IQ scores of the mothers and fathers is determined to be .44. The investigator wishes to determine whether there is a significant difference in the first two correlations cited. Then

$$z = \frac{5.196(0.20)}{\sqrt{0.1296 + 0.4096 - 0.1704 - (0.40)(-0.1936)}} = \frac{1.039}{0.6680} = 1.55$$

Given a two-tailed test at the .05 level, the finding is nonsignificant, indicating that there is no significant difference between (1) the correlation of the sons' IQ scores with their fathers', and (2) the correlation of the sons' IQ scores with their mothers'.

31.5   THE SPEARMAN RANK CORRELATION COEFFICIENT

When the Spearman rank correlation coefficient is calculated, each of the variables ranges from 1 to N. Suppose we arrange the X-ranks in order from 1 to N, and consider all possible pairings of the Y-ranks with these. There are $N!$ possible arrangements of the paired variables, and each of these is equally probable if the relationship between the two variables is a completely random one. For example, if N equals two, there are two possible combinations and two possible correlation coefficients, plus one and minus one. Each of these is equally probable, the probability being one-half. If we continued in this fashion, it would be a relatively simple though laborious process to develop the sampling distribution of the Spearman rank correlation coefficient for any sample size. Fortunately, this task has already been completed, and the results are recorded in Table 13 in the Appendix. A Spearman coefficient which equals or exceeds the tabled value is a significant one. For larger samples, the t-test used with the Pearson correlation coefficient may be used. A usable approximation is achieved with samples as small as 10 in size, and the approximation is excellent for samples as large as 30.

31.6   THE KENDALL TAU RANK CORRELATION COEFFICIENT

The probabilities associated with random values of the Kendall tau coefficient are derived in essentially the same fashion as those for the Spearman coefficient, and the two tests have precisely the same power. Table 14 in the Appendix lists critical values of the Kendall tau coefficient for testing the hypothesis of no relationship. A value of tau which equals or exceeds the tabled value is a significant one.

31.7   THE GOODMAN-KRUSKAL GAMMA COEFFICIENT,
THE PHI COEFFICIENT,
AND THE CONTINGENCY COEFFICIENT

The significance of these three correlation coefficients may be determined through the use of a chi-square test for a contingency table. In the case of the contingency coefficient, the calculation of the chi-square statistic is an intermediate step in the determination of the value of the correlation coefficient. The test of significance is discussed in some detail in Chapter 30.

In the case of the phi coefficient, the chi-square may be conveniently calculated from this relationship:

$$\chi^2 = N\phi^2$$

31.8 THE BISERIAL AND TETRACHORIC CORRELATION COEFFICIENTS

The sampling distribution of the biserial correlation coefficient is a function of the proportion of the subjects falling in each part of the dichotomy. This greatly complicates the problem of establishing a sampling distribution for the coefficient, and the author knows of no completely satisfactory test of its significance. Although the sampling distribution of the tetrachoric correlation coefficient is more readily established, the assumptions which underlie its use are not likely to be met in the practical research setting. For these reasons, tests of significance for these correlation coefficients will not be reported here; however, they will be found in some of the references listed at the end of the chapter. The use of a chi-square test (as in Section 30.3) may be appropriate for testing the significance of the tetrachoric coefficient.

EXERCISES

31.1 Refer to Exercise 12.1. Test the significance of the correlation coefficient at the .01 level.

31.2 Refer to Exercise 12.2. Test the significance of the correlation coefficient at the .05 level.

31.3 Refer to Exercise 12.3. Test the significance of the correlation coefficient at the .05 level.

31.4 Refer to Exercise 13.1. Test the significance of the correlation coefficient at the .05 level.

31.5 Refer to Exercise 13.4. Test the significance of the correlation coefficient at the .05 level.

SELECTED REFERENCES

BRADLEY, J. V., *Distribution-Free Statistical Tests*. Englewood Cliffs, N. J., Prentice-Hall, 1968.

CONOVER, W. J., *Practical Nonparametric Statistics*. New York: John Wiley, 1971.

GLASS, GENE V., and JULIAN C. STANLEY, *Statistical Methods in Education and Psychology*. Englewood Cliffs, N. J.: Prentice-Hall, 1970.

HENDRICKSON, GERRY F., JULIAN C. STANLEY, and JOHN R. HILLS, "Olkin's new formula for significance of r_{13} vs. r_{23} compared with Hotelling's method," *American Educational Research Journal*, **7**, 189–194 (1970).

HENDRICKSON, GERRY F., and JAMES R. COLLINS, "Note correcting the results in 'Olkin's new formula for the significance of r_{13} vs. r_{23} compared with Hotelling's method'," *American Educational Research Journal*, **7**, 639–641 (1970).

SIEGEL, SIDNEY, *Nonparametric Statistics for the Behavioral Sciences*. New York: McGraw-Hill, 1956.

WALKER, HELEN M., and JOSEPH LEV, *Statistical Inference*. New York: Holt, Rinehart and Winston, 1953.

WERT, JAMES E., CHARLES O. NEIDT, and J. STANLEY AHMANN, *Statistical Methods in Educational and Psychological Research*. New York: Appleton-Century-Crofts, 1954.

KOLMOGOROV TESTS OF GOODNESS OF FIT

32.1 FUNCTION

A Kolmogorov test of goodness of fit is used to determine whether an observed frequency distribution departs significantly from a hypothesized frequency distribution. Its use requires data on at least an ordinal scale and the assumption is made that the data are continuously distributed. It is the appropriate alternative to the chi-square test of goodness of fit if the data are continuously distributed and it may be used with smaller samples than the chi-square test. However, the Kolmogorov test loses much of its power if the assumption of continuity is violated. (The presence of numerous tie scores is indicative of violation of the assumption of continuity.) The use of grouped data with the Kolmogorov statistic is highly questionable (and unlikely to lead to rejection of the null hypothesis), whereas the use of grouped data is customary with the chi-square statistic. The Kolmogorov procedure is simple to apply and usually requires less computational effort than the chi-square.

32.2 RATIONALE

Consider the situation in which a random sample is drawn from a continuously distributed population whose distribution is known. If the relative cumulative frequency of the population and that of the sample are plotted on a single graph, the *deviation* of the sample curve from the population curve will be independent of the population distribution. Thus a single sampling distribution of this deviation suffices as a model for all samples of the same size, and it may be used for testing hypotheses about populations having any form of distribution. This sampling distribution has been determined for many different sample sizes and has been recorded in a form convenient for hypothesis testing in Table 15 of the Appendix.

The investigator need determine only the maximum deviation of the sample relative cumulative frequency distribution from the hypothesized relative cumulative frequency distribution. This maximum observed deviation is compared to the tabled value to determine if it is too large to be reasonably attributed to chance.

The assumption of continuity of distribution is one that is never met in practice—it is only approximated. As a result, the Kolmogorov test tends to be a conservative one, and in the practical research setting the probability of rejecting a true null hypothesis is likely to be somewhat smaller than the level of significance set by the investigator. This problem is intensified by grouping the data, and the investigator is cautioned to avoid this practice.

32.3 PROCEDURE

1. The application of a Kolmogorov goodness-of-fit test resembles that of a chi-square goodness-of-fit test. A theoretical distribution is either hypothesized or is known to exist in the completely random situation. The investigator must establish a hypothetical frequency distribution based upon whatever information he may have. This hypothesized distribution is translated into a relative cumulative frequency distribution with intervals that are directly comparable to sample intervals. Ordinarily, the author recommends that there be as many intervals as there are score levels.

2. A random sample is drawn from the population under study, and its relative cumulative frequency distribution is determined.

3. Each interval of the sample distribution is paired with the corresponding interval of the hypothesized distribution, and the difference between the two distributions is determined for each interval. Let D equal the maximum deviation.

4. D is compared to the tabled value for the appropriate sample size and level of significance. A two-tailed test is appropriate (D may be either negative or positive), and this is what the tabled values provide. If the calculated value of D equals or exceeds the tabled value, the null hypothesis is rejected, the finding is significant, and the population distribution is determined to differ from the hypothesized dis-

tribution at the specified level of significance. If the calculated value is smaller than the tabled value, the hypothesis of no difference is retained.

32.4 EXAMPLE

The experiment that served to illustrate the use of a chi-square test of goodness of fit (see Section 29.4) will also be used to illustrate the use of a Kolmogorov test of goodness of fit. This was the situation in which students were awarded percentile rank scores on the basis of national norms, and the teacher sought to determine whether the students were distributed in the same fashion as the norm group. Unfortunately, the data were grouped in ten intervals when reported, and this may handicap the Kolmogorov test.

1. H: The students are uniformly distributed.
   A: The students are not uniformly distributed.
2. Kolmogorov goodness-of-fit test, $N = 60$.
3. .05 level.
   R: $D \geq 0.18$.
4. The value of D may be quite readily determined by arranging the data as shown in the following table. The sample data are from Section 29.4.

Percentile rank	10	20	30	40	50	60	70	80	90	
Hypothesis rcf	.10	.20	.30	.40	.50	.60	.70	.80	.90	1.0
Sample rcf	.15	.32	.38	.42	.47	.55	.57	.68	.82	1.0
Difference	.05	.12	.08	.02	.03	.05	.13	.12	.08	

$D = .13$

5. Retain the null hypothesis. No significant difference between the two distributions has been determined to exist. The reader is reminded that the hypothesis was rejected when the chi-square test was used.

32.5 KOLMOGOROV TEST FOR GOODNESS OF FIT TO NORMAL

Let us suppose that the following collection of scores is a random sample from a much larger sample, and that the investigator desires to test the hypothesis that the parent population is normally distributed with mean of zero and standard deviation of one:

2.72,   2.01,   1.76,   1.29,   .67,   .10,   −.08,   −1.20,   −1.42,   −2.03

We will establish the hypothesized distribution with N equally probable intervals and present the data in tabular form as before:

z-score		−1.28	−.84	−.52	−.25	0	+.25	+.52	+.84	+1.28
Hypothesis rcf	.10	.20	.30	.40	.50	.60	.70	.80	.90	1.0
Sample rcf	.20	.30	.30	.30	.40	.50	.50	.60	.60	1.0
Difference	.10	.10	0	.10	.10	.10	.20	.20	.30	

The value of D is .30, and this is nonsignificant at the .05 level (R: $D \geq$ .41). Of course, a sample of size ten is hardly adequate to build a convincing case for normality of distribution.

The reader's attention is directed to the fact that there is no provision in the Kolmogorov test for estimating parameters from sample information. The hypothesis must completely specify the hypothesized distribution. In the case of the chi-square test for goodness of fit to normal, the investigator was privileged to estimate the population mean and standard deviation from sample information (resulting in the loss of two degrees of freedom).

32.6   DISCUSSION

The Kolmogorov maximum deviation test has been suggested as an alternative to the chi-square approximation of the multinomial for testing goodness of fit to a hypothesized distribution. The use of the chi-square statistic in goodness-of-fit tests usually necessitates rather large samples, while the Kolmogorov statistic may be used with samples of any size. The chi-square approximation of the multinomial may be used with either discrete or continuous data, while the Kolmogorov statistic requires data that are continuously distributed.

Ordinarily, when all of the assumptions are met and the statistician has a choice between two statistical tests, the more powerful of the two will be the one with the most rigorous assumptions. Thus, for a given sample size, one would anticipate that the Kolmogorov statistic would be superior to the chi-square when the data are continuously distributed. However, the Kolmogorov statistic suffers great loss of power if this assumption is violated, and this is quite likely to be the case in the practical research setting. Some experimentation on the part of the investigator may be necessary to determine which of the two tests is most appropriate for his data.

EXERCISES

32.1 Refer to Exercise 29.1. Substitute the Kolmogorov test for the chi-square.

32.2 Refer to Exercise 29.2. Substitute the Kolmogorov test for the chi-square.

32.3 An achievement test has been standardized with a large number of school children, and the scores are reported as percentile ranks based on the norm group. Use the Kolmogorov statistic to test the hypothesis that the following collection of scores is distributed in the same fashion as the norm group. Use the .05 level.

$$
\begin{array}{ccccc}
7 & 22 & 35 & 46 & 48 \\
49 & 51 & 52 & 56 & 62 \\
64 & 69 & 70 & 74 & 78 \\
81 & 82 & 88 & 90 & 94
\end{array}
$$

SELECTED REFERENCES

GOODMAN, L. A., "Kolmogorov-Smirnov Tests for Psychological Research," *Psychological Bulletin*, **51**, 160–168 (1954).

MASSEY, F. J., "The Kolmogorov-Smirnov test for goodness of fit," *Journal of the American Statistical Association*, **46**, 68–78 (1951).

KITTLESON, HOWARD M., and JOHN T. ROSCOE, "An empirical comparison of four chi-square and Kolmogorov models for testing goodness of fit to normal," paper presented at the annual convention of the American Educational Research Association, Chicago, April 1972.

SLAKTER, M. J., "A comparison of the Pearson chi-square and Kolmogorov goodness of fit tests with respect to validity. *Journal of the American Statistical Association*, **60**, 854–858 (1965).

33

THE KOLMOGOROV-SMIRNOV TWO-SAMPLE TEST

33.1 FUNCTION

The Kolmogorov-Smirnov two-sample test is used to determine whether two populations are distributed in the same fashion. Its use requires data on at least an ordinal scale, and the assumption is made that the data are continuously distributed. It is the appropriate alternative to the chi-square test of homogeneity for two samples if the data are continuously distributed, and it may be used with smaller samples than the chi-square test. Like the Kolmogorov test of goodness of fit, the Kolmogorov-Smirnov two-sample test loses much of its power if the assumption of continuity is violated. If the data are grouped, the chi-square statistic will usually prove more powerful. The Kolmogorov-Smirnov procedure is simple to apply and usually requires less computational effort than the chi-square.

33.2 RATIONALE

The Kolmogorov-Smirnov two-sample test is an extension of the Kolmogorov statistic to the two-sample situation. If two independent random samples are drawn from the same continuously distributed parent population, and if their cumulative frequencies (or relative cumulative frequencies) are plotted on a single graph, the difference between the two curves will be independent of the distribution of the population. The sampling distribution of this difference has been determined for many different sample sizes and has been recorded in a form convenient for hypothesis testing in Tables 16 and 17 of the Appendix. Table 16 is for use with small samples ($N \leq 40$)

of equal size, and Table 17 is for use with large samples ($N \geq 20$) without the restriction of equal size.

The investigator need determine only the maximum difference between the cumulative frequencies of the two samples and compare it to the tabulated value to see if it is too large to be reasonably attributed to chance.

The assumption of continuity of distribution is one that is never met in practice — it is only approximated. As a result, the Kolmogorov-Smirnov test tends to be a conservative one, and in the practical research setting, the probability of rejecting a true null hypothesis is likely to be somewhat smaller than the level of significance set by the investigator.

33.3 PROCEDURE FOR SMALL SAMPLES OF EQUAL SIZE

1. Two independent random samples of equal size are drawn by the investigator. These may be samples drawn from the same parent population and subjected to two different experimental treatments for determining the effects of the treatments, or they may be drawn from two different populations with the intent to determine whether the populations differ. Membership in one of the two samples constitutes the independent variable. The dependent variable is some continuously distributed criterion selected by the investigator.

2. The cumulative frequency distributions of the two samples are established in such a fashion that the intervals of the two samples are directly comparable. The investigator is cautioned to use as many intervals as practical. The difference in the cumulative frequencies is determined for each interval, and the largest of these differences is designated K_D.

3. K_D is compared to the tabled value for the appropriate sample size and level of significance. Table 16 of the Appendix is for samples of equal size up to $N = 40$, where N is the size of each sample. A one- or two-tailed test may be used. If the calculated value of K_D equals or exceeds the tabled value, the finding is significant, the null hypothesis is rejected, and the two distributions are determined to differ. The exact interpretation of a significant finding is dependent upon the nature of the criterion variable. If the calculated value is smaller than the tabled value, the null hypothesis is retained, and no significant difference in the two distributions has been established.

33.4 PROCEDURE FOR LARGE SAMPLES

1. Two independent samples are drawn by the investigator. These may be samples drawn from the same parent population and subjected to

two different experimental treatments, or they may be drawn from two different populations and compared on some criterion. In either event, a continuously distributed criterion measure must be available on each individual.

2. The relative cumulative frequency distributions of the two samples are established in such a fashion that the intervals of the two samples are directly comparable. The investigator is cautioned to use as many intervals as practical. The difference in the relative cumulative frequencies is determined for each interval, and the largest of these differences is designated D.

3. D is compared to the tabled value for the appropriate sample size and level of significance. Table 17 of the Appendix is for samples of size 20 or greater, and it is not necessary that the samples be of equal size. The table is two-tailed, but may be adapted for one-tailed use by being entered with twice the desired level of significance (use .10 in the table for a one-tailed test at the .05 level). If the calculated value of D equals or exceeds the tabled value, the finding is significant, and the null hypothesis is rejected. If the calculated value is smaller than the tabled value, the null hypothesis is retained.

33.5 EXAMPLE

The experiment that served to illustrate the use of the t-test for two independent samples (see Section 24.4) and the Mann-Whitney U-test (see Section 26.4) will also serve to illustrate the use of the Kolmogorov-Smirnov two-sample test. It should be noted that each of these three statistical analyses tests a different null hypothesis, but that each is a valid approach to the problem. To some extent, the selection of a statistical test is dependent upon the way in which the investigator sees the problem.

1. H: No difference in the distributions associated with the two diets.
   A: Different distributions are associated with the two diets. The difference in the two might result from either greater gains or greater variability associated with one of the diets.
2. Kolmogorov-Smirnov two-sample test, $n_a = n_b = 10$.
3. .05 level, two-tailed test.
   R: $K_D \geq 7$.
4. The raw scores will be used in establishing the two cumulative frequency distributions. The number of intervals will be set equal to the number of score levels.

Score	1	2	3	4	5	6	7	8	9	10	11
Sample A cf	1	2	3	5	7	7	7	8	10	10	10
Sample B cf	0	0	0	1	1	2	4	6	7	9	10
Difference	1	2	3	4	6	5	3	2	3	1	

$K_D = 6$

5. Retain the null hypothesis; no significant difference in the two distributions was found. Both the t-test for independent samples and the Mann-Whitney U-test gave significant results with these data. The fact that the Kolmogorov-Smirnov test was unable to detect the difference between the two samples might be due to the fact that it tests a null hypothesis with a broader range of alternatives. Generally a statistical test that may be rejected because of any one of several different kinds of departures from the sampling distribution will be less powerful than a statistical test that concentrates on a single alternative to the null hypothesis.

33.6 DISCUSSION

The Kolmogorov-Smirnov two-sample test has been suggested as an alternative to the chi-square test of homogeneity for testing whether there are significant differences in the distributions of two samples. The use of the chi-square statistic in tests of homogeneity usually necessitates rather large samples, while the Kolmogorov-Smirnov statistic may be used with samples of any size. The chi-square test may be used with either discrete or continuous data, while the Kolmogorov-Smirnov statistic requires data that are continuously distributed.

Under certain circumstances, the investigator may also regard the Kolmogorov-Smirnov two-sample test as an alternative to the t-test for two independent samples and the Mann-Whitney U-test.

The Kolmogorov-Smirnov two-sample test is a useful and versatile method of statistical analysis, but it suffers great loss in power if the assumption of continuity of distribution is violated. For this reason, it is not likely to be very useful where the data are grouped or where there are numberous tie scores. Some experimentation on the part of the investigator may be desirable to determine whether the Kolmogorov-Smirnov test provides an effective method of analyzing the data.

EXERCISES

33.1 Refer to Exercise 26.3. Substitute the Kolmogorov-Smirnov test for two independent samples for the Mann Whitney U-test.

33.2 Refer to Exercise 30.2. Substitute the Kolmogorov-Smirnov test for two independent samples for the chi-square test of independence.

SELECTED REFERENCES

CONOVER, W. J., *Practical Nonparametric Statistics*. New York: John Wiley, 1971.
NOETHER, G. E., *Elements of Nonparametric Statistics*. New York: John Wiley, 1967.

THE FISHER EXACT
PROBABILITY TEST

34.1 FUNCTION

The Fisher exact probability test is an extremely useful nonparametric test for use with two-by-two bivariate frequency tables. In contrast to the chi-square approximation of the multinomial, the Fisher test is based on exact probabilities and may be used with very small samples. Generally, the Fisher exact probability test is used to compare two independent samples on a dichotomous criterion. The two samples may be constituted by drawing them from a single population and subjecting them to experimentation, or by drawing them from two different populations for purposes of comparison.

34.2 RATIONALE

It is customary to let the letters A, B, C, and D represent the cell frequencies of a two-by-two bivariate frequency distribution:

A	B	$A + B$
C	D	$C + D$
$A + C$	$B + D$	

If the marginal frequencies are regarded as fixed, the probability of any particular arrangement of the cell frequencies may be determined from the hypergeometric distribution:

$$p = \frac{\left(\dfrac{A+C}{A}\right)\left(\dfrac{B+D}{B}\right)}{\left(\dfrac{N}{A+B}\right)} = \frac{(A+B)!(C+D)!(A+C)!(B+D)!}{N!A!B!C!D!}$$

In order to test a hypothesis with this distribution, it is necessary not only to determine the probability of occurrence of the particular frequency distribution observed, but also that of all more extreme distributions. The problem is essentially the same as that encountered in the use of the binomial probability distribution (see Section 19.4), in which the investigator sought to determine the probability that r was equal to or greater than some particular value. In the case of the hypergeometric distribution, the calculations get a little more involved; however, the distribution is tabled in a manner convenient for hypothesis testing in Table 16 of the Appendix.

Table 18 gives critical values of D (or C) corresponding to observed values of B (or A) when the marginal frequencies are fixed. If the calculated value of D (or C) is equal to or less than the tabled value, the finding is significant at the specified level. The table is one-tailed, but may be adapted to two-tailed tests by being entered with one-half the desired level of significance. In the case of two-tailed tests, the region of rejection must be established independently for each tail. Because of the discreteness of the data (fractional frequencies are not allowed), the tabled values are approximations, and any error is on the conservative side. Thus the critical value listed at the .05 level might also be listed as the critical value for the .025 level, and the probability of rejecting a true hypothesis might be on the order of .022.

34.3 PROCEDURE

1. All of the subjects in two independent random samples are assessed on a single dichotomous criterion. The data are arranged in a bivariate frequency table, and the cell frequencies are designated A, B, C, and D as outlined in Section 34.2.
2. The marginal frequencies, $A + B$ and $C + D$, are determined and the portion of Table 18 corresponding to these values is located.
3. The table is entered for the observed value of B, and the observed value of D is compared to the critical value under the desired level of significance. If the observed value of B is not listed, the table is entered for the observed value of A, and the observed value of C is compared to the critical value tabled under the desired level of significance.

4. If the observed value of D (or C) is equal to or smaller than the tabulated value, the null hypothesis is rejected. If the observed value exceeds the tabled value, the null hypothesis is retained.

5. The critical values listed in Table 18 are for a one-tailed test. A two-tailed test may be accomplished by entering the table with one-half the desired level of significance (use .025 in the table for a two-tailed test at the .05 level). If a two-tailed test is used, a separate determination of the region of rejection is required for each tail. The use of one- and two-tailed tests is illustrated in the following section.

34.4   EXAMPLES

Four examples will be given to illustrate the use of Table 18. In each case a median test will be used so that the logic of one- and two-tailed tests can be demonstrated. The .05 level of significance will be used with all examples.

EXAMPLE 1

	Group 1	Group 2	
Above median	$A = 7$	$B = 8$	$A + B = 15$
Below median	$C = 1$	$D = 14$	$C + D = 15$

One-tailed alternative: $\text{Mdn}_1 > \text{Mdn}_2$

$$R: \quad C \leq 1 \quad \text{(reject)}$$

One-tailed alternative: $\text{Mdn}_1 < \text{Mdn}_2$

$$R: \quad D \leq 2 \quad \text{(retain)}$$

Two-tailed alternative: $\text{Mdn}_1 \neq \text{Mdn}_2$

$$R: \quad C \leq 1, D \leq 1 \quad \text{(reject)}$$

EXAMPLE 2

	Group 1	Group 2	
Above median	$A = 6$	$B = 9$	$A + B = 15$
Below median	$C = 1$	$D = 14$	$C + D = 15$

One-tailed alternative: $\text{Mdn}_1 > \text{Mdn}_2$

$$R: \quad C \leq 1 \quad \text{(reject)}$$

One-tailed alternative: $Mdn_1 < Mdn_2$

$$R: \quad D \le 3 \quad \text{(retain)}$$

Two-tailed alternative: $Mdn_1 \ne Mdn_2$

$$R: \quad C \le 0, D \le 2 \quad \text{(retain)}$$

EXAMPLE 3

	Group 1	Group 2	
Above median	$A = 5$	$B = 10$	$A + B = 15$
Below median	$C = 12$	$D = 3$	$C + D = 15$

One-tailed alternative: $Mdn_1 > Mdn_2$

$$R: \quad C \le 0 \quad \text{(retain)}$$

One-tailed alternative: $Mdn_1 < Mdn_2$

$$R: \quad D \le 4 \quad \text{(reject)}$$

Two-tailed alternative: $Mdn_1 \ne Mdn_2$

$$R: \quad C \le 0, D \le 3 \quad \text{(reject)}$$

EXAMPLE 4

	Group 1	Group 2	
Above median	$A = 4$	$B = 11$	$A + B = 15$
Below median	$C = 0$	$D = 15$	$C + D = 15$

One-tailed alternative: $Mdn_1 > Mdn_2$

$$R: \quad C \le 0 \quad \text{(reject)}$$

One-tailed alternative: $Mdn_1 < Mdn_2$

$$R: \quad D \le 5 \quad \text{(retain)}$$

Two-tailed alternative: $Mdn_1 \ne Mdn_2$

$$R: \quad D \le 4 \quad \text{(retain)}$$

34.5 DISCUSSION

The Fisher exact probability test may be substituted for the chi-square test of independence when the data are cast into a two-by-two bivariate frequency table. Ordinarily, the Fisher test will be used only with those small samples provided for by Table 18 and the chi-square test will be used with larger samples. Occasionally, the Fisher test may be an appropriate alternative to the t-test for two independent samples or for the Mann-Whitney U-test.

EXERCISES

34.1 A random sample of 13 businesswomen is polled on whether they would favor a female president of the United States. They split, with 9 voting yes and 4 voting no. A random sample of 14 businessmen is polled on the same issue, and they split, with 3 voting yes and 11 voting no. Use the Fisher exact test to test the null hypothesis of no relationship between voting preference and sex of the respondent. Use a one-tailed test at the .05 level.

34.2 Repeat Exercise 34.1, substituting the chi-square test of independence for the Fisher exact test.

35

THE F-DISTRIBUTION

35.1 THE SAMPLING DISTRIBUTION OF THE F-RATIO

Consider the situation in which there are two normally distributed populations having the same variance. Let us draw a sample of size n_1 from the first population and a sample of size n_2 from the second, calculate an unbiased estimate of the common variance from each sample, then form the ratio:

$$F = \frac{S_1^2}{S_2^2} \quad \text{where } S_1^2 = \frac{SS_1}{n_1 - 1} \text{ and } S_2^2 = \frac{SS_2}{n_2 - 1}$$

If a very large number of samples are drawn in the fashion described (all samples from the first population of size n_1 and all from the second of size n_2), and the F-ratio is calculated for each pair of samples, the sampling distribution of these ratios (which was derived by Sir Ronald Fisher) is known as the F-distribution.

Like the t-distribution and the chi-square distribution, the F-distribution is a function of sample size, in this case, of the sizes of two samples. There is a family of F-curves, each determined by two parameters, the degrees of freedom associated with the numerator and the degrees of freedom associated with the denominator. F-curves are positively skewed, ranging from zero to infinity; those with large values of degrees of freedom tend toward symmetry. The distribution of the F-ratio for representative values of degrees of freedom is graphically illustrated in Figure 35.1.

Because there are two values of degrees of freedom associated with every F-ratio, F-tables tend to be bulky, and they are customarily tabled only for the right-hand tail and for a few commonly used levels of significance. Table 19 in the Appendix is in two parts, the first giving critical

286

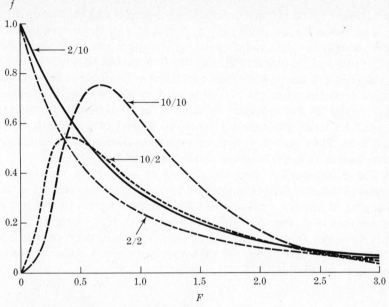

Figure 35.1   Representative F-distributions.

values of the F-ratio for the .05 level, the second for the .01 level. The degrees of freedom for the numerator (designated by df_1) are given across the top of the table and are used to determine which column is entered. The degrees of freedom for the denominator (designated by df_2) are given along the left side of the table and are used to determine which row is entered. To use the tables, one must choose the page with the desired level of significance, then find the point in the table where the column and row for the appropriate degrees of freedom intersect. For example, using the .01 level and 8, 20 degrees of freedom (degrees of freedom for the numerator is always given first), the tabled entry is 3.56. An F-ratio equal to or larger than the tabled value is significant.

As with the t-statistic, the table is not complete for all possible degrees of freedom, and the investigator will occasionally find that the desired degrees of freedom are not tabled. In this situation, he must discard degrees of freedom, that is, enter the table with fewer degrees of freedom than the problem merits. This procedure will necessitate a slightly larger statistic to achieve rejection.

35.2   THE RELATIONSHIPS AMONG THE COMMON PARAMETRIC STATISTICS

It has been previously noted that the t-distribution with degrees of freedom equal to infinity is identical to the normal distribution. Mathematicians would say that the normal distribution is a special case of the *t*-distribution.

The *F*-distribution, with degrees of freedom of the numerator equal to one, is equal to the square of the *t*-distribution having the same degrees of freedom as the denominator of the *F*. It should be noted here that the one-tailed *F* is the square of the two-tailed *t*. The logic of this will become more apparent when the *F*-ratio is used to test hypotheses about means in Chapter 36. For example, the tabled *t* for 60 degrees of freedom is 2.00 at the .05 level. The tabled *F* for 1, 60 degrees of freedom is 4.00 at the .05 level. Thus one may think of both the *t*-distribution and the normal distribution as special cases of the *F*-distribution.

The *F*-distribution, with the degrees of freedom of the denominator equal to infinity, may be multiplied by the degrees of freedom of the numerator to achieve a chi-square with the same degrees of freedom as the numerator of *F*. For example, the *F*-ratio with 10, ∞ degrees of freedom is 1.83 at the .05 level, and the chi-square with 10 degrees of freedom is 18.3 at the .05 level. Thus the chi-square distribution is also a special case of the *F*-distribution.

35.3   TESTING FOR HOMOGENEITY OF TWO VARIANCES

One of the assumptions underlying the *t*-test for two independent samples is that of homogeneity of variance — that the two samples are randomly drawn from populations having the same variance. The hypothesis of equal variances may be tested with the *F*-statistic.

Ordinarily, a two-tailed test is desired, and it is accomplished in the following fashion: The unbiased estimates of the variance are calculated for both samples, and the *F*-ratio is formed, the larger of these being in the numerator (the ratio should be equal to or larger than one). This ratio is compared to the tabled value for the appropriate degrees of freedom. The level of significance, however, is twice that specified by the table (the .01 table gives a two-tailed test at the .02 level).

Many authors prefer to use an alternate strategy for two-tailed tests, in which the left tail is defined by:

$$F'_{(df_1, df_2)} = \frac{1}{F_{(df_2, df_1)}}$$

That is, the *F* for the left tail is the reciprocal of the tabled *F* found by reversing the degrees of freedom. Suppose, for example, that a two-tailed test at the .02 level is desired, and that $n_1 = 16$ (so $df_1 = 15$) and $n_2 = 25$ (so $df_2 = 24$). Referring to Table 19 (the .01 level section), the right tail is defined by $F \geq 2.89$. The left tail is defined by:

$$F' \leq \frac{1}{3.18} = 0.314$$

When the two-tailed test is organized in this fashion, the F-ratio from the sample data is always formed with S_1^2 in the numerator and S_2^2 in the denominator, where S_1^2 is the variance estimate from population one and S_2^2 is the variance estimate from population two.

If a one-tailed test is desired, the investigator should specify which variance is expected to be the larger of the two before the variance estimates are calculated. The ratio is formed using the estimate for this hypothesized larger variance in the numerator. Then the table is entered in the usual fashion (the .01 table gives a one-tailed test at the .01 level). Of course, if the F-ratio formed in this fashion is less than one, the finding is nonsignificant.

Unfortunately, the statistical test described is very sensitive to departures from normality, and it should not be used with small samples unless the assumption of normality is well established. This is unlikely to be the case in the practical research setting.

35.4   THE HARTLEY TEST FOR HOMOGENEITY OF VARIANCES

The use of the F-ratio as outlined in Section 35.3 is limited to the two-sample situation. A number of other tests for homogeneity of variances have been suggested, and these are suitable for the k-sample (two or more samples) situation. The simplest of these is the Hartley F_{max} test.

The procedure for using the Hartley F_{max} test is as follows:

1. Calculate an unbiased estimate of the variance (S_j^2) for each sample.
2. Form the ratio of the largest variance estimate to the smallest variance estimate.

$$F_{max} = \frac{S_{max}^2}{S_{min}^2}$$

3. The statistic may be referred to Table 20 in the Appendix, where k is the number of samples and n is the common sample size (the table assumes equal sample sizes). The hypothesis of equal variances is rejected if the calculated value is equal to or greater than the tabled value.

Suppose, for example, that five samples ($k = 5$, $n = 10$) yield variance estimates of 24.08, 36.67, 51.55, 73.09, and 98.32. The calculated statistic is 4.08, and the tabled value is 7.11 at the .05 level. The hypothesis is retained — there is no significant difference in the variances.

The use of the Hartley F_{max} statistic is limited to the situation in which the sample sizes are equal. While a number of authors have suggested ways in which the statistic might be adapted for use with unequal sample sizes, the available evidence suggests that the Hartley F_{max} test should be restricted to use with samples of equal size.

The Hartley F_{max} test is intended for use with data drawn from normally distributed populations, and it is extremely sensitive to departures from normality. When sampling from platykurtic populations, the test is extremely conservative, and detection of a false hypothesis is unlikely. When sampling from leptokurtic or skewed populations, the procedure is extremely liberal, and rejection of a true hypothesis is much too likely. The Hartley F_{max} test should not be used unless the investigator has good reason to believe that he is sampling from normal populations.

35.5   THE COCHRAN TEST FOR HOMOGENEITY OF VARIANCES

The Cochran C test is almost as simple as the Hartley F_{max} test, and it offers a number of advantages. The procedure for using the Cochran C test is as follows:

1. Calculate an unbiased estimate of the variance (S_j^2) for each sample.
2. Form the ratio of the largest variance estimate to the sum of the variance estimates:

$$C = \frac{S_{max}^2}{\Sigma \, S_j^2}$$

3. The statistic may be referred to Table 21 in the Appendix, where k is the number of samples and n is the common sample size (the table assumes equal sample sizes). The hypothesis of equal variances is rejected if the calculated value is equal to or greater than the tabled value.

The procedure will be illustrated with the same example used to illustrate the Hartley F_{max} test. Again, five samples ($k = 5$, $n = 10$) yield variance estimates of 24.08, 36.67, 51.55, 73.09, and 98.32. The calculated statistic is 0.347 and the tabled value is 0.424. As before, the hypothesis is retained — there is no significant difference in the variances.

Like the Hartley F_{max} test, the Cochran C test is derived for the situation in which the sample sizes are equal. For most behavioral science applications, the Cochran C test appears to provide a reasonable approximation for unequal sample sizes if n is taken to be the average sample size.

The Cochran C test is also derived for use with data drawn from normally distributed populations. Like the Hartley F_{max} test the test is extremely conservative when sampling from platykurtic populations, and detection of a false hypothesis is unlikely. A good approximation is achieved, however, when sampling from leptokurtic or skewed populations of the sort commonly encountered in behavioral research.

Because the Hartley F_{max} test uses only the data yielded by two of the k-samples, it tends to lose power as the number of samples is increased. This loss begins to be apparent if the number of samples is five or more and is appreciable if the number of samples is ten or more. Under these circumstances, the Cochran C test will prove considerably more powerful than the Hartley F_{max} test.

EXERCISES

35.1 Given two samples with $n_1 = 22$, $n_2 = 25$, $S_1 = 33.70$, $S_2 = 98.40$.
   (a)   Determine the region of rejection for a two-tailed test of the hypothesis of equal variances. Use the .02 level.
   (b)   Calculate the F-ratio from the sample data.
   (c)   Test the hypothesis of equal variances.

35.2 Given three samples of equal size with $n = 10$, $S_1 = 18.07$, $S_2 = 22.14$, $S_3 = 89.88$.
   (a)   Determine the region of rejection for a Hartley F_{max} test at the .05 level.
   (b)   Calculate the value of F_{max} from the sample data.
   (c)   Test the hypothesis of equal variances.

35.3 Repeat Exercise 35.2 using the Cochran C test.

SELECTED REFERENCES

BOX, G. E. P., "Non-normality and tests on variance," *Biometrika*, **40**, 318–335 (1953).

COCHRAN, W. G., "The distribution of the largest of a set of variances," *Annals of Eugenics*, **11**, 47–52 (1941).

GLASS, G. V., "Testing homogeneity of variances," *American Educational Research Journal*, **3**, 187–190 (May 1966).

HARTLEY, H. O., "The maximum F-ratio as a short-cut test for homogeneity of variances," *Biometrika*, **37**, 308–312 (1950).

PEARSON, E. S., "Alternative tests of heterogeneity of variance: Some Monte Carlo results," *Biometrika*, **53**, 229–234 (1966).

VEITCH, WILLIAM R., "Homogeneity of variance: An empirical comparison of four statistical tests," unpublished Ph.D. dissertation, Kansas State University, 1972.

36

ONE-WAY ANALYSIS OF VARIANCE

36.1 FUNCTION

In all probability, the most useful experimental analysis considered in this text up to this point is the t-test for two independent samples. It provides what is generally the most powerful test of the hypothesis most frequently encountered in the behavioral sciences, that of equal means. However, the probability theory underlying the t-distribution is based upon the assumption that two (and only two) samples will be compared. If the number of comparisons is increased, the t-distribution no longer provides a valid test of the hypothesis of equal means. If the investigator uses the t-test to compare each pair of samples in a collection of three samples (a total of three comparisons instead of one), for example, the probability of rejecting a true hypothesis is much greater than the level of significance. The analysis of variance provides a statistical procedure that is appropriate for use with two *or more* samples. In the two-sample situation it is mathematically equivalent to the t-test.

One-way analysis of variance is used for testing the hypothesis that two or more independent samples were drawn from populations having the same mean. The samples may be constituted by drawing independent random samples from a single population, subjecting them to experimentation, then comparing them on a single criterion variable. Or, the samples may be randomly drawn from different populations, then compared on a single criterion to determine whether the various populations differ with respect to this criterion. The first example is, of course, experimental research; the second is *ex post facto*. The two situations are illustrated by Figures 36.1 and 36.2.

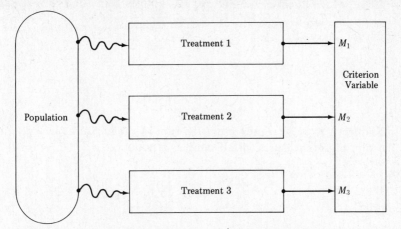

Figure 36.1 Simple randomized design — experimental research.

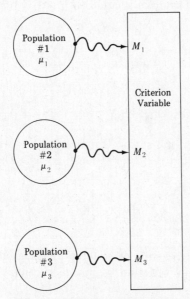

Figure 36.2 Simple randomized design — *ex post facto* research.

36.2 REVIEW OF DOUBLE CLASSIFICATION NOTATION

At this point, the student is advised to review the use of the summation operator and statistical notation for data subject to double classification, as in the situation where the subjects are divided into two or more experimental samples (see Sections 6.3, 6.4, and 7.3).

If we let k represent the number of samples and n_j be the number of subjects in any sample, the total number of subjects in all samples is equal to:

$$N = \sum_{j=1}^{k} n_j$$

Similarly, if we let T_j represent the sum of the measures in any sample, the sum of the measures for all samples is equal to:

$$T = \sum_{j=1}^{k} T_j = \sum_{j=1}^{k} \sum_{i=1}^{n_j} X_{ij}$$

And, if we let M_j represent the mean of any sample, the general mean for all measures in all samples is equal to:

$$M = \frac{1}{N} \sum_{j=1}^{k} n_j M_j = \frac{1}{N} \sum_{j=1}^{k} T_j = \frac{T}{N}$$

36.3 RATIONALE FOR ONE-WAY ANALYSIS OF VARIANCE

Let SS_t represent the total sum of squares for k samples, the sum of the squared deviations from the general mean for all subjects in all samples:

$$SS_t = \sum_{j=1}^{k} \sum_{i=1}^{n_j} (X_{ij} - M)^2 = \sum_{j=1}^{k} \sum_{i=1}^{n_j} X_{ij}^2 - \frac{T^2}{N}$$

Let SS_w represent the sum of squares *within* groups — a measure of dispersion based upon variation within each sample around its sample mean:

$$SS_w = \sum_{j=1}^{k} \sum_{i=1}^{n_j} (X_{ij} - M_j)^2 = \sum_{j=1}^{k} \sum_{i=1}^{n_j} X_{ij}^2 - \sum_{j=1}^{n_j} \frac{T_j^2}{n_j}$$

And let SS_b represent the sum of squares *between* groups — a measure of dispersion that reflects variation of the sample means from the general mean:

$$SS_b = \sum_{j=1}^{k} n_j(M_j - M)^2 = \sum_{j=1}^{k} \frac{T_j^2}{n_j} - \frac{T^2}{N}$$

With a little effort and some elementary algebra, it can be demonstrated that:

$$SS_t = SS_w + SS_b$$

This exercise is known as "partitioning" the sum of squares, and it permits the investigator to analyze the contributions that various factors make toward the total variability of the individuals participating in a given experiment.

The degrees of freedom for SS_t are equal to $N - 1$, there being N terms in the expression and the general mean of the population being estimated from sample data. The degrees of freedom for SS_w are equal to $N - k$, there being N terms in the expression and k group means being estimated from sample data. Finally, the degrees of freedom for SS_b are equal to $k - 1$, there being k terms in the expression and the general mean being estimated from sample data. It is immediately apparent that the degrees of freedom are partitioned when the sum of squares was partitioned:

$$N - 1 = N - k + k - 1$$

The sum of squares within groups (SS_w) may be divided by its degrees of freedom ($N - k$) to obtain the mean square within (MS_w). If the data from which MS_w is calculated come from independent random samples of normally distributed and equally variable populations, the MS_w provides an unbiased estimate of the common population variance — an estimate that reflects variation within each sample around its sample mean.

The sum of squares between groups (SS_b) may be divided by its degrees of freedom ($k - 1$) to obtain the mean square between (MS_b). If the data from which MS_b is calculated come from independent random samples of normally distributed and equally variable populations with equal means, the MS_b also provides an unbiased estimate of the common population variance — an estimate that reflects variation of the sample means from the general mean.

If the conditions stated in the two previous paragraphs are met, and if MS_w and MS_b are calculated from the same data, then these data provide independent estimates of the same common population variance, and the following expression is distributed as F:

$$F = \frac{MS_b}{MS_w} \quad \text{with } df = k - 1, N - k$$

If, however, the population means are unequal (but the other conditions are met), MS_b will tend to be larger than MS_w. The F-ratio thus calculated may be compared to the F-table to determine whether MS_b is significantly larger than MS_w. This procedure, known as the analysis of variance, provides a test of the hypothesis of equal means, which is mathematically equivalent to the t-test for independent samples in the two-sample situation. However, the analysis of variance is not limited to two samples; it may be used with as many samples as the investigator desires.

To be completely accurate, the null hypothesis tested by the simple analysis of variance is that of no significant difference between the parameters represented by the mean square between and the mean square within — with the one-tailed alternative that the mean square between is larger than the mean square within. A one-tailed test is achieved by always forming the F-ratio with the mean square between in the numerator and the mean square within in the denominator; this is compared to the right tail of the F-distribution as found in Table 19 in the Appendix.

36.4 PROCEDURE

1. Two or more random samples are independently drawn from the population under study, and these are subjected to different experimental treatments. The analysis of variance may also be used to analyze existing differences between different populations.

2. At the conclusion of the experiment, a single criterion measure is administered to all subjects in all samples. The sums of squares between groups, within groups, and for the total are calculated, along with their respective degrees of freedom. The sums of squares between groups and within groups are divided by their respective degrees of freedom to obtain the mean square between and the mean square within. The calculations needed to determine these values will be considered in great detail in the following section. It is customary to record these values in a summary table such as this:

Summary table for the analysis of variance			
Source of variation	df	Sums of squares	Mean squares
Between groups			
Within groups			
Total			XXXXX

3. The F-ratio is calculated as follows:

$$F = \frac{MS_b}{MS_w} \quad \text{with } df = k - 1, N - k$$

4. The calculated F is compared to the tabled F at the desired level of significance and with the appropriate number of degrees of freedom. If the calculated statistic equals or exceeds the tabled value, the hypothesis is rejected, and a significant difference in the means of the various samples is determined to exist. Unfortunately, if there are more than two samples, this procedure does not indicate exactly which means are significantly different from each other. This is a complex problem, the solution to which will be sought in a later chapter. If the calculated statistic is smaller than the tabled value, the null hypothesis is retained, and no significant difference in the means has been found.

36.5   COMPUTATIONAL PROCEDURES

The sum of squares between groups has previously been identified as:

$$SS_b = \sum_{j=1}^{k} \frac{T_j^2}{n_j} - \frac{T^2}{N}$$

The sum of squares within groups has been identified as:

$$SS_w = \sum_{j=1}^{k} \sum_{i=1}^{n_j} X_{ij}^2 - \sum_{j=1}^{k} \frac{T_j^2}{n_j}$$

And the sum of squares for total has been identified as:

$$SS_t = \sum_{j=1}^{k} \sum_{i=1}^{n_j} X_{ij}^2 - \frac{T^2}{N}$$

The reader's attention is directed to the fact that some of the terms are held in common by these three formulas. As a matter of fact, only three terms need be evaluated to complete the calculations.

Consider the first term of the sum of squares within, which is also the first term of the sum of squares for total:

$$\sum_{=1}^{k} \sum_{i=1}^{n_j} X_{ij}^2$$

To evalute this term, one need only (1) square each score in the entire collection, and (2) sum all of these squares.

Next, consider the second term of the sum of squares between, which is also the second term of the sum of squares for total:

$$\frac{T^2}{N}$$

To evaluate this term, one need only (1) sum all scores in the entire collection, (2) square this sum, and (3) divide by N. This term is sometimes referred to as the correction for the mean (CFM) — it reduces to zero and drops out of the expressions if deviation scores are used.

Finally, consider the first term of the sum of squares between, which is also the second term of the sum of squares within:

$$\sum_{j=1}^{k} \frac{T_j^2}{n_j}$$

To evaluate this term, (1) sum the scores in each group, (2) square the sum for each group, (3) divide each of these squares by the number of subjects in the respective group, and (4) sum for all groups.

The degrees of freedom for these sums of squares are $k - 1$ for the sum of squares between groups, $N - k$ for the sum of squares within groups, and $N - 1$ for the sum of squares for total. Of course, k is the number of groups, and N is the total number of subjects in all groups.

The mean square between is simply the sum of squares between groups divided by $k - 1$, and the mean square within is the sum of squares within divided by $N - k$.

36.6   EXAMPLE

The students in a psychology class are randomly assigned to three different study groups. Each of the three groups is given a separate reading assignment to be completed within a period of two weeks. The first group is assigned reading in psychoanalytic theory, the second in cognitive-field theory, and the third in stimulus-response theory. At the end of the two-week period, a ten-item multiple-choice examination is administered to the class — an examination deliberately structured so that it favors stimulus-response theory. The results of the examination are reported as follows:

Group 1		Group 2		Group 3	
X	X^2	X	X^2	X	X^2
9	81	6	36	9	81
8	64	6	36	8	64
8	64	5	25	7	49
6	36	4	16	7	49
5	25	4	16	7	49
4	16	3	9	6	36
4	16	2	4	5	25
4	16	2	4		
48	318	32	146	49	353

1. H: $\mu_1 = \mu_2 = \mu_3$ A: $\mu_1 \neq \mu_2 \neq \mu_3$.
2. Simple analysis of variance, $n_1 = 8$, $n_2 = 8$, $n_3 = 7$.
3. .05 level, $df = 2, 20$.
   R: $F \geq 3.49$.
4. $\Sigma\Sigma X^2 = 318 + 146 + 353 = 817$

$$\sum \frac{T_j^2}{n_j} = \frac{48^2}{8} + \frac{32^2}{8} + \frac{49^2}{7} = 759$$

$$\frac{T^2}{N} = \frac{129^2}{23} = 723.52$$

$$SS_b = 759 - 723.52 = 35.48$$

$$MS_b = \frac{35.48}{2} = 17.74$$

$$SS_W = 817 - 759 = 58$$

$$MS_W = \frac{58}{20} = 2.90$$

$$SS_t = 817 - 723.52 = 93.48$$

Summary table for the analysis of variance			
Source of variation	df	Sums of squares	Mean squares
Between groups	2	35.48	17.74
Within groups	20	58.00	2.90
Total	22	93.48	XXXXX

$$F = \frac{17.74}{2.90} = 6.12, df = 2, 20$$

5. Reject the null hypothesis. There is a significant difference in the means of the three samples. The mean of the first sample is 6, of the second sample is 4, and that of the third sample is 7. It is probably reasonable to conclude that the third group scored significantly

higher than the second group, but nothing can be stated on the basis of the analysis of variance with respect to whether the third group scored significantly higher than the first, or whether the first scored higher than the second.

36.7 DISCUSSION

The one-way analysis of variance provides one of the more popular statistical tests. It is a very powerful test that lends itself to a wide variety of research problems, and it is the basic unit of many more complex experimental designs.

The mathematical assumptions underlying the simple analysis of variance outlined in this chapter are essentially the same as those underlying the *t*-test for two independent samples. These assumptions are:

1. The criterion scores are statistically independent. This can be assured only by the process of randomization — either by the random selection of subjects from the populations under study in *ex post facto* research or by the random assignment of subjects to the various experimental treatments in experimental research. If this assumption is violated, it may profoundly influence the statistical analysis and invalidate the results.

2. The criterion scores are drawn from normally distributed populations. Ordinarily, the investigator has no way of knowing for sure precisely how the populations under study are distributed. If he is concerned about the possible violation of this assumption, the chi-square test for goodness of fit to normal may be used where $n_j = 20$ or more. For smaller samples, the Kolmogorov goodness-of-fit test may be used, but it is usually so feeble under these circumstances that non-normality may go undetected. When large samples are used, the Central Limit Theorem is effective, and the investigator need not be concerned about violation of the assumption of normality. It has also been established that the analysis is quite robust with respect to the assumption of normality if samples (even very small samples) of equal size are used. These findings do not apply to the situation in which small samples of unequal size are used, and the use of samples of equal size is highly recommended whenever the investigator has this option.

3. The criterion scores are drawn from populations having the same variance. This is known as the assumption of homogeneous variances. Ordinarily, the investigator has no way of knowing for sure what the population variances are, and there are some kinds of research (classroom learning, for example) which have a distinct tendency to produce

unequal variances. If a test for homogeneity of variances is desired, the Cochran test is recommended. However, it has been established that the analysis is quite robust with respect to the assumption of homogeneous variances if samples of equal size are used. Again, whenever the analysis of variance is used, there are compelling reasons for selecting samples of equal size. If samples of unequal size are used, and the assumption is violated, it may profoundly influence the statistical analysis and invalidate the results.

4. Some authors like to list a fourth assumption — that of equal population means. The mean square within provides an unbiased estimate of the common variance assumed in the preceding paragraph. The mean square between also provides an unbiased estimate of the common variance, provided that the population means are equal. If this is the case, the null hypothesis is true, and the expected value of the F-ratio is one. However, if the means are unequal, the mean square between tends to be inflated, and the expected value of the F-ratio is greater than one. If the F-ratio is sufficiently large to cause rejection of the null hypothesis, we would prefer to think that it is this fourth assumption which has been violated.

The Kruskal-Wallis test provides an excellent nonparametric alternative to one-way analysis of variance. The data must be ordinal and are assumed to be continuous; however, an effective tie correction is available if the assumption of continuity is violated. Under some circumstances, the investigator may prefer to use a chi-square test of homogeneity, especially the median test.

A number of suggestions has been made with respect to measures of association to be used with one-way analysis of variance, and several of these are discussed in the Selected References. The author is partial to the use of η^2 (lowercase Greek letter eta) for it is an extension of the r^2 interpretation given in Sections 12.5 and 24.5 and is equivalent to R^2 (the multiple correlation coefficient) to be discussed in those chapters treating multiple correlation. The value of η^2 (or R^2) is given by:

$$\eta^2 = \frac{SS_b}{SS_t}$$

Eta-square is that proportion of the variability of the dependent variable which is accounted for by the independent variable.

Whenever the investigator uses the analysis of variance with more than two groups, there is a question as to which means are significantly larger than which other means. This is the problem of *multiple comparisons*, which will be treated in a subsequent chapter. There are problems associated with these procedures and often no completely satisfactory solution exists. For this reason the reader should not regard the availability of the analysis of variance as an invitation to promiscuously add experimental treatments to a research design. His attention is also directed to the fact that the addition of irrelevant treatments to a research design tends to reduce the power of the overall test of the null hypothesis.

EXERCISES

36.1 Three groups of subjects selected at random participate in a psychological experiment. At the conclusion of the experiment, a test is administered, and the scores recorded are:

$$\text{Group I:}\quad 0, 1, 1, 2$$
$$\text{Group II:}\quad 1, 1, 2, 2$$
$$\text{Group III:}\quad 3, 3, 4, 4$$

(a) Determine the region of rejection for testing the hypothesis of equal means at the .05 level.
(b) Complete the summary table for the analysis of variance.
(c) Test the hypothesis of equal means.

36.2 A group of students are classified according to national heritage as belonging to Group A, B, C, or D. They are asked to respond to a rating scale that purports to assess the importance the individual places on formal education. A high score indicates that formal education is highly valued. The scores are recorded as follows:

$$\text{Group A:}\quad 0, 2, 5, 7$$
$$\text{Group B:}\quad 0, 1, 3, 5, 6$$
$$\text{Group C:}\quad 2, 4, 5, 5, 6, 9$$
$$\text{Group D:}\quad 1, 4, 6, 7, 8$$

(a) Determine the region of rejection for testing the hypothesis of equal means at the .05 level.
(b) Complete the summary table for the analysis of variance.
(c) Test the hypothesis of equal means.

36.3 Refer to Exercise 24.1. Substitute the analysis of variance for the t-test for independent samples and compare the results.

36.4 An experiment is conducted to determine which of four techniques for teaching spelling is most effective with first grade children. Thirty students in a single classroom are randomly assigned to four experimental treatments. At the conclusion of the experiment, a spelling test is administered to all the children. Their scores are reported below:

Sample 1: 0, 2, 5, 5, 6, 7, 9, 9
Sample 2: 3, 3, 3, 4, 5, 6, 6
Sample 3: 2, 4, 5, 7, 7, 8, 8, 9
Sample 4: 5, 7, 9, 9, 10, 12, 12

(a) Determine the region of rejection for testing the hypothesis of equal means at the .01 level.
(b) Construct the summary table for the analysis of variance.
(c) Test the hypothesis of equal means.

36.5 Given the following raw data:

Sample 1: 1, 1, 1, 1, 2, 2, 2, 2
Sample 2: 2, 2, 2, 2, 2, 2, 2, 2, 3, 3
Sample 3: 1, 2, 3, 3, 4, 4, 4, 5, 5, 5

(a) Determine the region of rejection for testing the hypothesis of equal means at the .01 level.
(b) Construct the summary table for the analysis of variance.
(c) Test the hypothesis of equal means.

36.6 Given the following raw data:

Sample 1: 1, 2, 3, 4, 4, 5, 5, 6, 7, 7, 7, 9
Sample 2: 0, 0, 1, 1, 3, 3, 3, 3, 4, 5, 6, 7
Sample 3: 0, 1, 1, 2, 2, 2, 2, 2, 2, 3, 3, 4
Sample 4: 1, 1, 2, 2, 2, 2, 2, 2, 2, 2, 3, 3

(a) Determine the region of rejection for testing the hypothesis of equal means at the .01 level.
(b) Construct the summary table for the analysis of variance.
(c) Test the hypothesis of equal means.

SELECTED REFERENCES

COCHRAN, W. G., "Some consequences when the assumptions for the analysis of variance are not satisfied," *Biometrics*, **3**, 22–38 (1947).

DONALDSON, THEODORE S., "Robustness of the *F*-test to errors of both kinds and the correlation between the numerator and denominator of the *F*-ratio," *American Statistical Association Journal*, **63**, 660–676 (June 1968).

GLASS, GENE V., and A. RALPH HAKSTIAN, "Measures of association in comparative experiments: Their development and interpretation," *American Educational Research Journal*, **6**, 403–414 (May 1969).

HSU, TSE-CHI, and L. S. FELDT, "The effect of limitation on the number of criterion score values on the significance level of the *F*-test," *American Educational Research Journal*, **6**, 515–27 (November 1969).

PECKHAM, PERCY D., GENE V. GLASS, and KENNETH D. HOPKINS, "The experimental unit in statistical analysis: Comparative experiments with intact groups." *Journal of Special Education*, **3**, 337–349 (1969).

SCHEFFE, HENRY, *The Analysis of Variance*. New York: John Wiley, 1959.

37

THE KRUSKAL-WALLIS TEST

37.1 FUNCTION

The Kruskal-Wallis test is a nonparametric alternative to the one-way analysis of variance. It may be used with either experimental or *ex post facto* research. Although the test statistic is calculated from the sums of ranks for the various samples, the interpretation is often that of a hypothesis of equal means, which is indeed the case if the distributions of the populations under study have the same shape. The Kruskal-Wallis test is appropriate for use in the k-sample (two or more samples) situation. In the two-sample situation, it is mathematically equivalent to the Mann-Whitney U-test. The Mann-Whitney U-test may be preferred in the two-sample situation, for it offers the option of a convenient one-tailed test.

The Kruskal-Wallis test requires data on at least an ordinal scale, and these data are assumed to come from a continuously distributed population. It does not require normality of distribution nor homogeneity of variance for the groups under study. The test is almost as powerful as the F-test in the analysis of variance under common research conditions, and the computations are relatively simple.

37.2 RATIONALE

The Kruskal-Wallis test is an extension of the Mann-Whitney U-test to the k-sample situation. Although a Kruskal-Wallis exact test is available, the

304

test statistic is ordinarily tabled for only very small sample sizes ($n_j = 5$ or less). Because of the author's prejudices with respect to the use of statistical analyses with samples this small (see Sections 20.5 and 21.3), the exact test will not be considered here. A chi-square approximation of the Kruskal-Wallis test is available which is comparable to the normal approximation of the Mann-Whitney U-test, and the procedure for its use will be discussed in the sections which follow. An excellent approximation is achieved with samples as small as $n_j = 10$. A tie correction factor is also available for use with the chi-square approximation. The derivation of these statistics is beyond the scope of this text but will be found in the Selected References.

37.3   PROCEDURE

1. Two or more random samples are independently drawn from the population under study, and these are subjected to different experimental treatments. The analysis may also be used to analyze existing differences among different populations.
2. At the conclusion of the experiment, a single criterion measure is administered to all subjects in all samples. The measures from the various samples are combined in a single-ordered series and ranked from 1 to N. Whenever large numbers of scores are to be converted to ranks, a frequency distribution should be constructed, and the rank for each score may be calculated from:

$$\text{rank} = cf - \frac{f}{2} + 0.5$$

3. Each rank order score is identified as belonging to a given sample, and the sum of ranks for each sample is calculated:

$$\sum_{i=1}^{n_j} R_{ij}$$

4. The chi-square approximation is calculated from:

$$x^2 = \frac{12}{N(N+1)} \sum_{j=1}^{k} \frac{(\Sigma R_{ij})^2}{n_j} - 3(N+1) \qquad \text{with } df = k - 1$$

5. In the event that tie scores are encountered, the use of the tie correction factor (CFT) is recommended. For each score level where ties exist, the quantity $T = f^3 - f$ should be calculated, where f is the frequency of the score interval. The values of T are summed for all score levels, and the tie correction factor is calculated from:

$$\text{CFT} = 1 - \frac{\Sigma T}{N^3 - N}$$

The chi-square corrected for ties is calculated by dividing the uncorrected chi-square by the tie correction factor:

$$\chi^2_{\text{cft}} = \frac{\chi^2}{\text{CFT}}$$

The value of the correction for ties will never exceed one, and the effect when ties are present is to inflate the value of the test statistic.

6. The chi-square statistic is compared to the tabled value (Table 10) with $df = k - 1$. If the calculated statistic is equal to or greater than the tabled value, the null hypothesis is rejected; the observed differences in the ranks include at least one significant difference. The precise interpretation attached to a significant finding is dependent upon the nature of the criterion variable. If the calculated chi-square is smaller than the tabled value, the null hypothesis is retained, and no significant difference in the ranks has been established.

37.4 EXAMPLE

The experiment that served to illustrate the one-way analysis of variance (see Section 36.6) will also be used to illustrate the use of the Kruskal-Wallis test.

1. H: No difference in the ranks for the three reading assignments.
   A: There is a difference in the ranks — at least one group scored higher than one other group.
2. Kruskal-Wallis test, $n_1 = 8$, $n_2 = 8$, $n_3 = 7$.
3. .05 level, $df = 2$.
   R: $\chi^2 \geq 5.99$.
4. The calculation of the sums of ranks is facilitated by the use of a table as follows:

Raw score	f	cf	Rank	T	Sample 1 f	Ranks	Sample 2 f	Ranks	Sample 3 f	Ranks
9	2	23	22.5	6	1	22.5			1	22.5
8	3	21	20.0	24	2	40.0			1	20.0
7	3	18	17.0	24					3	51.0
6	4	15	13.5	60	1	13.5	2	27.0	1	13.5
5	3	11	10.0	24	1	10.0	1	10.0	1	10.0
4	5	8	6.0	120	3	18.0	2	12.0		
3	1	3	3.0	0			1	3.0		
2	2	2	1.5	6			2	3.0		
	23			264	8	104.0	8	55.0	7	117.0

The calculation of the uncorrected chi-square is as follows:

$$\chi^2 = \frac{12}{23(24)} \left(\frac{10816}{8} + \frac{3025}{8} + \frac{13689}{7} \right) - 3(24) = 8.124$$

The tie correction factor (CFT) is calculated from:

$$CFT = 1 - \frac{264}{12,144} = 0.9783$$

The chi-square corrected for ties is:

$$\chi^2_{\text{cft}} = \frac{8.124}{0.9783} = 8.304$$

5. Reject the null hypothesis. There is a significant difference in the ranks of the three samples. The sum of ranks for the first sample is 104, for the second sample 55, and for the third sample 117. It is probably reasonable to conclude that the third group scored significantly higher than the second group, but nothing can be stated on the basis of the Kruskal-Wallis test with respect to whether the third group scored significantly higher than the first, or whether the first scored higher than the second.

37.5 DISCUSSION

The Kruskal-Wallis test is the appropriate nonparametric alternative to one-way analysis of variance under most circumstances where the assumptions of the analysis of variance are grossly violated. There may be circumstances, however, where some investigators will prefer a chi-square test of independence, especially a median test.

The Kruskal-Wallis test is almost as powerful as the analysis of variance (about 95 percent relative power with typical research data) and does not require homogeneity of variance nor normality of distribution. It does, however, have the assumption of continuity of distribution, and when large numbers of tie scores are encountered the test is much less powerful. The chi-square approximation of the Kruskal-Wallis test has been demonstrated to provide an excellent approximation with samples as small as ten in size, and an effective tie correction factor is available. Although the tie correction factor does not appear to alter the value of a given statistic appreciably, it has been demonstrated that its long-range effects are profound and its use is highly recommended.

EXERCISES

37.1 Refer to Exercise 36.5. Substitute the Kruskal-Wallis test for the analysis of variance. Use the tie correction factor.

37.2 Refer to Exercise 36.6. Substitute the Kruskal-Wallis test for the analysis of variance. Use the tie correction factor.

SELECTED REFERENCES

BRADLEY, J. V., *Distribution-Free Statistical Tests*. Englewood Cliffs, N. J.: Prentice-Hall, 1968.

CONOVER, W. J., *Practical Nonparametric Statistics*. New York: John Wiley, 1971.

KRUSKAL, W. H., "A nonparametric test for the several sample problem," *The Annals of Mathematical Statistics*, **23**, 252–540 (1952).

KRUSKAL, W. H., and W. A. WALLIS, "Use of ranks on one-criterion variance analysis," *Journal of the American Statistical Association*, **47**, 583–621(1952).

McNEIL, D. R., "Efficiency loss due to grouping in distribution-free tests." *Journal of the American Statistical Association*, **62**, 954–965 (1967).

SIEGEL, SIDNEY, *Nonparametric Statistics for the Behavioral Sciences*. New York: McGraw-Hill, 1956.

WOODS, DONALDSON G., "A comparison of the relative power and robustness of Mann-Whitney, Kruskal-Wallis, and analysis of variance tests employing discrete score point scales common to educational research," unpublished Ph.D. dissertation, Kansas State University, 1972.

MULTIPLE COMPARISONS IN THE ANALYSIS OF VARIANCE

38.1 THE PROBLEM OF MULTIPLE COMPARISONS AMONG MEANS

The analysis of variance has been presented as a statistical procedure for making comparisons of means from two or more random samples. When the overall hypothesis of equal means is rejected by the analysis of variance, this does not indicate that every sample mean differs significantly from every other sample mean. It is reasonable to assume, on the basis of rejection of the overall null hypothesis, that at least one sample mean differs significantly from one other sample mean. If all of the samples are of the same size, the investigator is justified in concluding that the largest mean is significantly larger than the smallest mean. However, the contribution of any sample to a test of significance is a function of sample size, and if either the largest mean or the smallest mean comes from a sample with fewer subjects than one of the other samples, one cannot be sure that the null hypothesis is rejected due to the difference between these two extreme means. Even if the investigator can identify the difference between the largest mean and the smallest as a significant one, he will usually want to know whether there are other significant differences among the sample means. Generally, the rejection of the overall hypothesis of equal means without identifying which means are significantly larger than which other means is not a very satisfactory conclusion to a research analysis.

Comparisons or contrasts among means may be simple pairwise comparisons, such as:

$$\mu_1 = \mu_2 \qquad \mu_1 = \mu_3 \qquad \mu_2 = \mu_3$$

Or they may be complex comparisons such as:

$$\mu_1 = \tfrac{1}{2}(\mu_2 + \mu_3)$$

38.2 *A POSTERIORI* VERSUS *A PRIORI* COMPARISONS

Multiple comparisons procedures tend to fall into two broad categories: (1) *a posteriori* (also called *post hoc* or *post mortem*) comparisons, and (2) *a priori* or planned comparisons.

Suppose, for example, that the investigator intends to make all possible simple pairwise comparisons among means. In the three-sample situation, he may wish to test the hypotheses:

$$\mu_1 = \mu_2 \qquad \mu_1 = \mu_3 \qquad \mu_2 = \mu_3$$

Whenever the investigator wishes to make all possible comparisons, or when he is unwilling to limit the number of comparisons in advance of the data collection, his comparisons are referred to as *a posteriori* tests. This is a common situation in behavioral research — it permits data snooping after the data have been collected and a significant F-ratio has been derived from the overall analysis of variance.

Sometimes the investigator plans his research in such a fashion that a limited number of comparisons is to be made, and these are specified prior to the data collection. For example, in the three-sample situation, he may wish to test only these two hypotheses:

$$\mu_1 = \mu_2 \qquad \mu_1 = \mu_3$$

Whenever the investigator is willing to limit the number of comparisons to be made and specify in advance which ones will be made, his comparisons are referred to as *a priori* or planned comparisons. As a general rule, statistical procedures designed for making *a priori* comparisons will be more powerful than those designed for making *a posteriori* comparisons.

The problem of multiple comparisons in the analysis of variance is a controversial topic among statisticians, and it is a problem that has no simple solution. The statistical literature contains a great variety of procedures for making multiple comparisons; these vary considerably with respect to the mathematical logic involved in their derivation and with respect to the situations in which they were intended to be used. The author has selected several such procedures for presentation in this chapter — others will be found in the Selected References. The multiple comparisons procedures recommended here tend to be those most popular with behavioral scientists; they fit a wide variety of research situations and (with one exception) they appear to be those whose use is most easily defended. The *a posteriori* procedures treated in the sections which follow are:

1. *Fisher's least significant difference (LSD)*. Most statisticians object to the use of this procedure, and it will not be recommended here. It does, however, provide a convenient introduction to the other multiple comparisons procedures.
2. *The Scheffé test*. This is the most general of the multiple-comparisons procedures, being suited for use under almost any circumstances. It is not limited to samples of equal size nor is it limited to simple pairwise comparisons. As a general rule, however, the other multiple comparisons procedures will be more powerful in those situations in which it is appropriate to use them.
3. *Tukey's honestly significant difference (HSD)*. This procedure is suitable for making all simple pairwise comparisons among means when the samples are of equal size. It is the preferred procedure for this particular situation.

The *a priori* procedures treated in this text are as follows:

1. *Orthogonal comparisons*. There are certain patterns of comparisons which meet the criterion of orthogonality (to be defined in Section 38.7). If the assumption of normality is met, orthogonal comparisons are statistically independent. If the criterion of orthogonality is met, this will be the preferred method of analysis.
2. *The Dunnett test*. Occasionally, the investigator will be interested only in comparing one sample to each of the other samples. This will usually be thought of as comparing a control group to two or more experimental groups. When the investigator is willing to place this restriction upon the comparisons to be made, the Dunnett procedure is recommended.

38.3   ERROR RATES FOR MULTIPLE COMPARISONS TECHNIQUES

The various multiple comparisons techniques described in the statistical literature often differ with respect to the strategy used in dealing with the rate of Type I errors. Some, for example, use an error rate per comparison. Such an error rate is defined as the probability that any one of a group of comparisons will be declared significant when the null hypothesis is true. This error rate appears to be justified only when the various comparisons are statistically independent (as with orthogonal comparisons). A more conservative approach is to use an experimentwise error rate. Such an error rate is defined as the probability that one or more Type I errors will be made in the comparisons derived from a given experiment. This is the error rate for Tukey's HSD, the Scheffé test, and Dunnett's test. It is assumed here that the value of alpha used for the various comparisons is the same as that for the test of the overall null hypothesis. For a more extensive discussion of error rates for multiple comparisons refer to the Selected References.

38.4 FISHER'S LEAST SIGNIFICANT DIFFERENCE

In the early history of the analysis of variance, Fisher suggested the use of individual t-tests to make multiple comparisons following rejection of the overall null hypothesis with the analysis of variance. This has come to be known as Fisher's Least Significant Difference (abbreviated LSD), for it provides the least protection against Type I errors of all the multiple comparisons techniques.

The usual procedure is to pool the sums of squares from all of the samples to arrive at an error term, and the t-test for the difference between any two means becomes:

$$t = \frac{M_A - M_B}{\sqrt{MS_w \left(\frac{1}{n_A} + \frac{1}{n_B}\right)}}$$

where MS_w is from the analysis of variance calculations, and the degrees of freedom for the t-test are the same as that for MS_w (where $df = N - k$). Either a one- or two-tailed test may be used.

The procedure is not limited to simple pairwise comparisons, but M_A and/or M_B in the expression above may be the weighted mean of two or more samples. Suppose, for example, that M_A is the mean of samples one and two combined, and that the samples are of equal size; then:

$$M_A = \frac{M_1 + M_2}{2} \quad \text{and} \quad n_A = n_1 + n_2$$

In the event that the samples are unequal in size, the use of weighted means is required and the calculations become:

$$M_A = \frac{n_1 M_1 + n_2 M_2}{n_1 + n_2} \quad \text{and} \quad n_A = n_1 + n_2 \quad \text{as before}$$

When all of the samples are of equal size, a critical difference (between any two means) may be calculated from:

$$d = t \sqrt{\frac{2 \, MS_w}{n}}$$

where t is the alpha-level two-tail t with $df = N - k$ from Table 7. A critical difference of this sort may be used for making simple pairwise comparisons; if the difference between any two means is equal to or greater than d, the difference is significant.

Although Fisher's LSD is relatively simple to carry out (the value of MS_w having been previously determined in the analysis of variance), most statisticians object to the use of this procedure. The reason for this is apparent when we realize that the t-test was derived for use in the situation in which two and only two samples are drawn. It is certainly invalid to draw more than two samples, select the two samples with the greatest difference between the means, and perform a t-test with these two samples. Under these circumstances, the probability of rejecting a true hypothesis is likely to be much greater than the level of significance specified by the investigator. Fisher's LSD, although sometimes encountered in the research literature, is not recommended here and is included only as a convenient introduction to the procedures which follow.

Occasionally, Fisher's LSD is reported as an F-ratio (with $df = 1$, $N - k$), which is simply the square of the t-statistic. This mathematically equivalent procedure is no more valid than the t-test.

38.5   THE SCHEFFÉ TEST FOR ALL POSSIBLE COMPARISONS

The Scheffé procedure for testing any and all possible comparisons between means has the important property that the probability of a Type I error for any comparison does not exceed the level of significance specified in the analysis of variance for the overall hypothesis. However, the probability of a Type I error for a given comparison may be considerably smaller than the level of significance set by the investigator. Like the analysis of variance, the Scheffé procedure is quite insensitive to departures from normality and homogeneity of the variances. The test statistic is quite simply calculated for any pair of means and it is referred to the same region of rejection as that specified for the test of the overall hypothesis of equal means. The formula is as follows:

$$F = \frac{(M_1 - M_2)^2}{MS_w \left(\dfrac{1}{n_1} + \dfrac{1}{n_2} \right)(k - 1)} \qquad \text{with } df = k - 1, \ N - k$$

The problem used to illustrate the simple analysis of variance (Section 36.6) will also be used to illustrate the Scheffé technique. The region of rejection was determined to be F equal to or greater than 3.49 in the previous illustration.

Comparing the means for Groups 1 and 2, we obtain:

$$F = \frac{(6 - 4)^2}{2.9(\frac{1}{8} + \frac{1}{8})(2)} = 2.76 \quad \text{(retain)}$$

Next, comparing the means for Groups 1 and 3 we obtain:

$$F = \frac{(7 - 6)^2}{2.9(\frac{1}{8} + \frac{1}{7})(2)} = 0.64 \quad \text{(retain)}$$

Finally, comparing the means for Groups 2 and 3, we obtain:

$$F = \frac{(7 - 4)^2}{2.9(\frac{1}{8} + \frac{1}{7})(2)} = 5.79 \quad \text{(reject)}$$

The conclusion is that Group 3 scored significantly higher than Group 2, but the differences between Group 1 and Group 2 and between Group 1 and Group 3 are both nonsignificant.

The Scheffé procedure is not limited to simple comparisons of pairs of means, but may be used to study more complex relationships. Suppose the investigator in the previous example wished to compare the mean of Group 3 to the mean of Groups 1 and 2 combined:

$$M_{1+2} = \frac{n_1 M_1 + n_2 M_2}{n_1 + n_2} = 5$$

$$F = \frac{(M_3 - M_{1+2})^2}{MS_w \left(\frac{1}{n_1 + n_2} + \frac{1}{n_3}\right)(k - 1)} = 3.37 \quad \text{(retain)}$$

If we compare the Scheffé test to the F-ratios discussed at the end of the previous section, we find that it is algebraically equivalent to these F-ratios divided by the quantity $k - 1$. It should be noted, however, that the Scheffé test does not have the same number of degrees of freedom for the numerator as these previous F-tests.

When the various samples are all of equal size, the Scheffé procedure may be used to arrive at a critical difference between the means, thereby saving some computational effort:

$$d = \sqrt{\frac{2(k - 1)(\text{tabled } F)(MS_w)}{n}}$$

The tabled F in this case is the region of rejection selected earlier, and the value of d is the minimum difference between any two means necessary for a significant finding.

Let us pretend, for purposes of illustration, that all of the samples in the illustration above had eight subjects. The critical difference is:

$$d = \sqrt{\frac{2(2)(3.49)(2.9)}{8}} = 2.25$$

The difference between Means 1 and 2 in the example is two (which is nonsignificant), the difference between Means 1 and 3 is one (which is nonsignificant), and the difference between Means 2 and 3 is three (which is significant). This is consistent with the earlier conclusions, even though the size of one sample was juggled for purposes of illustration.

The Scheffé test is the most flexible of the multiple-comparisons procedures available to the researcher — it fits a great variety of situations and it has received widespread use. Unfortunately, it is not at all uncommon to follow a significant test of the overall null hypothesis with the Scheffé procedure and find that the Scheffé does not detect any significant differences. The theory underlying the Scheffé test assures us that it will indeed find at least one significant difference under these circumstances; however, this significant difference need not be a simple pairwise comparison. The power of the Scheffé test is equal to that of the F-ratio for the test of the overall null hypothesis only with respect to the largest difference (which may be a simple pairwise comparison or some more complex comparison). As a general rule, the other multiple-comparisons techniques discussed will prove more powerful than the Scheffé test in those situations suited to their use, although the Scheffé is adapted for use in many situations in which the other techniques should not be used.

38.6   TUKEY'S HONESTLY SIGNIFICANT DIFFERENCE

Tukey's honestly significant difference (HSD) approach to the problem of multiple comparisons is intended for use in making any and all simple pairwise comparisons when the samples are of equal size. It makes use of the studentized range statistic, the distribution of which is a function of the maximum difference between two means in the k-sample situation. The test statistic is calculated from:

$$q = \frac{M_1 - M_2}{\sqrt{MS_w/n}}$$

The reader's attention is directed to the simple algebraic relationship between the formula used to determine the value of q and that used to determine the value of t with Fisher's LSD when the samples are of equal size. The value of q is equal to the t multiplied by the square root of 2.

The value of q calculated from the sample data may be compared to the tabled value (Table 22 in the Appendix), where k is the number of samples and $df = N - k$. The hypothesis is rejected if the calculated value equals or exceeds the tabled value.

The critical difference may be calculated from:

$$d = q\sqrt{MS_w/n}$$

where q is the appropriate value from Table 22. For the equal sample size situation, Tukey's HSD will ordinarily prove superior to the Scheffé test for simple pairwise comparisons. If the underlying assumptions are met, Tukey's HSD will yield at least one significant difference when the overall F-test is significant.

There are other multiple-comparisons techniques using the studentized range statistic; however, statisticians disagree with respect to their validity.

38.7   ORTHOGONAL COMPARISONS

Consider the three sample situation in which there are three simple pairwise comparisons:

$$\mu_1 = \mu_2 \qquad \mu_1 = \mu_3 \qquad \mu_2 = \mu_3$$

If the following two hypotheses are true:

$$\mu_1 = \mu_2 \qquad \tfrac{1}{2}(\mu_1 + \mu_2) = \mu_3$$

with a little algebra, it can be demonstrated that:

$$\mu_1 = \mu_2 = \mu_3$$

Thus there are only two algebraically independent comparisons in the three sample case. With some effort it can be demonstrated that for the k-sample situation the number of independent comparisons is equal to $k - 1$ (the number of degrees of freedom for the sum of squares between groups). Thus there is one independent contrast for each degree of freedom. The term "orthogonal comparisons" is used to designate comparisons which meet this criterion of independence. When applicable to the investigator's research, the method of orthogonal comparisons will usually be preferred over all of the other methods of making multiple comparisons.

A comparison (C_i) may be defined as any combination of all the means $(M_j\text{'s})$ from a given k-sample type problem such that each mean is multiplied by some constant (c_{ij}), and that the sum of these constants is equal to zero. That is:

$$\sum c_{ij} = 0 \quad \text{for a given } C_i$$

Consider the two examples from the three sample situation.

For the hypothesis:

$$\mu_1 = \mu_2, \text{ which is algebraically equivalent to } \mu_1 - \mu_2 = 0$$

The comparison (C_1) may be stated as:

$$C_1: \quad (1)(M_1) + (-1)(M_2) + (0)(M_3)$$

and the constants for the comparison are:

$$1 - 1 + 0 = 0$$

For the hypothesis:

$$\tfrac{1}{2}(\mu_1 + \mu_2) - \mu_3 = 0$$

The comparison (C_2) may be stated as:

$$C_2: \quad (\tfrac{1}{2})(M_1) + (\tfrac{1}{2})(M_2) + (-1)(M_3)$$

and the constants for the comparison are:

$$\tfrac{1}{2} + \tfrac{1}{2} - 1 = 0$$

Thus each of the two examples meets the criterion established for a comparison.

Two comparisons are orthogonal if the sum of products of their constants (the $c_{ij}\text{'s}$) is equal to zero. The rule is expressed algebraically for any two comparisons by:

$$\sum_{j=1}^{k} c_{1j}c_{2j} = 0$$

For the two comparisons considered above:

$$(1)(\tfrac{1}{2}) + (-1)(\tfrac{1}{2}) + (0)(-1) = 0$$

Thus the two comparisons meet the criterion of orthogonality.

Provided that the comparisons to be made (that is, the hypotheses to be tested) are restricted as prescribed above, the calculations may be completed in precisely the same fashion as those given for Fisher's LSD.

The procedure will be illustrated with the data used to illustrate the one-way analysis of variance in Section 36.6. An examination of the narrative accompanying the problem suggests that the investigator might be interested in testing the hypotheses:

$$\mu_1 - \mu_2 = 0 \quad \text{and} \quad \tfrac{1}{2}(\mu_1 + \mu_2) - \mu_3 = 0$$

The following information is taken from Section 36.6: $n_1 = 8$, $n_2 = 8$, $n_3 = 7$, $M_1 = 6$, $M_2 = 4$, $M_3 = 7$, $MS_w = 2.90$, and $dfw = 20$. The first comparison is made as follows:

$$t = \frac{6 - 4}{\sqrt{2.90(\tfrac{1}{8} + \tfrac{1}{8})}} = 2.35$$

With $df = 20$, the difference is significant at the .05 level. The second comparison is only slightly more involved. First, samples one and two must be combined, yielding $M_A = 5$ and $n_A = 16$. Then, the second t-statistic is calculated by:

$$t = \frac{5 - 7}{\sqrt{2.90(\tfrac{1}{16} + \tfrac{1}{7})}} = -2.59$$

Also with $df = 20$, the difference is significant at the .05 level. The student should probably learn to think of any contrast as the difference between two means, one being the combined mean of all those with positive constants and the other the combined mean of all those with negative constants. Due consideration should also be given to the existence of unequal sample sizes — the use of weighted means is mandatory.

The most popular pattern of orthogonal comparisons is that sometimes referred to as Helmert's contrasts. Coefficients for a variety of situations are given below.

For $k = 3$, $df = 2$.

$$\begin{array}{ccc} 1 & -\tfrac{1}{2} & -\tfrac{1}{2} \\ 0 & 1 & -1 \end{array}$$

For $k = 4$, $df = 3$.

$$
\begin{array}{cccc}
1 & -\tfrac{1}{3} & -\tfrac{1}{3} & -\tfrac{1}{3} \\
0 & 1 & -\tfrac{1}{2} & -\tfrac{1}{2} \\
0 & 0 & 1 & -1
\end{array}
$$

For $k = 5$, $df = 4$.

$$
\begin{array}{ccccc}
1 & -\tfrac{1}{4} & -\tfrac{1}{4} & -\tfrac{1}{4} & -\tfrac{1}{4} \\
0 & 1 & -\tfrac{1}{3} & -\tfrac{1}{3} & -\tfrac{1}{3} \\
0 & 0 & 1 & -\tfrac{1}{2} & -\tfrac{1}{2} \\
0 & 0 & 0 & 1 & -1
\end{array}
$$

By now the pattern is evident and the student should be able to extend it to any number of samples.

Orthogonal comparisons are not, however, limited to Helmert's contrasts. Suppose, for example, that a sociologist is studying the relationship between sex and race (two independent variables) and attitudes toward some pressing social issue (the dependent variable). Suppose also that he wishes to compare (C_1) white males to white females, (C_2) black males to black females, and (C_3) white males and females to black males and females. The coefficients would be as follows:

$$
\begin{array}{lcccc}
(C_1) & 1 & -1 & 0 & 0 \\
(C_2) & 0 & 0 & 1 & -1 \\
(C_3) & \tfrac{1}{2} & \tfrac{1}{2} & -\tfrac{1}{2} & -\tfrac{1}{2}
\end{array}
$$

The student should confirm that each pair of contrasts in these sets is indeed orthogonal.

The utility of orthogonal contrasts is dependent upon their being appropriately matched to the investigator's research. Such contrasts can be used legitimately only when planned in advance of the data collection, and when the hypotheses that are tested answer questions relevant to the research. Although the method fits only a special class of research problems, it is the preferred method for this class of problems.

Many authors prefer to approach the method of orthogonal comparisons by partitioning the sum of squares between into orthogonal components, each corresponding to one degree of freedom. The material which follows is presented primarily to demonstrate the equivalence of the two approaches.

The need for multiple comparisons arises only when the number of samples exceeds two and the number of degrees of freedom possessed by the sum of squares between groups exceeds one. Whenever the number of degrees of freedom for the sum of squares between groups exceeds one, it is possible to partition the sum of squares into independent component parts, each possessing one degree of freedom. Then, an F-ratio may be formed for each of these component parts by dividing its share of the sum of squares (which is also a mean square, possessing one degree of freedom) by the mean square within. These F-statistics will be the squares of the corresponding t-statistics calculated in the manner described in the preceding paragraphs.

The sum of squares for such a comparison may be determined from:

$$SS_{C_i} = \frac{(M_A - M_B)^2}{\dfrac{1}{n_A} + \dfrac{1}{n_B}}$$

where M_A, M_B, n_A, and n_B are defined as in Section 38.2.

In the example, the sum of squares for comparison one is:

$$SS_{C_1} = \frac{(6 - 4)^2}{\frac{1}{8} + \frac{1}{8}} = 16.00$$

And the sum of squares for comparison two is:

$$SS_{C_2} = \frac{(5 - 7)^2}{\frac{1}{16} + \frac{1}{7}} = 19.48$$

The sum of these is 35.48, the original sum of squares between groups.

The F-ratio for the first comparison is $F = 16.00/2.90 = 5.52$, with $df = 1, 20$. The F-ratio for the second comparison is $F = 19.48/2.90 = 6.72$, with $df = 1, 20$. These F-ratios are indeed the squares of the corresponding t-statistics achieved earlier.

38.8 THE DUNNETT TEST FOR MULTIPLE COMPARISONS TO A CONTROL GROUP

Dunnett has derived the sampling distribution for a t-statistic for use in comparing a control group to one or more experimental groups. Dunnett's t is calculated as follows:

$$t = \frac{M_1 - M_2}{\sqrt{MS_w \left(\dfrac{1}{n_1} + \dfrac{1}{n_2} \right)}} \qquad \text{with } df = N - k$$

The reader's attention is directed to the fact that the formula is the same as that used with Fisher's LSD. However, a different table of critical values is used. Provision is made for both one- and two-tailed tests in the distribution of Dunnett's t in Table 23. The problem used to illustrate the simple analysis of variance (Section 36.6) will also serve to illustrate the Dunnett test. A review of the experiment reveals that it is consistent to undertake a one-tailed test with the alternative that Group 3 will score higher than either Group 1 or Group 2. Referring to Table 18, we find that the region of rejection (.01 level, one-tailed, two experimental groups, $df = 20$) is determined to be t equal to or greater than 2.81.

Comparing Group 3 to Group 1, we obtain:

$$t = \frac{7 - 6}{\sqrt{2.9(\frac{1}{8} + \frac{1}{7})}} = 1.14 \qquad \text{(retain)}$$

Comparing Group 3 to Group 2, we obtain:

$$t = \frac{7 - 4}{\sqrt{2.9(\frac{1}{8} + \frac{1}{7})}} = 3.41 \qquad \text{(reject)}$$

Like the Scheffé procedure, the Dunnett procedure is conservative, and the probability of a Type I error for any comparison does not exceed the level of significance specified in the analysis of variance for the overall hypothesis.

As with Fisher's LSD, a critical difference may be used when the samples are of equal size:

$$d = t \sqrt{\frac{2\ MS_w}{n}}$$

where t is the appropriate value from Table 23.

EXERCISES

38.1 Refer to the data of Exercise 36.5.
  (a) Use a t-test (as in Sections 38.2 and 38.3) to compare Sample 2 to Sample 3. Test the hypothesis of equal means at the .01 level.
  (b) Use a t-test (as in Sections 38.2 and 38.3) to compare Sample 1 to Samples 2 and 3 combined. Test the hypothesis of equal means at the .01 level.

38.2 Refer to Exercise 38.1. Repeat the exercise, using the Scheffé test.

38.3 Refer to the data of Exercise 36.6.
  (a) Using the Scheffé procedure, calculate a critical difference between means. Use the .01 level.
  (b) Make all simple pairwise comparisons.

38.4 Refer to Exercise 38.3. Repeat the exercise, using the Tukey honestly significant difference.

38.5 Refer to the data of Exercises 36.5 and 38.1.
  (a) Use the Dunnett procedure to compare Sample 1 to Sample 2. Use the .01 level.
  (b) Use the Dunnett procedure to compare Sample 1 to Sample 3. Use the .01 level.

SELECTED REFERENCES

DUNNETT, C. W., "A multiple comparison procedure for comparing several treatments with a control," *Journal of the American Statistical Association*, **50**, 1096–1121 (1955).

GAMES, PAUL A., "Multiple Comparisons of Means," *American Educational Research Journal*, **8**, 531–564 (May 1971)

HOPKINS, K. D., and R. A. CHADBOURN, "A schema for proper utilization of multiple comparisons in research and a case study," *American Educational Research Journal*, **4**, 407–412 (November 1967).

KIRK, ROGER E., *Experimental Design Procedures for the Behavioral Sciences.* Belmont: Brooks/Cole, 1968.

MILLER, RUPERT G., *Simultaneous Statistical Inference.* New York: McGraw-Hill, 1966.

PETRINOVICH, LEWIS F., and CURTIS D. HARDYK, "Error rates for multiple comparison methods: Some evidence concerning the frequency of erroneous conclusions," *Psychological Bulletin*, **71**, 43–54 (1969).

RAMSEYER, GARY C., and TSE-KIA TCHENG, "The robustness of the studentized range statistic to violations of the normality and homogeneity of variance assumptions," *American Educational Research Journal*, **10**, 235–240 (Summer 1973).

RYAN, THOMAS A., "Comments on orthogonal components," *Psychological Bulletin*, **56**, 394–395 (1959).

RYAN, THOMAS A., "Multiple comparisons in psychological research," *Psychological Bulletin*, **59**, 26–47 (1962).

RYAN, THOMAS A., "The experiment as the unit for computing ratio of error," *Psychological Bulletin*, **59**, 301–304 (1962).

SCHEFFÉ, HENRY, "A method of judging all contrasts in the analysis of variance," *Biometrika*, **40**, 87–104 (1953).

SCHEFFÉ, HENRY, *The Analysis of Variance.* New York: John Wiley, 1959.

SPARKS, JACK N., "Expository notes on the problem of making multiple comparisons in a completely randomized design," *Journal of Experimental Education*, **31**, 343–349 (1963).

TUKEY, J. W., "Comparing individual means in the analysis of variance," *Biometrics*, **5**, 99–114 (1949).

WINER, B. J., *Statistical Principles in Experimental Design*, 2nd ed. New York: McGraw-Hill, 1971.

39

TWO-WAY ANALYSIS
OF VARIANCE
WITH ONE ENTRY
PER CELL

39.1 FUNCTION

The term, two-way analysis of variance, is applied to analyses of data
organized in a double-entry table, that is, with scores on the criterion
entered into cells organized into rows and columns. The columns in the
table will usually correspond to the different treatments as in the one-way
analysis of variance. The rows will usually correspond to (1) levels of a
controlled variable in a randomized blocks design, (2) subjects in a repeated
measurements design, or (3) the second factor in a factorial design. The
analysis of variance is not limited to designs in two dimensions; however,
the student will need to consult a text on experimental design for a discussion of three-way and higher-order designs.

Two-way analysis of variance with one entry per cell is the extension of
the t-test for two related samples to the k-sample situation. It is most often
encountered in repeated measurements designs and in a special class of
randomized blocks designs.

In experimental research, a repeated measurements design is one in which
the various experimental treatments are administered sequentially to each
subject. The criterion variable is administered to each subject following
each experimental treatment. The situation is illustrated for the two-
treatment case by Figure 39.1. Of course, when the analysis of variance is
used, repeated measurements designs are not limited to two treatments.
The use of repeated measurements designs in experimental research is
ordinarily limited to those situations in which the effects of all treatments
(with the possible exception of the last one) are known to be temporary

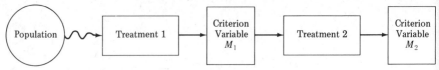

Figure 39.1 Repeated measurements design — experimental research.

and do not influence or confound the effects of subsequent treatments. A similar limitation is made with respect to the repeated administration of the criterion. If the effects of one measurement will influence scores on subsequent measurements, the use of the design is invalidated. There must be no carryover from treatment to treatment nor from measurement to measurement. The design has obvious limitations with respect to use when learning is taking place. In view of the fact that only one sample is used throughout, the degree of control of a wide range of variables is unique, and where applicable the repeated measurements design is likely to be superior to other designs. Repeated measurements designs are also used for *ex post facto* research in the field of measurement.

There is a special class of randomized blocks designs in which there is only one entry per cell. The matched-pairs design is an example of this. With the analysis of variance, however, the investigator is not limited to the use of matched-pairs, but may use matched-trios or higher-order grouping. Matched-pairs research is illustrated by Figure 39.2. Occasionally, the investigator is confronted with a situation in which groups rather than individuals are sampled. For example, the educational researcher may be limited to sampling of intact classes. Often, we see such data analyzed as though the investigator had randomly selected individual students rather than intact classes. A more legitimate approach to the analysis is to treat the mean of the intact group as the sampling unit instead of the score of the individual. The result will probably be a randomized blocks design with one entry per cell.

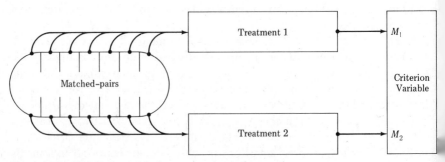

Figure 39.2 Matched-pairs design — experimental research.

39.2 RATIONALE

Data organized in double-entry tables with one entry per cell may be approached from two different perspectives with one-way analysis of variance. First, the existence of the row variable may be ignored, and the sum of squares for total (SS_t) partitioned into the sum of squares between columns (SS_c) and the sum of squares within columns. Second, the existence of the column variable may be ignored, and the sum of squares for total (SS_t) partitioned into the sum of squares between rows (SS_r) and the sum of squares within rows. The student should note that the sum of squares for total is the same for the two analyses described above, and that the procedures suggested are identical to those presented in Chapter 36 for the one-way analysis of variance.

The preferred procedure is to analyze the data in a two-way analysis of variance in which three sources of variation are seen to contribute to the total variation — that is, the sum of squares for total is partitioned into three additive components. The three sources of variation are (1) the sum of squares between rows (SS_r), which is calculated as noted above, (2) the sum of squares between columns (SS_c), which is also calculated as noted above, and (3) the sum of squares for row-by-column interaction (SS_{rc}), which is most conveniently calculated as a residual:

$$SS_{rc} = SS_t - SS_r - SS_c$$

It may be useful to think of interaction in this context as simply that variation that is unaccounted for by the independent effects of the row and column variables. Refer to Section 22.2 for a discussion of the nature of interaction. The actual calculations for the various sums of squares will be discussed in considerable detail in a later section of this chapter.

The number of degrees of freedom for SS_t can be demonstrated to be equal to $(rc - 1)$, that for SS_r equal to $(r - 1)$, that for SS_c equal to $(c - 1)$, and that for SS_{rc} equal to $(r - 1)(c - 1)$. The logic used to arrive at these values will be discussed in a later section. Thus the sum of squares for total has been partitioned into three independent components:

$$SS_t = SS_r + SS_c + SS_{rc}$$

And the degrees of freedom for total have likewise been partitioned into three independent components:

$$(rc - 1) = (r - 1) + (c - 1) + (r - 1)(c - 1)$$

SS_r, SS_c, and SS_{rc} may each be divided by their respective degrees of freedom to produce the corresponding mean square.

Provided that certain assumptions are met, the appropriate test for significance of the effects of the row variable is given by forming the ratio:

$$F = \frac{MS_r}{MS_{rc}} \quad \text{with } df = (r - 1), (r - 1)(c - 1)$$

Similarly, provided that certain assumptions are met, the appropriate test for significance of the effects of the column variable is given by forming the ratio:

$$F = \frac{MS_c}{MS_{rc}} \quad \text{with } df = (c - 1), (r - 1)(c - 1)$$

When there is only one entry per cell, there will be no sum of squares within and there will be no test of significance for the interaction effect. This can be a severe handicap in the analysis, and some authors recommend that the investigator consider the use of some strategy to produce two (or more) entries per cell. This is sometimes accomplished in a repeated measurements design by taking two measures where only one was required before. It may be accomplished in the randomized blocks design by placing two (or more) persons in each cell (reducing the number of rows). In either case, the method of analysis for two-way analysis of variance with more than one entry per cell is treated in Chapter 40.

39.3   PROCEDURE

1. When a repeated measurements design is called for, a single random sample is drawn from the parent population and sequentially subjected to treatment 1, criterion measure, treatment 2, criterion measure, and so on. When a randomized blocks design is called for, the available subjects are matched on the blocking variable, randomly assigned to the various experimental treatments in the manner described in Section 22.3, and administered the criterion at the end of the experiment.
2. The sum of squares for rows, for columns, for interaction, and for total are calculated and recorded in a summary table. Their degrees of freedom are similarly recorded. Finally, the mean squares for rows, columns, and for interaction are calculated and recorded in the summary table. The recommended form for the summary table is given by:

Summary table for the analysis of variance			
Source of variation	*df*	*Sums of squares*	*Mean squares*
Between rows			
Between columns			
Interaction			
Total			XXXXX

3. The hypothesis of equal row means is tested by the ratio:

$$F = \frac{MS_r}{MS_{rc}} \quad \text{with } df = (r - 1), (r - 1)(c - 1)$$

The *F*-ratio may be referred to Table 19 to determine whether or not it is statistically significant. We will make a practice of organizing the data so that the row variable represents the subjects in a repeated measurements design or the blocking variable in a randomized blocks design. The utility and validity of the test for equal row means will be discussed in the final section of this chapter.

4. The hypothesis of equal column means is tested by the ratio:

$$F = \frac{MS_c}{MS_{rc}} \quad \text{with } df = (c - 1), (r - 1)(c - 1)$$

The *F*-ratio may be referred to Table 19 in the Appendix to determine whether or not it is statistically significant. We will make a practice of organizing the data so that the column variable represents the treatment variable. The assumptions underlying the statistical test will be discussed in the final section of this chapter.

39.4 COMPUTATIONAL PROCEDURES

First, it will be necessary to extend our notational scheme to cover the situation in which score-type data are recorded in a double-entry table of this sort:

X_{11}	X_{12}	X_{13}
X_{21}	X_{22}	X_{23}
X_{31}	X_{32}	X_{33}
X_{41}	X_{42}	X_{43}

The notational scheme is as follows:

r = number of rows

c = number of columns

rc = product of r and c, which is the total number of measures or cells

X_{ij} = score in the ith row and jth column

$T_i. = \sum\limits_{j=1}^{c} X_{ij}$ = sum of scores for the ith row

$T._j = \sum\limits_{i=1}^{r} X_{ij}$ = sum of scores for the jth column

$T.. = T = \sum\sum X_{ij}$ = sum of all scores

$M_i. = \dfrac{T_i.}{c}$ = mean of the ith row

$M._j = \dfrac{T._j}{r}$ = mean of the jth column

$M.. = M = \dfrac{T}{rc}$ = general mean of all scores

The sum of squares for total (the sum of the squared deviations of all scores from the general mean) may be calculated from:

$$SS_t = \sum_{i=1}^{r}\sum_{j=1}^{c} (X_{ij} - M)^2 = \sum_{i=1}^{r}\sum_{j=1}^{c} X_{ij}^2 - \frac{T^2}{rc}$$

The number of degrees of freedom for the sum of squares for total is $(rc - 1)$, there being rc terms in the expression and the general mean being estimated from the sample data.

The sum of squares between rows may be calculated from:

$$SS_r = \frac{1}{c}\sum_{i=1}^{r} (M_i. - M)^2 = \frac{1}{c}\sum_{i=1}^{r} T_i.^2 - \frac{T^2}{rc}$$

The number of degrees of freedom for the sum of squares between rows is equal to $(r - 1)$, there being r terms in the expression and the general mean being estimated from the sample data.

The sum of squares between columns may be calculated from:

$$SS_c = \frac{1}{r}\sum_{j=1}^{c} (M._j - M)^2 = \frac{1}{r}\sum_{j=1}^{c} T._j^2 - \frac{T^2}{rc}$$

The number of degrees of freedom for the sum of squares between columns is equal to $(c - 1)$, there being c terms in the expression and the general mean being estimated from the sample data.

Finally, the sum of squares for interaction is calculated as a residual:

$$SS_{rc} = SS_t - SS_r - SS_c$$

The number of degrees of freedom for the sum of squares for interaction is equal to $(r - 1)(c - 1)$. Degrees of freedom for interaction may also be calculated as a residual by subtracting the degrees of freedom for rows and for columns from the degrees of freedom for total.

Note that some of the terms are held in common by these formulas. As a matter of fact, only four terms need be evaluated to complete the calculations.

Consider the first term of the sum of squares for total:

$$\sum_{i=1}^{r} \sum_{j=1}^{c} X_{ij}^2$$

To evaluate this term, one need only (1) square each score in the entire collection, and (2) sum all of these squares.

Next, consider the second term of the sum of squares between rows, between columns, and for total:

$$\frac{T^2}{rc}$$

To evaluate this term, one need only (1) sum all scores in the entire collection, (2) square this sum, and (3) divide by the product rc. This term is known also as the correction for the mean (CFM).

The first term of the sum of squares between rows is:

$$\frac{1}{c} \sum_{i=1}^{r} T_{i\cdot}^2$$

To evaluate this term, (1) sum all scores for each row, (2) square each sum, (3) sum for all rows, and (4) divide by c.

The first term of the sum of squares between columns is:

$$\frac{1}{r} \sum_{j=1}^{c} T_{\cdot j}^2$$

To evalute this term, (1) sum all scores for each column, (2) square each sum, (3) sum for all columns, and (4) divide by r.

39.5 EXAMPLE

A psychometrist is conducting exploratory research comparing three different techniques for measuring responses to a specified psychophysical stimulus. Each yields a score on a 0–4 scale. Five students are used as subjects in the research, and the identical stimulus is administered to each student in three separate trials. A different measurement technique is used on each of the three trials. With columns corresponding to the measurement techniques and rows to students, the results are as follows:

0	0	1
1	1	2
3	2	3
2	3	4
3	4	4

1. *H:* There is no difference in the average scores recorded on the three scales ($\mu._1 = \mu._2 = \mu._3$).
   A: There is a difference in the average scores recorded on the three scales.
2. Two-way analysis of variance with one entry per cell, three scales and five subjects.
3. .05 level, *R:* $F \geq 4.46$, $df = 2, 8$.
4. $\Sigma\Sigma X_{ij}^2 = 99.00$

$$\frac{T^2}{rc} = \frac{33^2}{15} = 72.60$$

$$\frac{1}{c}\sum T_{i\cdot}^2 = \frac{1}{3}(1 + 16 + 64 + 81 + 121) = 94.33$$

$$\frac{1}{r}\sum T_{\cdot j}^2 = \frac{1}{5}(81 + 100 + 196) = 75.40$$

$SS_t = 99.00 - 72.60 = 26.40$
$SS_r = 94.33 - 72.60 = 21.73$
$SS_c = 75.40 - 72.60 = 2.80$
$SS_{rc} = 26.40 - 21.73 - 2.80 = 1.87$

Summary table for the analysis of variance			
Source of variation	*df*	*Sums of squares*	*Mean squares*
Rows (students)	4	21.73	5.40
Columns (scales)	2	2.80	1.40
Interaction	8	1.87	0.23
Total	14	26.40	XXXX

$$F = \frac{1.40}{0.23} = 6.08 \qquad df = 2, 8$$

5. There is a significant difference in the average scores recorded on the three scales. They should not be regarded as equivalent and interchangeable measures.

39.6 FIXED AND RANDOM EFFECTS

The distinction between fixed and random effects was not a critical issue in one-way analysis of variance. However, this distinction is a critical issue in determining (1) the validity of two-way analysis of variance with one entry per cell, or (2) the choice of an error term in two-way analysis of variance with more than one entry per cell.

Consider the example in Section 39.5. The row variable is a random effect because the students presumably were chosen at random from a larger population of students. The column variable is a fixed effect because the measurement scales presumably were selected (or fixed) by the investigator in some systematic (as opposed to random) fashion.

A *fixed effect* is a factor in which (1) all levels of the factor are included in the analysis, or (2) only those levels of interest to the investigator are included in the analysis, or (3) some systematic (as opposed to random) process is used to select the factors used in the research. Most of the independent variables encountered in behavioral research are fixed effects.

A *random effect* is a factor in which the levels of the factor are randomly selected from a larger population of such factors. The most common example is the subjects in a repeated measurements design. Another example occurs when an educational researcher randomly selects a group of teachers or schools to participate in experimental research. The blocking variable in a randomized blocks design may be either a fixed or a random effect.

A special problem arises when the investigator does not use a random process in the selection of the levels of a factor, but his obvious intent is to generalize his findings from the levels selected to some larger population of the same. This is likely to be the case when the levels of a factor are classroom teachers, or schools, or counseling psychologists, or judges or graders. Such a factor should probably be considered a random effect.

Three possibilities exist in two-way analysis of variance:

1. Both independent variables may be fixed effects. This is a fixed effects model, sometimes called Model I. Fixed effects models are commonly encountered in two-way analysis of variance with more than one entry per cell — both factorial and randomized blocks designs are likely to be fixed effects models.

2. Both independent variables may be random effects. This is a random effects model, sometimes called Model II. A repeated measurements design in which subjects are one factor and graders or judges are the second factor would probably be classified as a random effects model. Random effects designs are much less common than fixed effects designs.

3. One independent variable may be a fixed effect and the other a random effect. This is a mixed effects model, sometimes called Model III. Repeated measurements designs tend to fit this pattern, and the example cited in Section 39.5 is illustrative of the mixed effects model.

39.7   DISCUSSION

Two-way analysis of variance with one entry per cell is an extension of one-way analysis of variance and has essentially the same underlying mathematical assumptions. The assumptions of normality of distribution and homogeneity of variances are extended to the rows (for the test of equal row means) and to the columns (for the test of equal column means). These assumptions may be ignored in most behavioral research using analyses of this sort — provided that there is no carryover from treatment-to treatment and from measure to measure.

The validity of two-way analysis of variance with one entry per cell is much dependent upon whether the independent variables are fixed or random effects. (1) There is no appropriate error term for the fixed effects model. The use of MS_{rc} as the error term provides a conservative test, and its use may be justified under some circumstances. (2) The use of MS_{rc} as the error term is appropriate for the random effects model. (3) In the mixed effects model, MS_{rc} is the appropriate error term for testing the significance of the fixed effect. The use of MS_{rc} as an error term for the random effect provides a conservative test and its use may be justified under some circumstances. The problem of identifying an appropriate error term may be simplified considerably by altering the design so that there are two (or more) entries per cell — the analysis under these circumstances is discussed in Chapter 40.

EXERCISES

39.1   A study is conducted in which the following sequence of events takes place: (1) Ten rats are randomly selected from an available colony of white rats. Each rat is confined in a revolving wire drum, and the amount of activity (based upon the number of revolutions in a specified time) is recorded on a

0–3 scale. (2) Each rat is administered an experimental drug, again placed in the revolving wire drum, and the amount of activity is again recorded on the 0–3 scale. (3) After the effects of the first drug have worn off, a second drug is administered, and the amount of activity is recorded a third time on the 0–3 scale. The results are as follows:

1	0	0
3	0	3
3	1	2
0	1	0
3	0	0
2	0	1
3	1	2
1	0	0
2	2	2
2	1	0

(a) Determine the region of rejection for the test of equal column means (the drug effect). Use the .05 level.
(b) Complete the summary table for the analysis of variance.
(c) Test the hypothesis of equal column means.

39.2 Given a three-by-four double-entry table with one entry per cell as follows:

3	4	3	5
0	2	2	2
0	0	1	2

(a) Determine the region of rejection for the test of equal row means. Use the .01 level.
(b) Determine the region of rejection for the test of equal column means. Use the .01 level.
(c) Complete the summary table for the analysis of variance.
(d) Test the hypothesis of equal row means.
(e) Test the hypothesis of equal column means.

SELECTED REFERENCES

DAYTON, C. MITCHELL, *The Design of Educational Experiments*. New York: Mc-Graw-Hill, 1970.

EDWARDS, ALLEN L., *Experimental Design in Psychological Research*, 3rd ed. New York: Holt, Rinehart and Winston, 1968.

KIRK, ROGER E., *Experimental Design Procedures for the Behavioral Sciences.* Belmont: Brooks/Cole, 1968.

LINDQUIST, E. F., *Design and Analysis of Experiments in Psychology and Education.* Boston: Houghton Mifflin, 1953.

MYERS, JEROME L., *Fundamentals of Experimental Design.* Boston: Allyn and Bacon, 1966.

SCHEFFÉ, HENRY, *The Analysis of Variance.* New York: John Wiley, 1959.

WINER, B. J., *Statistical Principles in Experimental Design*, 2nd ed. New York: McGraw-Hill, 1971.

40

TWO-WAY ANALYSIS OF VARIANCE

40.1 FUNCTION

Often, the most valuable research conducted by the behavioral scientist is that in which there are two or more independent variables (in addition to the dependent variable). In this chapter and those that follow, three different approaches to research analysis involving two or more independent variables will be considered: (1) two-way analysis of variance, (2) analysis of covariance, and (3) multiple regression. Traditionally, two-dimensional analysis of variance has been associated with experimental designs employing two discrete independent variables, the analysis of covariance with designs using one discrete and one or more continuous variables, and multiple regression with two or more continuous independent variables. Of course, the student is now aware that continuous data may be treated as discrete, and that it is often useful to transform discrete data into binary variables and to complete the analysis with procedures ordinarily reserved for continuously distributed data. For these reasons the various techniques considered in these chapters will often be interchangeable, and the selection of the statistical analysis will depend to some extent on how the investigator perceives his data and his objectives.

Two-way analysis of variance is an extension of simple analysis of variance to the situation in which there are two independent variables. It is customary to organize the data into a double-entry table, with one of the independent variables cast into columns and the other into rows. Scores on the dependent variable are entered in the cells of the table. Consider, for example, the following two-by-two table:

	Drug	No drug
Electroshock	$X_{111}, X_{112} \dots X_{11k}$	$X_{121}, X_{122} \dots X_{12k}$
No shock	$X_{211}, X_{212} \dots X_{21k}$	$X_{221}, X_{222} \dots X_{22k}$

The table represents a two-dimensional experimental design in which a psychiatrist is studying concomitantly the effects of a certain drug and electroshock therapy. Patients in cell 11 receive both the drug and shock treatments, those in cell 12 only the shock, those in cell 21 only the drug, and those in cell 22 neither the drug nor the shock treatment. Notice that all patients in column 1 receive the drug, while those in column 2 do not, and that all patients in row 1 receive the shock therapy, while those in row 2 do not. The investigator is enabled by this design to carry on simultaneously two experiments with the same group of subjects. In addition to the economies effected by carrying on two experiments with the same subjects at the same time, the investigator is enabled to study possible interaction between the row and column variables. Interaction is a treatment effect that is dependent upon the concomitant influence of two independent variables. In the example, a significant interaction would suggest that (1) the effects of the shock-treatment variable are not the same for those patients who received the drug as for those who did not, and that (2) the effects of the drug-treatment variable are not the same for those patients who received the shock treatment as for those who did not. A nonsignificant interaction would suggest that the drug and shock treatment effects are independent of each other.

Two-way analysis of variance may also be used to improve upon the simple analysis of variance through the use of stratified random samples. The educational researcher, for example, might find it desirable to stratify a population of schoolchildren, dividing it into three groups, those of high IQ, those of average IQ, and those of low IQ. Then independent random samples might be drawn from each of these three levels and cast into a three-by-two table for purposes of studying the effects of two methods of instruction:

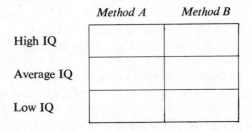

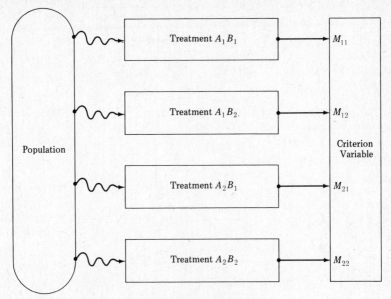

Figure 40.1 Factorial design — experimental research.

The first example cited, the psychiatric experiment, is a factorial design. The procedure is illustrated by Figure 40.1. The second example cited, the

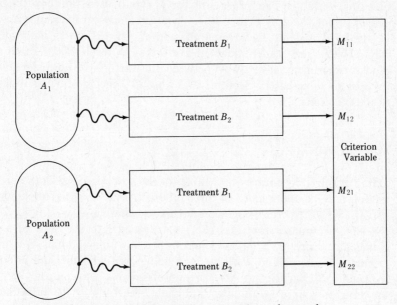

Figure 40.2 Randomized blocks design — experimental research.

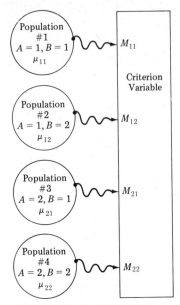

Figure 40.3 Factorial design — *ex post facto* research.

educational experiment, is a randomized blocks design with IQ as the blocking variable. The procedure for a two-by-two randomized blocks design is illustrated by Figure 40.2. Two-way analysis of variance may also be used for *ex post facto* research, and the procedure is illustrated by Figure 40.3. The student may find it useful to think of the randomized blocks experiment as midway between the factorial experiment and the *ex post facto* research. Sections 27.2 and 27.3 include further discussion of these designs.

40.2 RATIONALE

Data organized in double-entry tables of the sort described in the previous section may be approached in three different fashions with simple analysis of variance. First, the existence of the row variable may be ignored, and the sum of squares for total (SS_t) partitioned into the sum of squares between columns (SS_c) and the sum of squares within columns. Second, the existence of the column variable may be ignored, and the sum of squares for total (SS_t) partitioned into the sum of squares between rows (SS_r) and the sum of squares within rows. Finally, the cells in the table may be treated as distinct groups, and the sum of squares for total (SS_t) partitioned into the the sum of squares between cells (SS_{cells}) and the sum of squares within cells (SS_w), which we shall refer to simply as the *sum of squares within*. The student should note that the sum of squares for total is identical in the three analyses described above, and that the procedures suggested are identical to those presented in Chapter 36 on the one-way analysis of variance.

Suppose, however, that upon completing the calculations described above, the investigator determines that the sum of squares for cells is substantially larger than the sum of squares for columns plus the sum of squares for rows. In other words, the variability from cell to cell cannot be entirely accounted for by the effects of the row and column variables. This suggests the possibility of a third treatment effect, one that is due to interaction between the row and column variables. This is referred to as the *sum of squares for interaction (SS_{rc})*, and it is equal to the sum of squares for cells minus the sum of squares for columns and the sum of squares for rows:

$$SS_{rc} = SS_{cells} - SS_c - SS_r$$

The student may find it useful to think of this as partitioning the sum of squares for cells into its component parts.

The commonly accepted approach to two-way analysis of variance consists of partitioning the sum of squares for total into four parts: (1) sum of squares for columns, (2) sum of squares for rows, (3) sum of squares for interaction, and (4) sum of squares within. As in the simple analysis of variance, each of these sums of squares will be converted into a mean square by dividing each by its respective degrees of freedom. The actual computations will be treated in detail in a later section, where the student may demonstrate for himself that the total sum of squares has in fact been partitioned into four parts:

$$SS_t = SS_c + SS_r + SS_{rc} + SS_w$$

In similar fashion, one may demonstrate that the total number of degrees of freedom has been partitioned:

$$(N - 1) = (c - 1) + (r - 1) + (c - 1)(r - 1) + (N - rc)$$

where c is the number of columns and r the number of rows.

The test for significance of interaction is accomplished by forming the ratio:

$$F = \frac{MS_{rc}}{MS_w} \quad \text{with } df = (r - 1)(c - 1), \quad (N - rc)$$

Similarly, the test for significance of the column variable usually is accomplished by forming the ratio:

$$F = \frac{MS_c}{MS_w} \quad \text{with } df = (c - 1), \quad (N - rc)$$

This F-ratio provides a test of the hypothesis of equal column means, ignoring the existence of the rows. Such a test is sometimes referred to as a *test of main effects for columns.*

Finally, the test for significance of the row variable is usually accomplished by forming the ratio:

$$F = \frac{MS_r}{MS_w} \quad \text{with } df = (r - 1), \quad (N - rc)$$

This F-ratio provides a test of the hypothesis of equal row means, ignoring the existence of the columns. Such a test is sometimes referred to as a *test of main effects for rows.*

There are special circumstances in which MS_{rc} may be substituted for MS_w as the error term for the test of equal column means and/or equal row means.

Ordinarily, a significant interaction effect suggests that the tests of main effects for rows and columns outlined above are inappropriate. The presence of a significant interaction effect indicates that the row effects are not constant from column to column, and that the column effects are not constant from row to row. The accepted procedure in this situation is to break the total design into parts, performing a one-way analysis of variance for columns separately for each row in the table, and performing a one-way analysis of variance for rows separately for each column in the table. These tests are commonly referred to as *tests of simple effects for rows and columns.*

Underlying the derivation of the two-way analysis of variance presented in this chapter is a requirement of proportionality of cell frequencies from row to row and column to column. For a two-by-three double-entry table, this requirement may be stated algebraically by:

$$\frac{n_{11}}{n_{21}} = \frac{n_{12}}{n_{22}} = \frac{n_{13}}{n_{23}}$$

Often, this will mean that the investigator will choose to use random samples of the same size for the various cells; however, if he has a reason for doing so, the investigator might choose samples for a two-by-three design of unequal sizes such that:

$$\frac{10}{20} = \frac{15}{30} = \frac{20}{40}$$

Of course, this type of restriction precludes the use of available groups of odd sizes such as those commonly encountered in classroom research.

40.3 PROCEDURE

1. A double-entry table is constructed in the fashion described in Section 40.2, random samples are independently drawn from the population under study (one for each cell in the table), and these are subjected to experimentation. The analysis may also be used to analyze existing differences between different populations where these populations logically fit a double-entry table. The cell frequencies must meet the requirement of proportionality outlined in Section 40.2.
2. At the conclusion of the experiment, a single criterion measure is administered to all subjects in all samples. A summary table is constructed, and the various degrees of freedom, sums of squares, and mean squares needed to complete the analysis are calculated. The following summary table serves as an example:

Summary table for two-dimensional analysis of variance			
Source of variation	*df*	*Sums of squares*	*Mean squares*
(Between cells)	XX		XXXX
Between columns			
Between rows			
Interaction			
Within cells			
Total			XXXX

The student's attention is directed to the fact that while the sum of squares between cells does not enter into the final analysis, it is a useful intermediate step in the calculation of the sum of squares for interaction.

3. The significance of interaction is tested first by forming the ratio of MS_{rc} to MS_w. This is compared to the tabled F-ratio with the appropriate level of significance and degrees of freedom. If the interaction is significant, the effects of the row variable are not the same for each column and the effects of the column variable are not the same for each row. If the interaction is nonsignificant, the row and column variables may be considered independently.

4. If the interaction is significant, the analysis will be completed for the simple effects of the row and column variables. In the case of a two-by-three design, for example, there will be three simple analyses of variance with two groups each and two simple analyses with three groups each. If the interaction is nonsignificant, the main effects for columns may be tested by the ratio of MS_c to MS_w and the main effects for rows by the ratio of MS_r to MS_w. These, of course, are tests of the significance of difference between the column and row means, respectively.

40.4   COMPUTATIONAL PROCEDURES

The sum of squares between columns may be calculated from:

$$SS_c = \sum_{j=1}^{c} n_{.j}(M_{.j} - M)^2 = \sum_{j=1}^{c} \frac{T_{.j}^2}{n_{.j}} - \frac{T^2}{N}$$

where $n_{.j}$ is the number of subjects, $M_{.j}$ the mean, and $T_{.j}$ the sum of the scores in the jth column.

Similarly, the sum of squares between rows may be calculated from:

$$SS_r = \sum_{i=1}^{r} n_{i.}(M_{i.} - M)^2 = \sum_{i=1}^{r} \frac{T_{i.}^2}{n_{i.}} - \frac{T^2}{N}$$

where $n_{i.}$ is the number of subjects, $M_{i.}$ the mean, and $T_{i.}$ the sum of the scores in the ith row.

And the sum of squares between cells may be calculated from:

$$SS_{\text{cells}} = \sum_{i=1}^{r} \sum_{j=1}^{c} n_{ij}(M_{ij} - M)^2 = \sum_{i=1}^{r} \sum_{j=1}^{c} \frac{T_{ij}^2}{n_{ij}} - \frac{T^2}{N}$$

where n_{ij} is the number of subjects, M_{ij} the mean, and T_{ij} the sum of the scores in the ijth cell.

Both the sum of squares for interaction and the sum of squares within cells will be calculated as residuals:

$$SS_{rc} = SS_{\text{cells}} - SS_c - SS_r$$
$$SS_w = SS_t - SS_{\text{cells}}$$

Finally, the sum of squares for total may be calculated as follows:

$$SS_t = \sum_{i=1}^{r} \sum_{j=1}^{c} \sum_{k=1}^{n_{ij}} X_{ijk}^2 - \frac{T^2}{N}$$

The reader's attention is directed to the fact that some of the terms are held in common by these formulas. As a matter of fact, only five terms need be evaluated to complete the calculations.

Consider the first term of the sum of squares for total:

$$\sum_{i=1}^{r} \sum_{j=1}^{c} \sum_{k=1}^{n_{ij}} X_{ijk}^2$$

To evaluate this term, one need only (1) square each score in the entire collection and (2) sum all of these squares.

Next, consider the second term of the sum of squares between columns, between rows, between cells, and for total:

$$\frac{T^2}{N}$$

To evaluate this term, one need only (1) sum all scores in the entire collection, (2) square this sum, and (3) divide by N.

The first term of the sum of squares between columns is:

$$\sum_{j=1}^{c} \frac{T_{\cdot j}^2}{n_{\cdot j}}$$

To evaluate this term, (1) sum all scores for each column, (2) square each sum, (3) divide each of the squares by the number of subjects in the respective column, and (4) sum for all columns.

The first term of the sum of squares between rows is:

$$\sum_{i=1}^{r} \frac{T_{i\cdot}^2}{n_{i\cdot}}$$

To evaluate this term, (1) sum all scores for each row, (2) square each sum, (3) divide each of the squares by the number of subjects in the respective row, and (4) sum for all rows.

Finally, the first term of the sum of squares between cells is:

$$\sum_{i=1}^{r} \sum_{j=1}^{c} \frac{T_{ij}^2}{n_{ij}}$$

To evaluate this term, (1) sum all scores for each cell, (2) square each sum, (3) divide each of the squares by the number of subjects in the respective cell, and (4) sum for all cells.

The degrees of freedom for the various sums of squares are $(c - 1)$ for between columns, $(r - 1)$ for between rows, $(c - 1)(r - 1)$ for interaction, $(N - rc)$ for within, and $(N - 1)$ for total. Of course, c is the number of columns, r the number of rows, and N the number of subjects in the entire collection.

As always, a mean square is simply a sum of squares divided by its degrees of freedom.

40.5   EXAMPLE

The psychiatric experiment described in the earlier sections of the chapter will serve to illustrate the use of two-dimensional analysis of variance. Suppose that four independent random samples of size four are drawn and assigned to the four treatment combinations. At the conclusion of the experiment, a criterion of some sort is administered to the patients and their scores are recorded in the double-entry table as follows:

	Drug	No drug
Electroshock	2, 3, 3, 4	1, 2, 2, 3
No shock	0, 1, 2, 3	0, 1, 1, 2

1. $H1$:   There is no interaction between the row and column variables.
   $H2$:   The column means are equal.
   $H3$:   The row means are equal.
2. Two-dimensional 'analysis of variance, $n_{11} = n_{21} = n_{12} = n_{22} = 4$.
3. .05 level, R: $F \geq 4.75$, $df = 1$, 12 for all three hypotheses (the region of rejection will not always be the same for all three hypotheses in two-way analysis of variance).
4. $\Sigma\Sigma\Sigma X_{ijk}^2 = 76.00$
   $$\frac{T^2}{N} = \frac{30^2}{16} = 56.25$$

$$\sum \sum \frac{T_{ij}^2}{n_{ij}} = \frac{12^2}{4} + \frac{8^2}{4} + \frac{6^2}{4} + \frac{4^2}{4} = 65.00$$

$$\sum \frac{T_{\cdot j}^2}{n_{\cdot j}} = \frac{18^2}{8} + \frac{12^2}{8} = 58.50$$

$$\sum \frac{T_{i\cdot}^2}{n_{i\cdot}} = \frac{20^2}{8} + \frac{10^2}{8} = 62.50$$

$SS_{\text{cells}} = 65.00 - 56.25 = 8.75$

$SS_c = 58.50 - 56.25 = 2.25$

$SS_r = 62.50 - 56.25 = 6.25$

$SS_{rc} = 8.75 - (6.25 + 2.25) = .25$

$SS_w = 19.75 - 8.75 = 11.00$

$SS_t = 76.00 - 56.25 = 19.75$

Summary table for two-dimensional analysis of variance			
Source of variation	df	Sums of squares	Mean squares
(Between cells)	XX	(8.75)	XXXX
Between columns	1	2.25	2.25
Between rows	1	6.25	6.25
Interaction	1	.25	.25
Within cells	12	11.00	.917
Total	15	19.75	XXXX

Interaction: $F = \dfrac{0.25}{0.917} = .27$ (retain)

Column means: $F = \dfrac{2.25}{0.917} = 2.45$ (retain)

Row means: $F = \dfrac{6.25}{0.917} = 6.82$ (reject)

5. There is no significant interaction between the drug and shock treatments. There is no significant difference between the column means; apparently the drug is ineffective. There is a significant difference between the row means with those patients receiving electroshock therapy scoring higher on the criterion.

40.6 THE CHOICE OF AN ERROR TERM

In the two-way analysis of variance with more than one entry per cell, there will be a mean square within (MS_w), and this is the appropriate error term for testing the interaction effect. The choice of an error term for testing the row and column effects is dependent upon whether these variables are classified as random or fixed effects. The student may find it profitable to review the use of these terms in Section 39.6.

The most common situation encountered in behavioral research is that in which both independent variables are fixed effects (a fixed effects model, or Model I). Under these circumstances, MS_w is the appropriate error term for both the test of equal row means and the test of equal column means.

The situation in which both independent variables are random effects (a random effects model, or Model II) is much less common. When it occurs, MS_{rc} is the appropriate error term for both the test of equal row means and the test of equal column means.

It is not at all unusual for the behavioral scientist to encounter a situation in which one independent variable is a fixed effect and the other is a random effect (a mixed effects model, or Model III). Under these circumstances, MS_w is the appropriate error term for testing the random effects factor and MS_{rc} is the appropriate error term for testing the fixed effects factor.

Ordinarily, the use of MS_w as the error term where MS_{rc} is the more appropriate choice will result in a conservative test of the null hypothesis.

40.7   ANALYSIS OF VARIANCE WITH A NESTED VARIABLE

Consider the situation in which classroom learning research is conducted with two different learning strategies (B_1 and B_2). Four classroom teachers are to participate (A_1, A_2, A_3, and A_4). Ordinarily, a four-by-two analysis of variance would be preferred. Suppose, however, that each teacher can teach only one class and thus only one of the learning strategies. The situation is illustrated by Figure 40.4 and by this four-celled diagram:

	B_1	B_2
A_1	A_1B_1	
A_2	A_2B_2	
A_3		A_3B_2
A_4		A_4B_2

Notice that the columns are formed in the usual fashion, as in either factorial or randomized blocks designs; however, there is only one cell in each row — rows, as such, do not exist and are not subject to statistical analysis. The A-factor is said to be nested within the B-factor, and the analysis must take this fact into consideration. The validity of the example given above rests upon the assumption that students (not classrooms) are to be sampled (if classrooms were sampled, this would be a simple randomized design with $n_1 = n_2 = 2$).

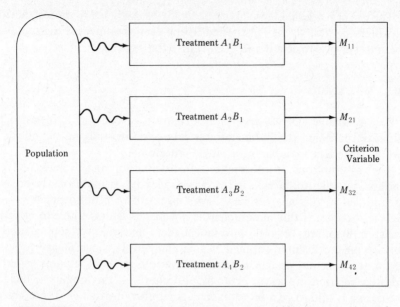

Figure 40.4 Design with nested factor (A nested within B) — experimental research.

The computational strategy for the analysis of variance with a nested variable is straightforward:

1. SS_t, SS_{cells}, and SS_w are calculated precisely as in two-way analysis of variance.
2. If the independent variable whose effects are to be studied is the column variable as in the example above, the sum of squares between groups (SS_b) is calculated precisely as SS_b in the one-way analysis or as SS_c in the two-way analysis.
3. The sum of squares for the nested factor (SS_{nf}) is most conveniently calculated as a residual:

$$SS_{nf} = SS_{\text{cells}} - SS_b = SS_t - SS_b - SS_w$$

The number of degrees of freedom allotted the nested factor is the number of cells minus the number of treatments (columns in the example).
4. If the nested factor is a fixed effect, the appropriate error term for the treatment effect is MS_w. If the nested factor is a random effect, the appropriate error term for the treatment effect is MS_{nf}. In either case, the error term for the nested factor is MS_w.

The use of nested designs should probably be restricted to those situations in which the number of cells nested within each treatment is the same and the number of subjects in each cell is also the same.

40.8 DISCUSSION

Two-way analysis of variance is an extension of one-way analysis to the situation in which the investigator is interested in studying the effects of two independent variables on a single criterion.

The assumptions of normality of distribution and homogeneity of variances applies to the cells in the two-way layout. In addition, the analysis as outlined in this chapter has the requirement of proportionality of the cell frequencies. If this requirement is not met, it is impossible to partition the sum of squares for total into independent and nonoverlapping sums of squares in the fashion outlined in this chapter. It is customary to speak of an experimental design in which the various sums of squares have this additive relationship as an orthogonal design — if this quality is lacking, the design is said to be nonorthogonal. A number of strategies for analyzing data that do not meet the requirement of orthogonality have been suggested, and these are treated in the Selected References. The author is partial to the use of multiple regression in these circumstances, and this strategy will be outlined subsequently.

The ideas used in developing two-way analysis of variance in this chapter may be extended to three-way and higher order designs and to designs in which organization of the data is much more complex than that suggested here. These, however, are beyond the scope of this book — they are reserved for a textbook on experimental design.

EXERCISES

40.1 A group of freshman college students is classified according to sex and according to whether they graduated from an urban or rural high school. Their grades (as recorded on a four-point scale) in a first course in public speaking are recorded in the double-entry table below:

	Urban H. S.	Rural H. S.
Males	0, 1, 2, 2, 3	0, 0, 1, 2, 3
Females	1, 2, 3, 4, 4	1, 2, 2, 2, 3

Construct a summary table for two-way analysis of variance, and:
(a) Test the hypothesis of no interaction between the row and column variables. Use the .05 level.

(b) Test the main effects for the column variable. Is there a significant difference in the performance of the urban and rural students?

(c) Test the main effects for the row variable. Is there a significant difference in the performance of the male and female students?

40.2 Given the following raw data from a three-by-two table:

Sample 1,1:   2, 6, 6, 8, 8
Sample 1,2:   5, 6, 9, 9, 11
Sample 2,1:   1, 3, 3, 6, 7
Sample 2,2:   4, 8, 8, 10, 10
Sample 3,1:   0, 0, 2, 2, 6
Sample 3,2:   0, 0, 2, 2, 6

Complete a summary table for two-way analysis of variance. Assume that the row variable is a random effect and that the column variable is a fixed effect.

(a) Test the hypothesis of no interaction at the .01 level.

(b) Test the hypothesis of equal column means at the .01 level.

(c) Test the hypothesis of equal row means at the .01 level.

40.3 Given the following raw data:

0, 1	0, 2	1, 3
1, 1	1, 3	0, 3
0, 1	1, 1	1, 2
0, 2	1, 1	2, 3

(a) Complete the summary table for two-way analysis of variance.

(b) Test the hypothesis of interaction at the .05 level.

(c) Test the hypothesis of equal row means at the .05 level. (Assume a fixed effects model.)

(d) Test the hypothesis of equal column means at the .05 level.

40.4 Refer to the data of Exercise 40.3. Assume that the cells are nested within the columns.

(a) Complete the summary table for the analysis of variance.

(b) Test the hypothesis of equal column means at the .05 level. (Assume a fixed effects model.)

SELECTED REFERENCES

ANDERSON, R. L., "Some remarks on the design and analysis of factorial experiments," in *Contributions to Probability and Statistics* (Ingram Olkin, et al., eds.). Stanford: Stanford University Press, 1960, pp. 35–56.

BINDER, ARNOLD, "The choice of an error term in analysis of variance designs," *Psychometrika*, **20**, 29–50 (March 1955).

DAYTON, C. MITCHELL, *The Design of Educational Experiments*. New York: McGraw-Hill, 1970.

EDWARDS, ALLEN L., *Experimental Design in Psychological Research*, 3rd ed. New York: Holt, Rinehart and Winston, 1968.

KIRK, ROGER E., *Experimental Design Procedures for the Behavioral Sciences*. Belmont: Brooks/Cole, 1968.

LINDQUIST, E. F., *Design and Analysis of Experiments in Psychology and Education*. Boston: Houghton Mifflin, 1953.

LUBIN, ARDIE, "The interpretation of significant interaction," *Educational and Psychological Measurement*, **21**, 807–817 (1961).

MARASCUILO, L. A., and J. R. LEVIN, "Appropriate *post hoc* comparisons for interaction and nested hypotheses in analysis of variance designs: The elimination of Type IV errors," *American Educational Research Journal*, **7**, 397–421 (1970).

MILLMAN, JASON, and GENE V. GLASS, "Rules of thumb for writing the anova table," *Journal of Educational Measurement*, **4**, 41–51 (Summer 1967).

MYERS, JEROME L., *Fundamentals of Experimental Design*. Boston: Allyn and Bacon, 1966.

SCHEFFÉ, HENRY, *The Analysis of Variance*. New York: John Wiley, 1959.

STEINHORST, R. KIRK, and C. DEAN MILLER, "Disproportionality of cell frequencies in psychological and educational experiments multiple classification," *Educational and Psychological Measurement*, **29**, 799–811 (1969).

WILLIAMS, JOHN D., "Two way fixed effects analysis of variance with disproportionate cell frequencies," *Multivariate Behavioral Research*, **7**, 67–83 (January 1972).

WINER, B. J., *Statistical Principles in Experimental Design*, 2nd ed. New York: McGraw-Hill, 1971.

41

THE ANALYSIS OF COVARIANCE

41.1 FUNCTION

The success of an experiment and the ability to detect significant differences in the criterion variable are often determined by the ability of the investigator to control one or more variables that influence the criterion. In Chapters 25 and 27, the possibility of controlling variables through the use of matched pairs or repeated measurements designs was discussed; in Chapter 40, the possibility of controlling variables through the use of stratified samples was considered. Each of these techniques uses experimental procedures to control the influence of these variables on the criterion, and each introduces special problems with respect to the conduct of the experiment. Random samples stratified into two or three levels on some variable related to the criterion may be used with two-dimensional analysis of variance to grossly control the influence of the stratification variable; however, the requirement of proportionality of cell frequencies restricts the use of this design in some research. The use of repeated measurements designs permits simultaneous control of many variables, but this design is inapplicable in the vast majority of behavioral experiments due to the carryover from treatment to treatment and from measure to measure. Occasionally, the use of matched pairs can be substituted to achieve the same end, but with less precision. The analysis of covariance is a blending of regression and the analysis of variance, which permits *statistical* rather than *experimental* control of variables. The result is equivalent to matching the various experimental groups with respect to the variable or variables being controlled.

The use of analysis of covariance ordinarily involves a pretest (the variable to be controlled) and a posttest (the criterion) that are known to be correlated. IQ scores, achievement test scores, and previous course grades are often used as pretest measures in educational research. In some circumstances, it may be completely appropriate to use the same instrument for both pretest and posttest. Occasionally, the variable to be controlled is a stable trait that is unlikely to be influenced by experimental manipulation (sex or IQ, for example), and the actual recording of the measure may take place during or following the experiment.

Whenever two measures are correlated, one can be used to predict scores on the other. To the extent that performance on the posttest can be predicted from performance on the pretest, this performance cannot be attributed to the experimental activities. The analysis of covariance consists essentially of determining that a proportion of the variance of the criterion existed prior to the experiment and this proportion is eliminated from the final analysis. It should be immediately apparent that two substantial benefits accrue from such a procedure: (1) any variable that influences the variation of the criterion variable may be controlled, and (2) the error variance in the analysis is substantially reduced. The situation is illustrated in Figure 41.1.

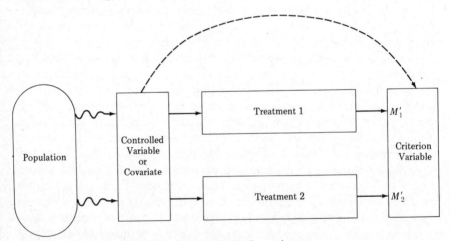

Figure 41.1 Experimental design with analysis of covariance.

Like one-way analysis of variance, the analysis of covariance may be used with any number of samples. It may be substituted for one-way analysis of variance whenever a suitable pretest is available. In fact, much educational research can be improved by introducing the analysis of covariance *after* the experiment has been completed. One need only find in the students' files some measure such as IQ, achievement test score, or previous grade that can be paired with the criterion score for the analysis.

41.2 RATIONALE

The essential task in statistical analysis is to account for the variability of the criterion. Variance has been defined as the mean of the squared deviations from the mean, and it may be calculated from the formula:

$$\sigma^2 = \frac{SS}{N}$$

In the analysis of variance, the total sum of squares is partitioned into its component parts in an attempt to identify the sources of variability of the criterion. The r-square interpretation of correlation is a similar procedure.

The analysis of covariance is concerned with concomitant variation in the criterion variable and a variable whose relationship to the criterion is to be controlled. Covariance may be defined as the mean of the products (for paired variables) of the deviations from the mean, and it may be calculated from the formula:

$$\text{cov} = \frac{SP}{N}$$

It should be noted that covariance may be either negative or positive; variance, of course, is always a positive quantity.

The analysis of covariance involves the partitioning of the total sum of products in a fashion similar to the partitioning of the total sum of squares in the analysis of variance. The total sum of products may be calculated from:

$$SP_t = \sum_{j=1}^{k} \sum_{i=1}^{n_j} X_{ij} Y_{ij} - \frac{T_x T_y}{N}$$

where T_x and T_y are the sums of the respective X- and Y-measures for all groups.

The sum of products between groups may be calculated from:

$$SP_b = \sum_{j=1}^{k} \frac{T_{xj} T_{yj}}{n_j} - \frac{T_x T_y}{N}$$

where T_{xj} and T_{yj} are the sums of the X- and Y-measures for the jth group.

The sum of products within groups may be calculated as a residual:

$$SP_w = SP_t - SP_b$$

We will follow the custom of letting X represent the predictor (controlled) variable and Y the predicted (criterion) variable. The student is reminded that there is a third variable of great importance, the independent variable whose effects on the criterion are to be studied, this variable being represented by membership in one of the experimental groups. If we introduce the assumption that the slope of the regression line (predicting Y from X) is the same for the various experimental groups, the best estimate of this common slope may be obtained from:

$$b_w = \frac{SP_w}{SS_{wx}}$$

where SS_{wx} is the sum of squares within for the X-measures.

The predicted criterion score (in deviation form) for any individual is:

$$\hat{y} = b_w x$$

This predicted score is the portion of the criterion measure that may be determined from a knowledge of the variable to be controlled. It may be subtracted from the criterion score to obtain an adjusted criterion score:

$$y' = y - b_w x \quad \text{or} \quad Y' = Y - b_w x$$

It is these adjusted scores upon which the final analysis is based. Statistical inferences are drawn with respect to adjusted group means, which are calculated in the following manner:

$$\overline{Y}'_j = \overline{Y}_j - b_w(\overline{X}_j - M_x)$$

where M_x is the general mean of the X-measures.

To test the hypothesis of equal adjusted means, it is necessary to obtain adjusted sums of squares and mean squares for the criterion variable. The adjusted total sum of squares may be calculated from:

$$SS'_{ty} = SS_{ty} - \frac{(SP_t)^2}{SS_{tx}}$$

where SS_{ty} is total sum of squares for Y.

Similarly, the adjusted sum of squares within groups may be determined from:

$$SS'_{wy} = SS_{wy} - \frac{(SP_w)^2}{SS_{wx}}$$

where SS_{wy} is sum of squares within for Y.

Finally, the adjusted sum of squares between groups may be calculated as a residual:

$$SS'_{by} = SS'_{ty} - SS'_{wy}$$

As in one-way analysis of variance, the number of degrees of freedom for the adjusted sum of squares between groups is $(k - 1)$. However, one degree of freedom is lost by imposing the restriction that the deviations be computed from the common within groups regression line, and the number of degrees of freedom for the adjusted sum of squares within groups is $(N - k - 1)$, and for the adjusted total sum of squares is $(N - 2)$. The adjusted mean squares between and within are, of course, obtained by dividing the sums of squares by their respective degrees of freedom. The test of the hypothesis of equal adjusted means is obtained from:

$$F = \frac{MS'_{by}}{MS'_{wy}} \quad \text{with } df = (k - 1), \quad (N - k - 1)$$

41.3   PROCEDURE

1. Two or more random samples are independently drawn from the population under study, and these are subjected to different experimental treatments. The analysis of covariance may also be used to analyze existing differences between different populations.
2. Prior to experimentation, a variable that is believed to be correlated with the criterion variable is selected for purposes of experimental control. A measure on this variable is recorded for each subject and later paired with the subject's measure on the criterion.
3. At the conclusion of the experiment, a single criterion measure is administered to all subjects in all samples. The sums of squares between groups, within groups, and for total are calculated, along with their respective degrees of freedom. These calculations are completed for both the criterion and the control variable and the calculations are precisely the same as those encountered earlier in one-way analysis of variance.
4. The sums of products between groups, within groups, and for total are calculated in the manner outlined in Section 41.2; step-by-step procedures for calculating these quantities are given in Section 41.4. Adjusted sums of squares, degrees of freedom, and mean squares are calculated for the criterion, using the formulas given in Section 41.2. It is customary to record all of the sums of squares, sums of products, mean squares, and degrees of freedom in a summary table such as this:

Summary table for the analysis of covariance							
Source	df	SS_x	SP	SS_y	df'	SS'_y	MS'_y
Between							
Within							
Total							XXX

5. The F-ratio is calculated from:

$$F = \frac{MS'_{by}}{MS'_{wy}} \quad \text{with } df = (k - 1), \quad (N - k - 1)$$

The adjusted criterion means are calculated from:

$$\overline{Y}'_j = \overline{Y}_j - b_w(\overline{X}_j - M_x)$$

6. The calculated F is compared to the tabled F at the desired level of significance and the appropriate number of degrees of freedom. If the calculated statistic equals or exceeds the tabled value, the null hypothesis is rejected, and a significant difference in the adjusted means of the various samples is determined to exist. If the calculated statistic is smaller than the tabled value, the null hypothesis is retained, and no significant difference in the adjusted means has been found.

41.4 COMPUTATIONAL PROCEDURES

It was noted earlier that the sums of squares and degrees of freedom for the X- and Y-measures are calculated in precisely the same fashion as encountered in the simple analysis of variance. These computations have been explained in great detail in a previous chapter (see Section 36.4).

The computations involved in the partitioning of the sum of products will be treated in detail in this section. The total sum of products has been identified as:

$$SP_t = \sum_{j=1}^{k} \sum_{i=1}^{n_j} X_{ij} Y_{ij} - \frac{T_x T_y}{N}$$

The formula for the sum of products between groups has been given as:

$$SP_b = \sum_{j=1}^{k} \frac{T_{xj} T_{yj}}{n_j} - \frac{T_x T_y}{N}$$

And the sum of products within groups may be calculated as a residual:

$$SP_w = SP_t - SP_b$$

Notice that only three terms need be evaluated to complete the calculations. Consider the first term of the total sum of products:

$$\sum_{j=1}^{k} \sum_{i=1}^{n_j} X_{ij}Y_{ij}$$

To evaluate this term, one need only (1) pair each X-score in the entire collection with its corresponding Y-score and take the product of each pair, and (2) sum all of these products.

Next, consider the second term of the total sum of products and of the sum of products between:

$$\frac{T_x T_y}{N}$$

To evaluate this term, (1) take the sum of all of the X-measures in the entire collection, and do the same for the Y-measures, (2) take the product of these two sums, and (3) divide by N.

Finally, to evaluate the first term of the sum of products between:

$$\sum_{j=1}^{k} \frac{T_{xj}T_{yj}}{n_j}$$

(1) take the sum of the X-measures for each group in the collection, and do the same for the Y-measures, (2) pair each X-sum with the Y-sum for the same group, and take the product for each group, (3) divide each product by the number of subjects in the group, and (4) sum for all groups.

The adjusted sums of squares, degrees of freedom, and mean squares may be calculated as outlined in Section 41.2.

41.5 EXAMPLE

Consider the data in the following table. Group membership constitutes the independent variable whose effects are to be studied, the X-score is the independent variable whose effects are to be controlled, and the Y-score is the dependent variable.

		Group I		
X	Y	X^2	Y^2	XY
0	3	0	9	0
1	5	1	25	5
3	5	9	25	15
4	6	16	36	24
4	8	16	64	32
6	9	36	81	54
18	36	78	240	130

		Group II		
X	Y	X^2	Y^2	XY
2	2	4	4	4
3	4	9	16	12
5	4	25	16	20
6	5	36	25	30
6	7	36	49	42
8	8	64	64	64
30	30	174	174	172

1. *H*: The two populations means are equal when the effects of the X-variable are controlled.
2. Analysis of covariance, $n_1 = n_2 = 6$.
3. .05 level, $df = 1, 9$.
   R: $F \geq 5.12$.
4. $SS_{bx} = 204 - 192 = 12$ $\qquad$ $SS_{by} = 366 - 363 = 3$
   $SS_{wx} = 252 - 204 = 48$ $\qquad$ $SS_{wy} = 414 - 366 = 48$
   $SS_{tx} = 252 - 192 = 60$ $\qquad$ $SS_{ty} = 414 - 363 = 51$
   $SP_b = 258 - 264 = -6$ $\qquad$ $SS'_{by} = 26.93 - 7.67 = 19.26$
   $SP_w = 302 - 258 = 44$ $\qquad$ $SS'_{wy} = 48.00 - 40.33 = 7.67$
   $SP_t = 302 - 264 = 38$ $\qquad$ $SS'_{ty} = 51.00 - 24.07 = 26.93$

Summary table for the analysis of covariance							
Source	DF	SS_x	SP	SS_y	DF'	SS'_y	MS'_y
Between	1	12	−6	3	1	19.26	19.26
Within	10	48	44	48	9	7.67	.85
Total	11	60	38	51	10	26.93	XXX

$$F = \frac{19.26}{0.85} = 22.6, \ df = 1, 9$$

Means:

$$\bar{X}_1 = 3 \qquad \bar{Y}_1 = 6 \qquad \bar{Y}'_1 = 6 - .92(3 - 4) = 6.92$$
$$\bar{X}_2 = 5 \qquad \bar{Y}_2 = 5 \qquad \bar{Y}'_2 = 5 - .92(5 - 4) = 4.08$$

5. Reject the null hypothesis. The adjusted mean for Group I is significantly larger than that of Group II. Notice that a simple analysis of variance performed on the unadjusted Y-measures would have yielded a nonsignificant finding ($F = .6$).

41.6 DISCUSSION

The analysis of covariance is one of the most valuable tools of statistical inference. It is especially appropriate in human learning research, where, to a large extent, the subjects' performance on the criterion may be determined before the experiment is undertaken. It is convenient to use and it fits a wide variety of research situations.

The advantage of the analysis of covariance over simple analysis of variance is dependent upon the relationship between the pretest and the posttest. If the correlation between the two variables is less than about 0.30, the benefit from covariance analysis is likely to be negligible. In the analysis as outlined in this chapter it is assumed that the relationship between the two variables is linear — to the extent that this is not the case the control of the pretest will be diminished. Absolute control of the effects of the pretest is dependent upon the trait being measured with perfect reliability — to the extent that this is not the case the control of the pretest will be imperfect.

It is extremely important that the scores on the pretest are not influenced by the experimental treatments. If this is not the case, the test for equal means is likely to be meaningless. The investigator may guard against this problem by recording the X-scores prior to the random assignment of the subjects to the experimental treatments.

The assumptions underlying the analysis of covariance are essentially the same as those for the analysis of variance, with the added assumption of homogeneity of regression. This assumption requires that the slope of the regression line (predicting Y from X) be the same within each of the populations under study. This is the common within-groups regression coefficient that was estimated by b_w. The analysis of covariance appears to be fairly robust with respect to the assumption of homogeneity of regression — if the assumption is violated, the test for equal means tends to be conservative. If a test for homogeneity of regression is desired, it may be developed by partitioning the adjusted sum of squares within into two parts:

$$SS_{wr} = SS_{wy} - \sum_{j=1}^{k} \frac{SP_j^2}{SS_{xj}} \quad \text{with } df = (N - 2k)$$

$$SS_{br} = SS'_{wy} - SS_{wr} \quad \text{with } df = (k - 1)$$

These sums of squares may be converted into mean squares by dividing by their respective degrees of freedom, and an F-ratio may be calculated from:

$$F = \frac{MS_{br}}{MS_{wr}} \quad \text{with } df = (k - 1), (N - 2k)$$

The test of the null hypothesis (no difference in the slopes of the regression lines) is accomplished by comparing the calculated F to the tabled value in the usual fashion. With a little effort, the student should be able to demonstrate that, for the example given in Section 41.5, $SS_{wr} = 7.67$, $SS_{br} = 0.0$, and $F = 0.0$.

It is also assumed that in the populations under study (these are hypothetical populations in experimental research), all persons having the same X-score are normally distributed with respect to their Y-scores, and that the Y-scores are equally variable for different values of X. These assumptions were encountered earlier in Section 14.7.

Generally, if the X-scores are known to be normally distributed and are recorded prior to the administration of the experimental treatments, and if the subjects are assigned at random to the experimental treatments, the analysis of covariance has been demonstrated to be quite robust with respect to the other assumptions.

The analysis of covariance may be extended to situations requiring the control of more than one variable or to the situation in which the regression is curvilinear through the use of multiple regression techniques, which are discussed in subsequent chapters.

EXERCISES

41.1   Consider the data in the following table:

	Control group		Experimental group	
SUBJECT	PRETEST	POSTTEST	PRETEST	POSTTEST
1	3	4	2	6
2	4	6	1	5
3	0	3	2	5
4	1	3	3	6
5	2	4		
Totals	10	20	8	22
Means	2	4	2	5.5

(a) Test the hypothesis of equal means for the control and experimental groups. Use the .05 level.

(b) Calculate the adjusted means for the two groups.

41.2   Refer to the data of Exercise 40.1. Convert sex to a binary variable by awarding a score of one to each male and a score of zero to each female. Complete an analysis of covariance summary table using urban versus rural group membership as the independent variable whose effects are to be studied, sex as the variable whose effects are to be controlled, and course grade as the dependent variable.

(a)   Test the hypothesis of equal means for the urban and rural groups. Use the .05 level.

(b)   Calculate the adjusted means for the two groups.

41.3   An experiment is conducted in which a pretest is administered to an available group of 90 subjects, which are then randomly assigned to three samples of equal size, subjected to experimentation, after which a posttest is administered. The means were as follows: $\bar{X}_1 = 47.4$, $\bar{X}_2 = 50.2$, $\bar{X}_3 = 52.4$, $\bar{Y}_1 = 20.0$, $\bar{Y}_2 = 21.8$, and $\bar{Y}_3 = 24.2$. Other facts are summarized in the table:

Summary table for the analysis of covariance

Source	DF	SS_x	SP	SS_y	DF'	SS'_y	MS'_y
Between		376.8	373.5	266.4			
Within		2560.0	4096.0	6682.6			
Total		2936.8	4469.5	6949.0			

(a)   Complete the summary table for the analysis of covariance.

(b)   Test the hypothesis of equal means. Use the .01 level.

(c)   Calculate the adjusted means for the three samples.

SELECTED REFERENCES

ATIQULLAH, M., "The robustness of the covariance analysis of a one-way classification," *Biometrika*, **51**, 83–92 (December 1964).

BILLEWICZ, W. Z., "The efficiency of matched samples: An empirical investigation," *Biometrics*, **21**, 623–644 (September 1965).

ELASHOFF, JANET D., "Analysis of covariance: A delicate instrument," *American Educational Research Journal*, **6**, 383–401 (May 1969).

FELDT, LEONARD S., "A comparison of the precision of three experimental designs employing a concomitant variable," *Psychometrika*, **23**, 335–353 (December 1958).

SCHEFFÉ, HENRY, *The Analysis of Variance*. New York: John Wiley, 1959.

42

MULTIPLE REGRESSION

42.1 INTRODUCTION

Human behavior is an exceedingly complex affair, and the behavioral scientist will ordinarily expect to find a multitude of factors influencing any given action. Stated statistically, variation in a given dependent variable is usually a function of concomitant variation in many independent variables acting simultaneously. We have noted that academic achievement is in part determined by the variable commonly referred to as intelligence. Thus the educational researcher should be able to predict student grades in a college composition class from intelligence test scores. He should also be able to predict these same grades from other variables, such as rank in high school graduating class, grade in high school composition class, or score on a college entrance examination. Each of these variables is an attempt to assess a different facet of the student's total personality and each could be expected to make some contribution toward future academic performance. The techniques of multiple regression enable the behavioral scientist to use his knowledge of two or more independent variables to predict scores on a single dependent variable with greater success than is possible with a knowledge of a single independent variable. In the example cited above, the investigator should be able to do a better job of predicting performance in the college class from a knowledge of performance in the corresponding high school class plus intelligence test score than from either of these variables taken alone.

Generally, the different kinds of variables encountered by the behavioral scientist will be correlated with one another, and to the extent that two variables are correlated, they may be said to measure the same thing. Thus there will be some overlap in the contribution of two-predictor variables in the typical multiple regression situation. Referring again to the example above, let us suppose that grade in the high school class accounts for 40 percent of the variance in college class grade and that intelligence score accounts for 30 percent of the variance in the college grade. When high school grade and intelligence score are combined in a multiple regression equation, it is highly improbable that the investigator will be able to account for 70 percent of the variance in the college grade, but he can anticipate that he will be able to account for something between 40 and 70 percent. Because of this intercorrelation between the predictor variables, it is often expedient to use the best combination of two or three predictors, and there is little advantage to be gained from using additional predictor variables. For this reason, and because the computations become increasingly laborious with added variables, only two- and three-predictor regression equations will be treated in this chapter.

42.2   THE TWO-PREDICTOR CASE

The general form of the regression equation for raw scores in the two-predictor case is:

$$\hat{Y} = b_1X_1 + b_2X_2 + c$$

The same relationship may be expressed in deviation score form by:

$$\hat{y} = b_1x_1 + b_2x_2$$

As with the single-predictor case, the least squares criterion will be used to determine the appropriate values of b_1, b_2, and c. The derivation of the least squares solution requires the use of differential calculus; however, the problem has already been solved and we shall be able to determine the prediction equation through the use of simple algebra. In the two-predictor situation using raw scores, there are three unknowns (b_1 b_2, and c) and three equations that must be solved simultaneously:

$$\Sigma X_1Y = b_1 \Sigma X_1^2 + b_2 \Sigma X_1X_2 + c \Sigma X_1$$
$$\Sigma X_2Y = b_1 \Sigma X_1X_2 + b_2 \Sigma X_2^2 + c \Sigma X_2$$
$$\Sigma Y = b_1 \Sigma X_1 + b_2 \Sigma X_2 + Nc$$

These are sometimes referred to as *conditional* or *normal* equations. The problem may be simplified somewhat by using deviation scores. Then there are only two unknowns (b_1 and b_2), and two equations to be solved simultaneously:

$$\Sigma\, x_1 y = b_1 \Sigma\, x_1^2 + b_2 \Sigma\, x_1 x_2$$
$$\Sigma\, x_2 y = b_1 \Sigma\, x_1 x_2 + b_2 \Sigma\, x_2^2$$

The student is reminded that the values of the regression coefficients (b_1 and b_2) are the same whether deviation or raw scores are used. For this reason some computational effort may be saved if the conditional equations are solved using deviation scores even though the end product is a regression equation using raw scores.

For the student who is unprepared to solve equations simultaneously, the author has derived the following formulas for b_1, b_2, and c:

$$b_1 = \frac{(\Sigma\, x_1 y)(\Sigma\, x_2^2) - (\Sigma\, x_1 x_2)(\Sigma\, x_2 y)}{(\Sigma\, x_1^2)(\Sigma\, x_2^2) - (\Sigma\, x_1 x_2)^2}$$

$$b_2 = \frac{(\Sigma\, x_1^2)(\Sigma\, x_2 y) - (\Sigma\, x_1 x_2)(\Sigma\, x_1 y)}{(\Sigma\, x_1^2)(\Sigma\, x_2^2) - (\Sigma\, x_1 x_2)^2}$$

$$c = \frac{\Sigma\, Y - b_1 \Sigma\, X_1 - b_2 \Sigma\, X_2}{N}$$

where:

$$\Sigma\, x_1^2 = \Sigma\, X_1^2 - \frac{(\Sigma\, X_1)^2}{N}$$

$$\Sigma\, x_2^2 = \Sigma\, X_2^2 - \frac{(\Sigma\, X_2)^2}{N}$$

$$\Sigma\, x_1 x_2 = \Sigma\, X_1 X_2 - \frac{(\Sigma\, X_1)(\Sigma\, X_2)}{N}$$

$$\Sigma\, x_1 y = \Sigma\, X_1 Y - \frac{(\Sigma\, X_1)(\Sigma\, Y)}{N}$$

$$\Sigma\, x_2 y = \Sigma\, X_2 Y - \frac{(\Sigma\, X_2)(\Sigma\, Y)}{N}$$

Consider the data of Table 42.1. The dependent variable (Y) and the first independent variable (X_1) are those used to illustrate the single-predictor situation (see Section 14.6). A second independent variable (X_2) in binary form has been added to illustrate the two-predictor situation. The least squares solution for the regression equation follows:

$$b_1 = \frac{(53)(2.5) - (10)(8)}{(90)(2.5) - (10)^2} = .42$$

$$b_2 = \frac{(90)(8) - (10)(53)}{(90)(2.5) - (10)^2} = 1.52$$

$$c = \frac{50 - .42(50) - 1.52(5)}{10} = 2.14$$

$$\hat{Y} = .42X_1 + 1.52X_2 + 2.14$$

Table 42.1   Data for the two-predictor example

Y	X_1	X_2	Y^2	X_1^2	X_2^2	X_1Y	X_2Y	X_1X_2
2	1	0	4	1	0	2	0	0
3	2	0	9	4	0	6	0	0
2	2	0	4	4	0	4	0	0
5	3	1	25	9	1	15	5	3
5	4	0	25	16	0	20	0	0
7	5	1	49	25	1	35	7	5
5	6	0	25	36	0	30	0	0
6	8	1	36	64	1	48	6	8
7	9	1	49	81	1	63	7	9
8	10	1	64	100	1	80	8	10
50	50	5	290	340	5	303	33	35

42.3   THE THREE-PREDICTOR CASE

The general form of the regression equation for raw scores in the three-predictor case is:

$$\hat{Y} = b_1X_1 + b_2X_2 + b_3X_3 + c$$

The same relationship may be expressed in deviation score form by:

$$\hat{y} = b_1x_1 + b_2x_2 + b_3x_3$$

As with the single- and two-predictor cases, the least squares criterion will be used to determine the appropriate values of b_1, b_2, b_3, and c. The conditional equations are:

$$\Sigma X_1Y = b_1 \Sigma X_1^2 + b_2 \Sigma X_1X_2 + b_3 \Sigma X_1X_3 + c \Sigma X_1$$
$$\Sigma X_2Y = b_1 \Sigma X_1X_2 + b_2 \Sigma X_2^2 + b_3 \Sigma X_2X_3 + c \Sigma X_2$$
$$\Sigma X_3Y = b_1 \Sigma X_1X_3 + b_2 \Sigma X_2X_3 + b_3 \Sigma X_3^2 + c \Sigma X_3$$
$$\Sigma Y = b_1 \Sigma X_1 + b_2 \Sigma X_2 + b_3 \Sigma X_3 + Nc$$

These equations may be stated in deviation score form, eliminating one unknown (c) and one equation:

$$\Sigma \ x_1 y = b_1 \ \Sigma \ x_1^2 + b_2 \ \Sigma \ x_1 x_2 + b_3 \ \Sigma \ x_1 x_3$$
$$\Sigma \ x_2 y = b_1 \ \Sigma \ x_1 x_2 + b_2 \ \Sigma \ x_2^2 + b_3 \ \Sigma \ x_2 x_3$$
$$\Sigma \ x_3 y = b_1 \ \Sigma \ x_1 x_3 + b_2 \ \Sigma \ x_2 x_3 + b_3 \ \Sigma \ x_3^2$$

It is a relatively simple matter to extend the pattern that is evident in the two- and three-predictor cases to four or more predictors. However, the calculations become increasingly laborious and will probably be undertaken only when an electronic computer is available.

42.4   CURVILINEAR REGRESSION

Some of the relationships encountered by the behavioral scientist are curvilinear and the use of linear regression equations in these situations will grossly underestimate the degree of relationship and introduce unnecessary prediction error. The most commonly encountered type of curvilinear regression calls for a quadratic regression equation, an equation that yields a curve with a single bend in it when plotted graphically. Such a relationship is graphically illustrated in Figure 12.6.

The general form of the equation for quadratic regression is:

$$\hat{Y} = b_1 X + b_2 X^2 + c$$

If we represent X in this expression by X_1 and generate a dummy variable X_2 that is equal to X^2, the quadratic regression equation takes on the form of the two-predictor linear regression equation:

$$\hat{Y} = b_1 X_1 + b_2 X_2 + c \qquad \text{where } X_2 = X_1^2$$

Treated in this fashion, the conditional equations for the quadratic situation are the same as those for the two-predictor situation (see Section 42.2). We will use this approach to determine the values of b_1, b_2, and c.

The data recorded in scattergram form in Figure 12.3 suggest a slight curvilinear relationship and will be used to illustrate quadratic regression. These same data were used earlier to illustrate simple linear regression (see Section 14.6).

The data of Figure 12.3, along with the variable X_2, which has been generated equal to X_1^2, have been recorded in Table 42.2. The least squares solution for quadratic regression follows:

$$b_1 = \frac{(53)(11388) - (990)(553)}{(90)(11388) - (990)^2} = 1.25$$

$$b_2 = \frac{(90)(553) - (990)(53)}{(90)(11388) - (990)^2} = -.06$$

$$c = \frac{50 - 62.58 + 20.48}{10} = .79$$

$$\hat{Y} = 1.25X - .06X^2 + .79$$

The graph of this equation, along with the original data from which it was generated, is reported in Figure 42.1.

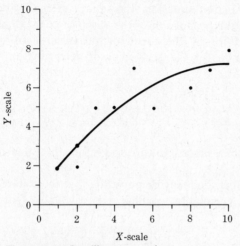

Figure 42.1 Curvilinear regression.

Table 42.2 Data for the quadratic example

Y	X_1	X_2	Y^2	X_1^2	X_2^2	X_1Y	X_2Y	X_1X_2
2	1	1	4	1	1	2	2	1
3	2	4	9	4	16	6	12	8
2	2	4	4	4	16	4	8	8
5	3	9	25	9	81	15	45	27
5	4	16	25	16	256	20	80	64
7	5	25	49	25	625	35	175	125
5	6	36	25	36	1296	30	180	216
6	8	64	36	64	4096	48	384	512
7	9	81	49	81	6561	63	567	729
8	10	100	64	100	10000	80	800	1000
50	50	340	290	340	22948	303	2253	2690

42.5 THE MULTIPLE CORRELATION COEFFICIENT

The Pearson product moment correlation coefficient has proved extremely useful as an index of relationship between two variables, which we customarily designate X and Y, with X the independent variable and Y the dependent variable. In the case of simple linear regression, $\hat{Y}$ is a linear function of X; therefore, the Pearson correlation between $\hat{Y}$ and Y is precisely the same as that between X and Y. This suggests the possibility of developing a multiple correlation coefficient that is the correlation between $\hat{Y}$ and Y. This multiple correlation coefficient shall be represented by R.

The usefulness of the squared correlation coefficient, also referred to as the coefficient of determination, has already been noted. As in the simple linear prediction situation, the square of the multiple correlation coefficient (R^2) is the proportion of the variance of the dependent variable that is accounted for by the predictor variables. Thus one may derive a general expression for the coefficient of determination that fits all regression models:

$$R^2 = \frac{\Sigma (\hat{Y} - \bar{Y})^2}{\Sigma (Y - \bar{Y})^2}$$

Because the $\hat{Y}$-values regress toward the mean, the value of R^2 will never exceed unity.

With a little effort, one may demonstrate that the square of the Pearson product moment correlation coefficient is calculable from the following expression:

$$r^2 = \frac{b \, \Sigma \, xy}{\Sigma \, y^2}$$

A similar expression may be derived that provides a convenient means of calculating the value of R^2 for the two-predictor and quadratic regression models:

$$R^2 = \frac{b_1 \, \Sigma \, x_1y + b_2 \, \Sigma \, x_2y}{\Sigma \, y^2}$$

The idea may be extended to the three-predictor model by:

$$R^2 = \frac{b_1 \, \Sigma \, x_1y + b_2 \, \Sigma \, x_2y + b_3 \, \Sigma \, x_3y}{\Sigma \, y^2}$$

For purposes of illustration, the values of the coefficient of determination will be calculated for the single-predictor, the two-predictor, and the quadratic regression examples used earlier. For this single-predictor case:

$$r^2 = \frac{(.59)(53)}{(40)} = .78$$

The investigator was able to account for 78 percent of the variance of the dependent variable from a knowledge of the single independent variable. For the two-predictor case:

$$R^2 = \frac{(.42)(53) + (1.52)(8)}{(40)} = .89$$

The additional information gained from the second independent variable enabled the investigator to account for an additional 11 percent of the variance of the dependent variable. Finally, for the quadratic model:

$$R^2 = \frac{(1.25)(53) - (.06)(553)}{(40)} = .83$$

Notice the small difference (only 5 percent) between the simple linear and the quadratic models. This suggests that the relationship between the two variables is almost linear, and this is apparent from examining Figure 42.1.

When the coefficient of determination is known, it is a relatively simple matter to calculate the standard error of estimate:

$$SE_{est} = S_y \sqrt{1 - R^2} \qquad \text{where } S_y = \sqrt{\frac{SS_y}{df}}$$

The number of degrees of freedom for S_y is equal to $N - 2$ for the single-predictor model, $N - 3$ for the two-predictor and quadratic models, and $N - 4$ for the three-predictor model. For a discussion of the uses of the standard error of estimate and of the assumptions underlying its use, the reader is directed to Section 14.7.

EXERCISES

42.1 Given the following collection of scores, in which the X_1-variable is a score on an academic rating scale, X_2 is IQ score, and Y is grade received in a certain course of study.

Y	X_1	X_2
4	2	120
3	1	110
2	0	100
0	0	90
1	0	80

(a) Determine the regression equation for the two-predictor model, using X_1 and X_2 as independent variables.

    (b)   Calculate the coefficient of determination.

    (c)   Calculate the standard error of estimate.

42.2 (a)   Determine the regression equation for the single-predictor model, using X_1 as the independent variable.

    (b)   Calculate the coefficient of determination.

    (c)   Calculate the standard error of estimate.

    (d)   Determine the regression equation for the quadratic model, using X_1 as the independent variable.

    (e)   Calculate the coefficient of determination.

    (f)   Calculate the standard error of estimate.

    (g)   Plot the regression lines for both the linear and the quadratic regression equations.

42.3 Given the following data:

X_1	X_2	Y	X_1^2	X_2^2	Y^2	X_1Y	X_2Y	X_1X_2
2	1	8	4	1	64	16	8	2
3	2	7	9	4	49	21	14	6
3	2	8	9	4	64	24	16	6
4	3	5	16	9	25	20	15	12
5	4	4	25	16	16	20	16	20
5	5	5	25	25	25	25	25	25
5	6	3	25	36	9	15	18	30
7	8	5	49	64	25	35	40	56
8	9	3	64	81	9	24	27	72
8	10	2	64	100	4	16	20	80
50	50	50	290	340	290	216	199	309

    (a)   Determine the regression equation for the two-predictor case.

    (b)   Calculate the coefficient of determination.

    (c)   Calculate the standard error of estimate.

SELECTED REFERENCES

DARLINGTON, RICHARD B., "Multiple regression in psychological research and practice," *Psychological Bulletin*, **69**, 161–182 (1968).

EZEKIAL, M., and K. E. FOX, *Methods of Correlation and Regression Analysis*, 3rd ed. New York: John Wiley, 1959.

LORD, FREDERIC M., "Nomograph for computing multiple correlation coefficients," *Journal of the American Statistical Association*, **50**, 1073–1077 (December 1955).

ROZEBOOM, WILLIAM W., *Foundations of the Theory of Prediction*. Homewood, Ill.: The Dorsey Press, 1966.

WERT, JAMES E., CHARLES O. NEIDT, and J. STANLEY AHMANN, *Statistical Methods in Educational and Psychological Research*. New York: Appleton-Century-Crofts, 1954.

THE ANALYSIS OF REGRESSION

43.1 INTRODUCTION

The analysis of regression provides a test of the hypothesis of no relationship between the independent and the dependent variables in a regression equation. In the single-predictor situation, the analysis of regression is equivalent to the test of the Pearson correlation coefficient cited in Section 31.2. In the multiple regression situation, the analysis of regression provides a test of the hypothesis of no relationship between the combined independent variables and the dependent variable — it may be thought of as a test of the hypothesis of no relationship between the Y-scores and the $\hat{Y}$-scores.

In the analysis of regression, the total sum of squares (SS_t) for the dependent variable is partitioned into a sum of squares for regression (SS_{reg}) and a sum of squares for error, or sum of squares for residuals (SS_{res}) as it is often called. This is equivalent to the partitioning of the total sum of squares in the simple analysis of variance; the sum of squares for regression corresponds to the sum of squares between groups, and the sum of squares for residuals corresponds to the sum of squares within groups. These are divided by their respective degrees of freedom to obtain the mean square for regression (MS_{reg}) and the mean square for residuals (MS_{res}); the F-ratio is formed with the residual mean square in the denominator.

The coefficient of determination (R^2) has been previously identified as that proportion of the variance of the dependent variables which is accounted for by the predictor variable or variables. Stated mathematically:

$$R^2 = \frac{SS_{reg}}{SS_t} \quad \text{or} \quad SS_{reg} = R^2 \, \Sigma \, y^2$$

Similarly, the coefficient of nondetermination $(1 - R^2)$ is that proportion of the variance of the dependent variable which is not accounted for by the predictor variables:

$$(1 - R^2) = 1 - \frac{SS_{reg}}{SS_t} = \frac{SS_{res}}{SS_t} \quad \text{or} \quad SS_{res} = (1 - R^2) \, \Sigma \, y^2$$

Certain key relationships in the analysis of regression will be demonstrated with data contrived by the author to accomplish this with the greatest possible simplicity. Table 43.1 gives the raw data plus the sums of squares and sum of products needed to derive the regression equation and the coefficient of determination.

Table 43.1   Raw data for the regression equation

X_i	x_i	x_i^2	Y_i	y_i	y_i^2	$x_i y_i$
2	−3	9	2	−3	9	+9
4	−1	1	6	+1	1	−1
4	−1	1	6	+1	1	−1
5	0	0	5	0	0	0
7	+2	4	3	−2	4	−4
8	+3	9	8	+3	9	+9
30	0	24	30	0	24	+12

The column sums from Table 43.1 reveal that $SS_x = 24$, $SS_y = 24$, and $SP = 12$. With a little effort, the student should be able to confirm that the least squares regression equation is $\hat{Y}_i = .5X_i + 2.5$, and that $r = .50$, and $r^2 = .25$.

Table 43.2 presents in detail an analysis of the predicted scores. The reader is reminded of the following relationships which were used to derive the table:

$$y_i = Y_i - \bar{Y} \qquad \hat{y}_i = \hat{Y}_i - \bar{Y} \qquad e_i = Y_i - \hat{Y}_i$$

Table 43.2   Detailed analysis of the predicted scores

Y_i	y_i	y_i^2	$\hat{Y}_i$	$\hat{y}_i$	$\hat{y}_i^2$	e_i	e_i^2
2	−3	9	3.5	−1.5	2.25	−1.5	2.25
6	+1	1	4.5	−0.5	0.25	+1.5	2.25
6	+1	1	4.5	−0.5	0.25	+1.5	2.25
5	0	0	5.0	0.0	0.00	0.0	0.00
3	−2	4	6.0	+1.0	1.00	−3.0	9.00
8	+3	9	6.5	+1.5	2.25	+1.5	2.25
30	0	24	30.0	0.0	6.00	0.0	18.00

An examination of the column sums reveals that:

$$SS_{reg} = \Sigma \, (\hat{Y}_i - \bar{Y})^2 = \Sigma \, \hat{y}_i^2 = r^2 \Sigma \, y_i^2 = 6.00$$
$$SS_{res} = \Sigma \, (Y_i - \hat{Y})^2 = \Sigma \, e_i^2 = (1 - r^2) \Sigma \, y_i^2 = 18.00$$
$$SS_t = \Sigma \, (Y_i - \bar{Y})^2 = \Sigma \, y_i^2 = \Sigma \, \hat{y}_i^2 + \Sigma \, e_i^2 = 24.00$$

The student should become thoroughly familiar with the concepts and relationships expressed here. The relationships revealed in Table 43.2 and the mathematical expressions immediately above hold for multiple regression as well as the one-predictor case.

43.2 ANALYSIS OF REGRESSION WITH A SINGLE PREDICTOR

In the interest of simplicity, the formulas used in the analysis of regression are presented in deviation form. If the student has any doubts about the symbols used in these formulas, he should refer to Chapter 42. The formulas for the single-predictor case are:

$$SS_{reg} = \Sigma \, (\hat{Y} - \bar{Y})^2 = \Sigma \, \hat{y}^2 = b \, \Sigma \, xy \quad \text{with } df = 1$$
$$SS_{res} = \Sigma \, (Y - \hat{Y})^2 = \Sigma \, e^2 = SS_t - SS_{reg} \quad \text{with } df = N - 2$$
$$SS_t = \Sigma \, (Y - \bar{Y})^2 = \Sigma \, y^2 \quad \text{with } df = N - 1$$

$$MS_{reg} = \frac{SS_{reg}}{1}$$

$$MS_{res} = \frac{SS_{res}}{N - 2}$$

$$F = \frac{MS_{reg}}{MS_{res}} \quad \text{with } df = 1, (N - 2)$$

The single-predictor case is illustrated using the data of Section 14.6:

$$SS_{reg} = b \, \Sigma \, xy = 0.59(53) = 31.27$$
$$SS_{res} = SS_t - SS_{reg} = 40.00 - 31.27 = 8.73$$
$$MS_{reg} = 31.27$$
$$MS_{res} = \frac{8.73}{8} = 1.09$$

$$F = \frac{31.27}{1.09} = 28.7 \quad \text{with } df = 1, 8$$

The calculated F-ratio is significant at the .01 level, suggesting a relationship between the Y-scores and the $\hat{Y}$-scores. The investigator would conclude that the independent variable does in fact account for a portion of the variance of the dependent variable.

Occasionally, it will be convenient to calculate the F-ratio from this mathematically equivalent formula:

$$F = \frac{r^2}{\dfrac{1-r^2}{N-2}} = \frac{r^2(N-2)}{1-r^2}$$

A summary table for the analysis of regression may be constructed in the same fashion as for the analysis of variance. For the example:

Summary table for the analysis of regression			
Sources	DF	Sums of squares	Mean squares
Regression	1	31.27	31.27
Residuals	8	8.73	1.09
Totals	9	40.00	XXXXX

43.3   ANALYSIS OF REGRESSION WITH MULTIPLE PREDICTORS

The formulas for the analysis of regression with two predictors are as follows:

$$SS_{reg} = \Sigma\,(\hat{Y} - \bar{Y})^2 = b_1\,\Sigma\,x_1 y + b_2\,\Sigma\,x_2 y \qquad \text{with } df = 2$$

$$SS_{res} = \Sigma\,(Y - \hat{Y})^2 = SS_t - SS_{reg} \qquad \text{with } df = N - 3$$

$$SS_t = \Sigma\,(Y - \bar{Y})^2 = \Sigma\,y^2 \qquad \text{with } df = N - 1$$

$$MS_{reg} = \frac{SS_{reg}}{2}$$

$$MS_{res} = \frac{SS_{res}}{N-3}$$

$$F = \frac{MS_{reg}}{MS_{res}} \qquad \text{with } df = 2,\,(N-3)$$

The two-predictor case is illustrated using the data of Section 42.2:

$$SS_{reg} = 0.42(53) + 1.52(8) = 34.42$$

$$SS_{res} = 40.00 - 34.42 = 5.58$$

$$MS_{reg} = \frac{34.42}{2} = 17.21$$

$$MS_{res} = \frac{5.58}{7} = 0.80$$

$$F = \frac{17.21}{0.80} = 21.5 \quad \text{with } df = 2, 7$$

The calculated F-ratio is significant at the .01 level, suggesting a relationship between the Y-scores and the $\hat{Y}$-scores. The investigator would conclude that the two independent variables combined in a least squares regression equation do in fact account for a portion of the variance of the dependent variable.

As before, the F-ratio may be calculated from a knowledge of the coefficient of determination:

$$F = \frac{\dfrac{R^2}{2}}{\dfrac{1 - R^2}{N - 3}} = \frac{(N - 3)R^2}{2(1 - R^2)}$$

The formulas developed for use with two predictors may also be used for the single-predictor quadratic case. In this situation X_2 is set equal to X^2 in the fashion set forth in Section 42.4.

The formulas for the analysis of regression with three predictors are as follows:

$$SS_{reg} = b_1 \Sigma x_1 y + b_2 \Sigma x_2 y + b_3 \Sigma x_3 y \qquad \text{with } df = 3$$
$$SS_{res} = SS_t - SS_{reg} \qquad \text{with } df = N - 4$$
$$SS_t = \Sigma y^2 \qquad \text{with } df = N - 1$$

The mean squares and the F-ratio are formed as before. By now the pattern should be apparent and the student should be able to extend the analysis of regression to any number of predictors.

43.4   COMPARING THE EFFECTIVENESS OF TWO PREDICTION EQUATIONS

The examples cited in the previous sections were drawn from the same data, the two-predictor case using the independent and the dependent variables from the single-predictor case plus one additional independent variable. The addition of a second predictor variable in this fashion cannot account for less of the variance than the single-predictor equation, but there may be a question as to whether it significantly increases the amount of variance accounted for. Stated differently, the addition of a second independent variable in the prediction equation cannot increase the error of prediction, but a test of significance may be desirable to determine whether the two-predictor equation produces significantly less error than the single-predictor equation. The test of significance to be outlined may be used to compare regression equations with any number of predictors.

The prediction equation using all of the predictors will be referred to as the full model (*fm*) and the equation with fewer predictors as the restricted model (*rm*). An adjusted sum of squares for regression will be calculated, which is the difference between the sums of squares for regression of the full and the restricted models:

$$SS_{reg(\text{diff})} = SS_{reg(fm)} - SS_{reg(rm)}$$

The number of degrees of freedom for this difference sum of squares is the difference between the degrees of freedom for the full model and that for the restricted model. If the restricted model has one less predictor than the full model (as in our example), the difference sum of squares will have one degree of freedom. The error term for the test of significance of difference between the two models will be the mean square for residuals of the full model:

$$F = \frac{MS_{reg(\text{diff})}}{MS_{res(fm)}}$$

Comparing the one- and two-predictor models used as examples in the previous sections:

$$SS_{reg(\text{diff})} = 34.42 - 31.27 = 3.15 \qquad \text{with } df = 1$$

$$F = \frac{3.15}{0.80} = 3.9 \qquad \text{with } df = 1, 7$$

The calculated F-ratio is nonsignificant, indicating that the two-predictor equation is not significantly better than the single-predictor equation. In a practical setting, this would probably result in the adoption of the single-predictor equation.

If the coefficients of determination are known, the F-ratio may be calculated from:

$$F = \frac{\dfrac{R^2_{fm} - R^2_{rm}}{u - v}}{\dfrac{1 - R^2_{fm}}{N - u - 1}} = \frac{(N - u - 1)(R^2_{fm} - R^2_{rm})}{(u - v)(1 - R^2_{fm})}$$

where u is the number of predictors in the full model and v the number of predictors in the restricted model.

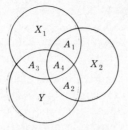

Figure 43.1 Variances held in common in the two-predictor case.

The situation is illustrated pictorially in Figure 43.1. The areas of the circles represent the variances of the three variables X_1, X_2, and Y. The areas labeled A_1, A_2, A_3, and A_4 represent the proportions of the variances which are held in common by the variables. The r-square interpretation of these areas may be expressed by:

$$r_{12}^2 = A_1 + A_4 = \text{variance common to } X_1 \text{ and } X_2$$
$$r_{1y}^2 = A_3 + A_4 = \text{variance common to } X_1 \text{ and } Y$$
$$r_{2y}^2 = A_2 + A_4 = \text{variance common to } X_2 \text{ and } Y$$
$$R_{fm}^2 = A_2 + A_3 + A_4 \quad \text{(predicting from both } X_1 \text{ and } X_2)$$
$$R_{rm}^2 = A_3 + A_4 \quad \text{(predicting from } X_1 \text{ only)}$$

The hypothesis of no difference between the full and restricted models $(\rho_{fm} - \rho_{rm} = 0)$ asks the question: Is A_2 a significant proportion of the variance of Y? The student's attention is directed to the fact that A_2 is that proportion of the variance of Y which is *uniquely* held in common with X_2. A common error is to phrase the hypothesis and complete the calculations in such a fashion that A_4 is included in the hypothesis test.

43.5 DISCUSSION

Given a prediction equation in the form:

$$\hat{Y}_i = b_1 X_1 + b_2 X_2 + c$$

which may be generalized to any number of predictor variables, and given the definition of an error of prediction such that:

$$e_i = Y_i - \hat{Y}_i$$

The value of Y_i (as opposed to $\hat{Y}_i$) is given by:

$$Y_i = b_1X_1 + b_2X_2 + c + e_i$$

The tests of significance reported in this chapter are dependent upon the following assumptions: (1) the errors are independent — this can only be assured by drawing the subjects at random from a parent population; (2) the distribution of the errors is normal for all combinations of the predictor variables; and (3) the errors are equally variable for all combinations of the predictor variables. Because of the simple additive relationships in the prediction equations, these assumptions may be transferred from the errors to the predicted scores.

The use of these hypothesis testing techniques is recommended only when the ratio of the number of subjects to the number of predictor variables is quite high (at least ten to one). Some peculiar phenomena occur when the number of subjects is too small. For example, there may be a significant relationship between X_1 and Y, but when a two-predictor equation is established by adding X_2, the relationship may become non-significant due to the loss of one degree of freedom. When the total number of variables equals or exceeds the number of subjects, all R-squares will be equal to one.

EXERCISES

43.1 An admissions officer in a small college is interested in studying some factors which influence performance during the freshman year of college. He selects a sample of 30 sophomore students for whom the following variables are a matter of record: (1) percentile rank in high school graduating class, (2) score on a college admissions test, and (3) grade-point-average in the freshman year. The regression model using both high school percentile rank and admissions test score as predictors of grade-point-average yields an R-square of 0.6188. The regression model using percentile rank only as a predictor yields an R-square of 0.4877. The regression model using the admissions test only as a predictor yields an R-square of 0.3544.

(a) Test the hypothesis of no relationship for the two-predictor model. Use the .01 level.

(b) Test the hypothesis of no difference between the two-predictor equation and the one-predictor equation using high school percentile rank. Use the .01 level.

(c) Test the hypothesis of no difference between the two-predictor equation and the one-predictor equation using admissions test score. Use the .01 level.

43.2   A psychologist is interested in studying the relationship between two different estimates of human intelligence. Both measures are recorded on a sample of 20 freshman psychology students. The quadratic model yields an R-square of 0.8133. The linear model yields an R-square of 0.6845.

(a)   Test the hypothesis of no relationship for the quadratic model. Use the .01 level.

(b)   Test the hypothesis of no difference between the two models. Use the .01 level.

(c)   What is the nature of the relationship between the two variables?

43.3   A high school counselor is seeking to predict whether or not entering freshmen in a ghetto neighborhood will finish high school. The following variables are recorded on a group of 100 entering freshmen: (1) IQ, (2) sex, (3) score on a socioeconomic rating scale. Five years later, the criterion is recorded as a binary (1 = finished high school, 0 = dropped out). A regression model using all three predictors yields an R-square of 0.7333. A regression model using only IQ and sex yields an R-square of 0.5987. A regression model using only IQ and socioeconomic rating yields an R-square of 0.4102. A regression model using only sex and socioeconomic rating yields an R-square of 0.5124.

(a)   Does IQ contribute significantly (.01 level) to the three-predictor model?

(b)   Does sex contribute significantly (.01 level) to the three-predictor model?

(c)   Does socioeconomic rating contribute significantly (.01 level) to the three-predictor model?

SELECTED REFERENCES

EZEKIAL, M., and K. E. FOX, *Methods of Correlation and Regression Analysis*, 3rd ed. New York: John Wiley, 1959.

MOOD, ALEXANDER M., "Partitioning variance in multiple regression analyses as a tool for developing learning models," *American Educational Research Journal*, **8**, 191–202 (March 1971).

OVERALL, JOHN E., and C. JAMES KLETT, *Applied Multivariate Analysis*. New York: McGraw-Hill, 1972.

PUGH, RICHARD C., "The partitioning of criterion score variance accounted for in multiple correlation," *American Educational Research Journal*, **5**, 639–646 (November 1968).

WERT, JAMES E., CHARLES O. NEIDT, and J. STANLEY AHMANN, *Statistical Methods in Educational and Psychological Research*. New York: Appleton-Century-Crofts, 1954.

44

THE MULTIPLE REGRESSION APPROACH TO ONE-WAY ANALYSIS

44.1 INTRODUCTION

The mathematical equivalence of the analysis of variance and the analysis of regression has long been known to mathematical statisticians. The mathematical statistician is likely to perceive the analysis of variance as a special case of regression analysis, which was derived to facilitate computation in a precomputer age. He is likely to describe these related phenomena in terms of a "general linear model" which very closely resembles the multiple regression equations previously discussed. It is only in recent years, however, that behavioral scientists have become aware of this relationship and of the numerous advantages which derive from approaching certain research problems with multiple regression analysis as opposed to the traditional analysis of variance. All of the parametric statistical analyses presented in this text can be duplicated with multiple regression analysis and most of these will be considered in later chapters.

There is little practical advantage in substituting the analysis of regression for one-way analysis of variance — this chapter serves primarily as an introduction to the chapters which follow. Substitution of regression analysis for two-way analysis of variance with one entry per cell is ordinarily a messy process, computationally inefficient, and conceptually difficult — it is discussed in the Selected References. The computational procedures are outlined in the next chapter, but the author does not recommend it as an alternative to the traditional analysis of variance. The multiple regression approach has distinct advantages in two-way (and higher order) analysis of variance with more than one entry per cell — the most striking advantage being the ease with which two-way analysis with disproportional cell frequencies is conceptualized and carried out. The multiple regression approach also has distinct advantages with respect to the analysis of covariance with multiple covariates and/or curvilinear regression.

The computational strategies involved will ordinarily require the availability of an electronic computer and a multiple regression computer program such as that contained in *The Funstat Package*. However, the examples and exercises given in this text may be worked with paper-and-pencil computational strategies, and many practical problems can be conveniently solved with an electronic desk calculator of the sort that is commonly available in department stores and discount houses.

44.2 ONE-WAY ANALYSIS WITH MULTIPLE REGRESSION

Consider the data from a very simple analysis of variance type problem:

Group	Scores	n_j	T_j	M_j	ΣX^2
I	0, 1, 2	3	3	1	5
II	0, 2, 4	3	6	2	20
III	2, 3, 4	3	9	3	29

$$\Sigma\Sigma X^2 = 5 + 20 + 29 = 54$$

$$\frac{T^2}{N} = \frac{18^2}{9} = 36$$

$$\frac{\Sigma T_j^2}{n_j} = \frac{9}{3} + \frac{36}{3} + \frac{81}{3} = 42$$

$$SS_b = 42 - 36 = 6$$

$$S_w = 54 - 42 = 12$$

$$SS_t = 54 - 36 = 18$$

Summary table for the analysis of variance			
Sources	df	Sums of squares	Mean squares
Between	2	6	3
Within	6	12	2
Totals	8	18	XXX

$$F = \frac{3}{2} = 1.5 \quad \text{with } df = 2, 6$$

This type of problem may be translated into a least squares regression equation, and the analysis of regression yields the same result as the analysis of variance. Two-predictor variables, X_1 and X_2, are generated to represent group membership (the independent variable in the analysis of variance). Members of Group I are awarded scores of one on X_1 and zero on X_2; members of Group II are awarded scores of zero on X_1 and one on X_2; members of Group III are awarded scores of zero on both X_1 and X_2. The data are recorded in Table 44.1 in a form convenient for completion of the analysis of regression.

Table 44.1   Data for the one-way analysis

Y	X_1	X_2	Y^2	X_1^2	X_2^2	X_1Y	X_2Y	X_1X_2
0	1	0	0	1	0	0	0	0
1	1	0	1	1	0	1	0	0
2	1	0	4	1	0	2	0	0
0	0	1	0	0	1	0	0	0
2	0	1	4	0	1	0	2	0
4	0	1	16	0	1	0	4	0
2	0	0	4	0	0	0	0	0
3	0	0	9	0	0	0	0	0
4	0	0	16	0	0	0	0	0
18	3	3	54	3	3	3	6	0

The solution of the least squares regression equation is as follows:

$$\Sigma x_1^2 = 3 - \frac{9}{9} = 2$$

$$\Sigma x_2^2 = 3 - \frac{9}{9} = 2$$

$$\Sigma y^2 = 54 - \frac{324}{9} = 18$$

$$\Sigma x_1 y = 3 - \frac{54}{9} = -3$$

$$\Sigma x_2 y = 6 - \frac{54}{9} = 0$$

$$\Sigma x_1 x_2 = 0 - \frac{9}{9} = -1$$

$$b_1 = \frac{(-3)(2) - (-1)(0)}{(2)(2) - (-1)^2} = \frac{-6}{3} = -2$$

$$b_2 = \frac{(2)(0) - (-1)(-3)}{(2)(2) - (-1)^2} = \frac{-3}{3} = -1$$

$$c = \frac{18 - 3(-2) - 1(-3)}{9} = \frac{27}{9} = 3$$

$$\hat{Y} = -2X_1 - X_2 + 3$$

To predict the score of any individual, it is only necessary to know to which of the three groups he belongs:

For members of Group I: $\hat{Y} = -2 - 0 + 3 = 1$
For members of Group II: $\hat{Y} = -0 - 1 + 3 = 2$
For members of Group III: $\hat{Y} = -0 - 0 + 3 = 3$

Notice that the predicted score for any individual is the mean for his group on the dependent variable.

A summary table for the analysis of regression may be completed as follows:

$$SS_{reg} = (-2)(-3) + (-1)(0) = 6 \quad \text{with } df = 2$$
$$SS_{res} = 18 - 6 = 12 \quad \text{with } df = 6$$

Summary table for the analysis of regression			
Sources	df	Sums of squares	Mean squares
Regression	2	6	3
Residuals	6	12	2
Totals	8	18	XXX

$$F = \frac{3}{2} = 1.5 \quad \text{with } df = 2, 6$$

The results, of course, are precisely the same as those achieved with the analysis of variance completed earlier. In other words, the analysis of regression completed in this fashion provides a test of the hypothesis of equal means. These procedures may be extended to multidimensional designs and to the analysis of covariance. When a computer is available, the use of regression analysis permits the investigator to develop complex designs using large numbers of variables. This is often desirable in research involving human subjects because of the complexity of human behavior and its determinants.

Many investigators using regression analysis as an alternative to the analysis of variance would have generated a third predictor, X_3, for this problem. Members of Groups I and II are awarded zeros on X_3, while members of Group III are awarded scores of one on X_3. Because X_3 contains no information not already included in X_1 and X_2, it does not alter the predicted scores or the analysis of regression nor does it add another degree of freedom to the sum of squares for regression. The addition of this type of variable results in an infinity of solutions which meet the least squares criterion. The original regression equation suggests one of these:

$$\hat{Y} = -2X_1 - X_2 + 0X_3 + 3$$

With a little effort, the student may be able to demonstrate that these also meet the least squares criterion:

$$Y = 0X_1 + X_2 + 2X_3 + 1$$
$$Y = X_1 + 0X_2 + X_3 + 2$$

Another interesting solution which meets the least squares criterion uses the group means as regression coefficients and eliminates the need for the constant:

$$\hat{Y} = X_1 + 2X_2 + 3X_3$$

Each of these regression equations yields the group mean as the predicted score for each individual in the group.

Table 44.2   Detailed analysis of the predicted scores

Y	y	y^2	$\hat{Y}$	$\hat{y}$	$\hat{y}^2$	e	e^2
0	−2	4	1	−1	1	−1	1
1	−1	1	1	−1	1	0	0
2	0	0	1	−1	1	+1	1
0	−2	4	2	0	0	−2	4
2	0	0	2	0	0	0	0
4	+2	4	2	0	0	+2	4
2	0	0	3	+1	1	−1	1
3	+1	1	3	+1	1	0	0
4	+2	4	3	+1	1	+1	1
18	0	18	18	0	6	0	12

Table 44.2 presents in detail an analysis of the predicted scores. The reader is reminded of the following relationships, which were used to complete the table:

$$y_i = Y_i - \bar{Y} \qquad \hat{y}_i = \hat{Y}_i - \bar{Y} \qquad e_i = Y_i - \hat{Y}_i$$

A number of interesting relationships is apparent from even a casual examination of the entries in the table. Special note should be taken of the column totals, which include the sums of squares for regression, residuals, and total.

44.3 R-SQUARES AND SUMS OF SQUARES

Typically, the computer output from a multiple regression analysis reports R-squares rather than sums of squares. The sums of squares for the one-way analysis may be calculated from a knowledge of the R-square for the full model and the sum of squares for total as follows:

$$SS_b = SS_{reg} = (R^2)(SS_t)$$
$$SS_w = SS_{res} = (1 - R^2)(SS_t)$$

Often, the computer output will report the standard deviation of the criterion rather than the sum of squares for total. The sum of squares for total may be calculated from:

$$SS_t = N\sigma^2 = (N - 1)S^2$$

Of course, the user will need to know whether the standard deviation is reported as σ or S.

The use of R^2 as a measure of association is discussed in Sections 36.7 and 42.5. In the situation under discussion, it is precisely the same as η^2.

44.4 MULTIPLE COMPARISONS WITH MULTIPLE REGRESSION

Given a full model for the three group situation with parameters (β_j's) replacing statistics (b_j's) in the regression equation, we obtain

$$\hat{Y} = \beta_1 X_1 + \beta_2 X_2 + \beta_3 X_3 + c$$

The reader's attention is directed to the data in Table 44.3, which are the same as those given earlier in the chapter, but now with variables added

Table 44.3 Data for multiple comparisons

Y	X_1	X_2	X_3	X_4	X_5	X_6	X_7	X_8
0	1	0	0	1	1	0	1	0
1	1	0	0	1	1	0	1	0
2	1	0	0	1	1	0	1	0
0	0	1	0	1	0	1	0	1
2	0	1	0	1	0	1	0	1
4	0	1	0	1	0	1	0	1
2	0	0	1	0	1	1	$\frac{1}{2}$	$\frac{1}{2}$
3	0	0	1	0	1	1	$\frac{1}{2}$	$\frac{1}{2}$
4	0	0	1	0	1	1	$\frac{1}{2}$	$\frac{1}{2}$

which were generated specifically for purposes of making multiple comparisons. Notice that the sum of $X_1 + X_2 + X_3$ is a constant for all subjects ($X_1 + X_2 + X_3 = 1$). This indicates that there is a redundant variable in the expression above and the investigator would probably prefer to derive the regression equation in the form:

$$\hat{Y} = b_1X_1 + b_2X_2 + c$$

Of course, the reader is already aware that the three group analysis of variance has only two degrees of freedom and thus should be duplicated by a two-predictor regression equation. When problems of this sort are worked on a computer, the reader should be aware that many computer programs will refuse to function if the redundant variable is included in the equation. There are other programs, however, which function very nicely despite the redundant variable and give the appropriate least squares solution with the correct number of degrees of freedom. The multiple regression program in *The Funstat Package* fits in this latter category.

The test of the overall hypothesis of equal means may be stated as:

$$\beta_1 = \beta_2 = \beta_3$$

This hypothesis may be superimposed upon the full model as an algebraic restriction and the restricted model is produced in the form:

$$\hat{Y} = \beta_1(X_1 + X_2 + X_3) + c$$

Again, the reader's attention is directed to Table 44.3 and the fact that the sum of $X_1 + X_2 + X_3$ is a constant. If a regression equation corresponding to the restricted model above were derived, R^2_{rm} would most certainly equal zero. For this reason there is no need to derive the restricted model to test the overall hypothesis of equal means.

The hypotheses for simple pairwise comparisons take the form:

$$\beta_1 = \beta_2 \qquad \beta_1 = \beta_3 \qquad \beta_2 = \beta_3$$

And the restricted model for the first of these takes the form:

$$\hat{Y} = \beta_1(X_1 + X_2) + \beta_3X_3 + c$$

The customary approach to the test of the hypothesis is to generate a new variable:

$$X_4 = X_1 + X_2$$

for each subject. And the restricted model takes on the form:

$$\hat{Y} = \beta_1 X_4 + \beta_3 X_3 + c$$

Notice, however, that the sum of $X_3 + X_4$ is a constant, signifying that there is a redundant variable in the expression. Of course, we should anticipate that this restricted model, which is to be compared to a full model having two predictors and two degrees of freedom, would have only one predictor and one degree of freedom. Thus one variable may be eliminated, and the investigator would probably prefer to derive the regression equation for the restricted model in the form:

$$\hat{Y} = bX_3 + c$$

This restricted model may be compared to the full model in the usual test of the difference between two regression equations for testing the hypothesis $\beta_1 = \beta_2$. Similarly, the variables X_5 and X_6 were generated for purposes of making the other simple, pairwise comparisons. They will not be needed, however, for it is evident that the hypothesis:

$$\beta_1 = \beta_3$$

may be tested by comparing the full model to the restricted model:

$$\hat{Y} = bX_2 + c$$

And the hypothesis:

$$\beta_2 = \beta_3$$

may be tested by comparing the full model to the restricted model:

$$\hat{Y} = bX_1 + c$$

If the number of samples exceeds three, however, it will be necessary to go through the process of placing restrictions on the full model and the generation of new variables as described.

The hypothesis for a more complex comparison may take the form:

$$\tfrac{1}{2}(\beta_1 + \beta_2) = \beta_3$$

And the restricted model takes the form:

$$\hat{Y} = \beta_1 X_1 + \beta_2 X_2 + \tfrac{1}{2}(\beta_1 + \beta_2)X_3 + c$$
$$= \beta_1(X_1 + \tfrac{1}{2}X_3) + \beta_2(X_2 + \tfrac{1}{2}X_3) + c$$

The customary procedure is to generate two new variables:

$$X_7 = X_1 + \tfrac{1}{2}X_3$$
$$X_8 = X_2 + \tfrac{1}{2}X_3$$

And the form of the restricted model becomes:

$$\hat{Y} = \beta_1 X_7 + \beta_2 X_8 + c$$

Again, the reader's attention is directed to Table 44.2 and the fact that the sum of $X_7 + X_8$ is a constant, thereby signifying the presence of a redundant variable. As before, we should have anticipated that this restricted model would have only one predictor and one degree of freedom. Thus the investigator would probably prefer to derive the equation:

$$\hat{Y} = bX_7 + c$$

This may be compared to the full model as before for the test of the null hypothesis.

   The method of comparing two regression models outlined in this text yields an F-ratio. The various multiple comparisons techniques outlined in Chapter 38 may be derived from this F-ratio in the following fashion:

1.   *Fisher's LSD.* Simply take the square root of the F-ratio. It is assumed here that the hypothesis being tested yields an F-ratio with one degree of freedom for the numerator.
2.   *Orthogonal comparisons.* Restrict the hypotheses in the fashion outlined in Section 38.3. The hypotheses may be tested either with the F-ratio or with the t-statistic, which is the square root of the F.
3.   *The Scheffé test.* Simply divide the F-ratio from the comparison of the two regression models by the quantity $k - 1$ to get the F-ratio for the Scheffé test. The Scheffé F will have $k - 1$ rather than one degree of freedom for the numerator.
4.   *Tukey's HSD.* Restrict the hypotheses to simple, pairwise comparisons from samples of equal size. Under these circumstances, $q = \sqrt{2F}$. The calculated value of q may be compared to the tabled value in Table 22.
5.   *The Dunnett test.* Restrict the hypotheses in the fashion outlined in Section 38.6. Take the square root of the F-ratio and compare to the tabled value in Table 23.

44.5 DISCUSSION

The computational procedures outlined in this chapter are mathematically equivalent to procedures treated in Chapters 36 and 38. They have the same underlying mathematical assumptions, and comments with respect to robustness and related topics apply as well to this chapter.

Generally, the computational procedures outlined in this chapter will be used only when an electronic computer is available. The calculations are conveniently and efficiently completed on a computer. Otherwise, the traditional analysis of variance approach will probably be preferred.

The multiple regression approach offers few, if any, advantages over the traditional analysis of variance approach with one-way analysis (some authors feel it is conceptually more appealing). However, the multiple regression approach does have some striking advantages with more complex analyses, which will be pointed out in the following chapters.

EXERCISES

44.1 Refer to the data of Exercise 36.1. Repeat the exercise using the analysis of regression.

44.2 Refer to the data of the example given in Sections 36.6, 38.3, 38.4, and 38.6. When the author set up this problem for the computer, he defined the following variables:

$$X_1 = 1 \text{ if Group I, zero otherwise}$$
$$X_2 = 1 \text{ if Group II, zero otherwise}$$
$$X_3 = 1 \text{ if Group III, zero otherwise}$$

The following R-squares were calculated on the computer:

(1) Predicting from X_1 and X_2, R-square $= 0.3795$.
(2) Predicting from X_1, R-square $= 0.0201$.
(3) Predicting from X_2, R-square $= 0.3396$.
(4) Predicting from X_3, R-square $= 0.2084$.

Using the procedures outlined in this chapter and in Chapter 43, select the appropriate full and restricted models and calculate the F-ratio for testing the following hypotheses:

(a) The test of the overall hypothesis of equal means.
(b) The test of equal means for Groups I and II.
(c) The test of equal means for Groups I and III.
(d) The test of equal means for Groups II and III.

(*Note:* Your answers to (b), (c), and (d) should be equal to the square of the t-statistic for Fisher's LSD.)

44.3 Refer to the answers for Exercise 44.2. Convert the answers for (b), (c), and (d) to F-ratios for the Scheffé test.

SELECTED REFERENCES

BOTTENBERG, ROBERT A., and JOE H. WARD, JR., *Applied Multiple Linear Regression.* (PRL-TDR-63-6) Lackland Air Force Base, Texas, 1963.

COHEN, JACOB, "Multiple regression as a general data-analytic system," *Psychological Bulletin,* **70,** 426–443 (1968).

FREESE, FRANK, *Linear regression methods for forest research.* U. S. Forest Service Research Paper, FPL 17, December 1964.

HURST, REX L., "Qualitative variables in regression analysis," *American Educational Research Journal,* **7,** 541–552 (November 1970).

KELLY, FRANCIS J., et. al., *Research Design in the Behavioral Sciences: Multiple Regression Approach.* Carbondale: Southern Illinois University Press, 1969.

JENNINGS, EARL, "Fixed Effects Analysis of Variance," *Multivariate Behavioral Research,* **2,** 95–108 (January 1967).

MENDENHALL, WILLIAM, *Introduction to Linear Models and the Design and Analysis of Experiments.* Belmont: Wadsworth, 1968.

ROSCOE, JOHN T., and HOWARD M. KITTLESON, "A computational strategy for inversion of correlation matrices having linear dependencies," *Journal of Experimental Education,* **41,** 51–53 (Winter 1972).

ROSCOE, JOHN T., *The Funstat Package in Fortran IV.* New York: Holt, Rinehart and Winston, 1973.

SMITH, BRANDON B., "Applied Multiple Regression: A General Research Strategy," *Journal of Industrial Education,* **7,** 21–19 (Fall 1969).

SUITS, DANIEL B., "Use of dummy variables in regression equations," *Journal of the American Statistical Association,* **52,** 548–551 (December 1957).

WALBERG, HERBERT J., "Generalized Regression Models in Educational Research," *American Educational Research Journal,* **8,** 71–91 (January 1971).

WARD, JOE H., JR., and EARL JENNINGS, *Introduction to Linear Models.* Englewood Cliffs, N. J.: Prentice-Hall, 1973.

WILLIAMS, JOHN D., "A regression approach to experimental design," *Journal of Experimental Education,* **39,** 86–90 (Fall 1970).

WILLIAMS, JOHN D., "A multiple regression approach to multiple comparisons for comparing several treatments with a control," *Journal of Experimental Education,* **39,** 94–96 (Spring 1971).

WILLIAMS, JOHN D., "Multiple comparisons in a regression approach," *Psychological Reports,* **30,** 639–647 (1972).

45

THE MULTIPLE REGRESSION
APPROACH TO
TWO-WAY ANALYSIS

45.1 INTRODUCTION

There are many compelling reasons why the analysis of variance should be
conducted with samples of equal size. It is with samples of equal size that
the investigator is most likely to achieve maximum power and robustness
plus conceptual and computational simplicity. These principles hold for
both one-way and two-way analyses, although there are additional reasons
for using equal sample sizes with two-way analyses. Rarely are situations
encountered in two-way analysis in which there is a conscious justification
for the use of unequal but proportional cell frequencies. Although the
traditional two-way analysis of variance procedure is suitable for use with
the proportional cell frequencies case, the investigator should carefully
weigh the possible additional advantages of using equal cell frequencies
before discarding this possibility and going to unequal but proportional cell
frequencies. There are numerous situations, however, in which the investigator has little option but to use sample sizes that are not only unequal
but disproportional as well. In experimental research where the sample
sizes are under the control of the investigator at the beginning of the
experiment, they are often no longer under his control at the end — a rat
dies, a child gets chicken pox, a sophomore drops out of college. In *ex
post facto* research, the sample sizes may be random variables or they may
be ultimately determined by the proportion of the subjects who choose
to respond.

   A number of suggestions which are treated in some detail has been made
for handling the analysis of data with disproportional cell frequencies.

Some have suggested that the investigator randomly discard subjects from the larger samples until all samples are of equal size; others have suggested that various techniques be used to estimate the missing scores and that these estimates be used in the analysis. Several strategies have been suggested for accommodating the analysis of variance for use with disproportional cell frequencies, but these tend to be approximations of the exact test which is presented in this chapter and will not be considered here.

The strategy outlined in this chapter will get precisely the same results as the traditional two-way analysis of variance if the cell frequencies are either equal or proportional. The reader should be aware that there are some multiple regression approaches to two-way analysis that do not have this consistency and they should be avoided. The strategy outlined here is mathematically equivalent to a procedure called the *method of fitting constants* by some authors, although this author's computational scheme is much simpler than that of some other authors and the reader may experience some difficulty in demonstrating their mathematical equivalence.

The computational strategies discussed in this chapter will ordinarily require the availability of an electronic computer and a multiple regression computer program such as that contained in *The Funstat Package*. However, the examples and exercises given in this text may be worked with paper-and-pencil computational strategies and many practical problems can be solved with an electronic desk calculator.

45.2 COMPUTATIONAL STRATEGY WITH EXAMPLE

Prior to undertaking the regression analysis, it will be necessary to organize the data in such a way that multiple regression equations can be derived for predicting the criterion from a knowledge of cell membership, row membership, and column membership. The data used to illustrate the traditional two-way analysis (see Section 40.5) have been so organized in Table 45.1. Computer programs designed for using multiple regression as an alternative to analysis of variance ordinarily have some provision for generating the binary predictor variables.

The first step in the analysis is to determine the R-square for the least squares regression equation for predicting the criterion from cell membership. In the two-by-two analysis, when the number of cells is four and the number of degrees of freedom for cells is 3, the equation will take the form:

$$\hat{Y} = b_1X_1 + b_2X_2 + b_3X_3 + c$$

Although some authors prefer to include the redundant variable (X_4), we will omit it here. The student who is adept at solving simultaneous equations

Table 45.1 Data for two-way analysis with multiple regression

Y	X_1	X_2	X_3	X_4	X_5	X_6	X_7	X_8
2	1	0	0	0	1	0	1	0
3	1	0	0	0	1	0	1	0
3	1	0	0	0	1	0	1	0
4	1	0	0	0	1	0	1	0
1	0	1	0	0	1	0	0	1
2	0	1	0	0	1	0	0	1
2	0	1	0	0	1	0	0	1
3	0	1	0	0	1	0	0	1
0	0	0	1	0	0	1	1	0
1	0	0	1	0	0	1	1	0
2	0	0	1	0	0	1	1	0
3	0	0	1	0	0	1	1	0
0	0	0	0	1	0	1	0	1
1	0	0	0	1	0	1	0	1
1	0	0	0	1	0	1	0	1
2	0	0	0	1	0	1	0	1

where:

$X_1 = 1$ if cell$_{11}$, zero otherwise
$X_2 = 1$ if cell$_{12}$, zero otherwise
$X_3 = 1$ if cell$_{21}$, zero otherwise
$X_4 = 1$ if cell$_{22}$, zero otherwise
$X_5 = X_1 + X_2 = 1$ if row 1, zero otherwise
$X_6 = X_3 + X_4 = 1$ if row 2, zero otherwise
$X_7 = X_1 + X_3 = 1$ if column 1, zero otherwise
$X_8 = X_2 + X_4 = 1$ if column 2, zero otherwise

should be able to demonstrate that the solution for the example is:

$$\hat{Y} = 2.0X_1 + X_2 + 0.5X_3 + 1.0$$

This equation yields what we shall refer to as $RSQ(1)$:

$$RSQ(1) = 0.4430$$

$RSQ(1)$ shall be used as the full model for all of our hypotheses. Some authors prefer to derive this model with the mathematically equivalent procedure of using as predictors the products of the row and column variables. That is, X_1 (representing cell$_{11}$) is the product of X_5 (representing row 1) and X_7 (representing column 1); X_2 is the product of X_5 and X_8; X_3 is the product of X_6 and X_7; and so on.

The second step in the analysis is to determine the R-square for the least squares regression equation predicting the criterion from row and column membership. In the two-by-two analysis, in which there are two rows and two columns with a total of two degrees of freedom, the equation will take the form:

$$\hat{Y} = b_1X_5 + b_2X_7 + c$$

With a little effort, the student should be able to demonstrate that the solution for the example is:

$$\hat{Y} = 1.25X_5 + 0.75X_7 + 0.875$$

This equation yields what we shall refer to as $RSQ(2)$:

$$RSQ(2) = 0.4304$$

$RSQ(2)$ shall be used as the restricted model for the interaction test.

The third step in the analysis is to determine the R-square for the least squares regression equation for predicting the criterion from row membership. In the two-by-two analysis, in which there are only two rows and one degree of freedom for rows, the equation will take the form:

$$\hat{Y} = bX_5 + c$$

The student should have no difficulty demonstrating that the solution for the example is:

$$\hat{Y} = 1.25X_5 + 1.25$$

This equation yields what we shall refer to as $RSQ(3)$:

$$RSQ(3) = 0.3164$$

$RSQ(3)$ will be used to derive a restricted model for the test for equal column means.

The fourth step in the analysis is to determine the R-square for the least squares regression equation predicting the criterion from column membership. In the two-by-two analysis, in which there are only two columns and one degree of freedom for columns, the equation will take the form:

$$\hat{Y} = bX_7 + c$$

The student should have no difficulty demonstrating that the solution for the example is:

$$\hat{Y} = 0.75X_7 + 1.5$$

This equation yields what we shall refer to as $RSQ(4)$:

$$RSQ(4) = 0.1139$$

$RSQ(4)$ will be used to derive a restricted model for the test for equal row means.

The test for interaction is completed simply by comparing $RSQ(1)$ and $RSQ(2)$ in the usual fashion for comparing two regression equations (see Section 43.4). For the example:

$$F = \frac{12(0.4430 - 0.4304)}{(1 - 0.4430)} = 0.27 \quad \text{with } df = 1, 12$$

This, of course, is precisely the same answer as that achieved with the traditional analysis of variance (Section 40.5).

The restricted model for the test of equal row means is:

$$R^2_{rm\text{-}rows} = RSQ(1) - RSQ(2) + RSQ(4) = 0.1265$$

and the test of the hypothesis is accomplished with:

$$F = \frac{12(0.4430 - 0.1265)}{(1 - 0.4430)} = 6.82 \quad \text{with } df = 1, 12$$

Again, this is the same answer as achieved with the traditional analysis of variance approach to these data.

The restricted model for the test of equal column means is:

$$R^2_{rm\text{-}cols} = RSQ(1) - RSQ(2) + RSQ(3) = 0.3290$$

And the test of the hypothesis is accomplished with:

$$F = \frac{12(0.4430 - 0.3290)}{(1 - 0.4430)} = 2.45 \quad \text{with } df = 1, 12$$

Again, the answer is the same as that achieved with the traditional analysis of variance approach to these data.

45.3 R-SQUARES AND SUMS OF SQUARES

Typically, the computer output from a multiple regression analysis reports R-squares rather than sums of squares. The sums of squares for the two-way analysis may be calculated from a knowledge of the R-squares and the sum of squares for total as follows:

$$SS_{\text{cells}} = (R^2_{fm})(SS_t) = [RSQ(1)][SS_t]$$
$$SS_r = (R^2_{fm} - R^2_{rm\text{-rows}})(SS_t) = [RSQ(2) - RSQ(4)][SS_t]$$
$$SS_c = (R^2_{fm} - R^2_{rm\text{-cols}})(SS_t) = [RSQ(2) - RSQ(3)][SS_t]$$
$$SS_{rc} = (R^2_{fm} - R^2_{rm\text{-}rc})(SS_t) = [RSQ(1) - RSQ(2)][SS_t]$$
$$SS_w = (1 - R^2_{fm})(SS_t) = [1 - RSQ(1)][SS_t]$$

In the event that the computer output does not report the sum of squares for total but does report the standard deviation for the criterion, the sum of squares for total may be calculated from:

$$SS_t = N\sigma^2 = (N - 1)S^2$$

The use of R-square as a measure of association is discussed in Sections 36.7 and 42.5. The usage reported previously generalizes to the two-way analysis, and the investigator may interpret his findings in terms of that proportion of the criterion variability which is accounted for by row membership, column membership, and by row-by-column interaction.

When the cell frequencies are equal or proportional, the computational strategies given previously will provide precisely the same sums of squares as the traditional analysis of variance. However, when the cell frequencies are disproportional, the various sums of squares cease to be orthogonal and they will not sum to the sum of squares for total.

45.4 MULTIPLE REGRESSION WITH A NESTED VARIABLE

It is a relatively simple matter to extend the computational strategies discussed thus far to the analysis of data with a nested variable. The data should be organized in a fashion similar to that in Table 45.1, although the row variables may be eliminated (assuming cells nested within columns).

The first step in the analysis is to determine the R-square for predicting the criterion from cell membership. This is done in precisely the same manner as in Section 45.3 and the result will be called $RSQ(1)$ as before. $RSQ(1)$ will serve as the full model for both hypothesis tests to follow.

The second step in the analysis is to determine the R-square for predicting the criterion from column membership. This is done in precisely the same manner as in Section 45.3 where it was called $RSQ(4)$. As there are only two equations to be derived for the analysis with nested variable, we will refer to this as $RSQ(2)$.

The restricted model for the treatment effect may be calculated from:

$$R^2_{rm\text{-cols}} = RSQ(1) - RSQ(2)$$

The restricted model for the nested factor is simply $RSQ(2)$.

There are some problems associated with the use of nested variables, where the number of cells nested within each treatment is unequal or where the cell frequencies are unequal. Refer to the Selected References for a discussion of the issues associated with the use of nested variables in statistical analysis.

45.5   TWO-WAY ANALYSIS WITH ONE ENTRY PER CELL

The data should be organized in a fashion similar to that in Table 45.1, although the cell variables may be eliminated as there is no within cells variability. A regression equation derived from a knowledge of cell membership would most assuredly yield an R-square of one as the number of predictor variables is equal to the total number of degrees of freedom available.

The first step in the analysis is to determine the R-square for predicting the criterion from row and column membership. If a repeated measurements design exists with 30 subjects and five treatments, the number of predictors will be $29 + 4 = 33$. Call this $RSQ(1)$, although it corresponds to $RSQ(2)$ in the example given in Section 45.2. $RSQ(1)$ will serve as the full model for both of the hypothesis tests which follow.

The second step in the analysis is to determine the R-square for predicting the criterion from row membership. Call this $RSQ(2)$. $RSQ(2)$ will be used as the restricted model for testing the hypothesis of equal column means.

The third step in the analysis is to determine the R-square for predicting the criterion from column membership. Call this $RSQ(3)$. $RSQ(3)$ will be used as the restricted model for testing the hypothesis of equal row means (if there is any interest in testing this hypothesis).

As a general rule, the traditional analysis of variance approach will involve considerably less effort than the multiple regression approach for data of this sort. The repeated measurements design with a substantial number of subjects is especially clumsy to handle with multiple regression.

45.6   DISCUSSION

The computational procedures outlined in this chapter are mathematically equivalent to procedures treated in Chapters 39 and 40. They have the same underlying mathematical assumptions, and comments with respect to robustness and related topics apply as well to this chapter.

The multiple regression approach to two-way analysis with more than one entry per cell offers distinct advantages over the traditional analysis of variance approach. The most obvious advantage is the lifting of the restriction to proportional cell frequencies. The multiple regression approach also provides a convenient strategy for more complex analyses involving both added factors and covariates. These topics will be introduced in the following chapter.

The procedures outlined here for two-way analysis with more than one entry per cell are applicable only to the fixed effects model. Although it is possible to modify these procedures to handle the analysis of random and mixed effects models, the author knows of no simple strategy for accomplishing this. These models are omitted from this text, but the reader will find them treated in some detail in the Selected References.

The use of the multiple regression approach to two-way analysis with only one entry per cell tends to be rather clumsy due to the large number of predictor variables required. The traditional analysis of variance approach will ordinarily be preferred.

Generally, the computational procedures outlined in this chapter will be used only when an electronic computer is available. Some small practical research problems may be completed on a desk calculator, but most will be more conveniently and efficiently handled on a large computer.

EXERCISES

45.1   Refer to the data of Exercise 40.1. The author has determined that the R-square for the full model (predicting the criterion from cell membership) is 0.2518.

(a)   Determine the value of R-square for the restricted model for testing the significance of interaction.

(b)   Determine the F-ratio and its degrees of freedom for testing the interaction effect.

(c)   Determine the value of R-square for the model in which the criterion is predicted from row membership.

(d)   Determine the value of R-square for the model in which the criterion is predicted from column membership.

(e)   Determine the value of R-square for the restricted model for testing the significance of the row effect.

(f) Determine the F-ratio and its degrees of freedom for testing the row effect.

(g) Determine the value of R-square for the restricted model for testing the significance of the column effect.

(h) Determine the F-ratio and its degrees of freedom for testing the column effect.

45.2 Given a two-by-three factorial analysis with fixed effects, the data are as follows:

Cell 11: 0, 1, 3, 4
Cell 12: 0, 1, 1, 2, 2, 3, 3, 4
Cell 13: 5, 5, 6, 8
Cell 21: 4, 6
Cell 22: 8, 10, 12, 14
Cell 23: 5, 13

The author has derived the following R-squares:

(1) Predicting from cells: 0.7784.
(2) Predicting from rows and columns: 0.6486.
(3) Predicting from rows: 0.5189.
(4) Predicting from columns: 0.1297.

(a) Determine the value of R-square for the full model.

(b) Determine the value of R-square for the restricted model for testing the significance of interaction.

(c) Determine the value of R-square for the restricted model for testing the significance of the row effect.

(d) Determine the value of R-square for the restricted model for testing the significance of the column effect.

(e) Determine the F-ratio and its degrees of freedom for testing the interaction effect.

(f) Determine the F-ratio and its degrees of freedom for testing the row effect.

(g) Determine the F-ratio and its degrees of freedom for testing the column effect.

SELECTED REFERENCES

BOTTENBERG, ROBERT A., and JOE H. WARD, JR., *Applied Multiple Linear Regression*. (PRL-TDR-63-6) Lackland Air Force Base, Texas, 1963.

COHEN, JACOB, "Multiple Regression as a General Data-Analytic System," *Psychological Bulletin*, **70**, 426–443 (1968).

HARVEY, HARVEY R., *Least-Squares Analysis of Data with Unequal Subclass Numbers*. (ARS 20-8) Agricutural Research Service, U.S. Department of Agriculture, April 1966.

JENNINGS, EARL, "Fixed Effects Analysis of Variance," *Multivariate Behavioral Research*, **2**, 95–108 (January 1967).

MENDENHALL, WILLIAM, *Introduction to Linear Models and the Design and Analysis of Experiments*. Belmont: Wadsworth, 1968.

OVERALL, J. E., and D. K. SPIEGEL, "Concerning least squares analysis of experimental data," *Psychological Bulletin*, **72**, 311–322 (1969).

OVERALL, JOHN E., and C. JAMES KLETT, *Applied Multivariate Analysis*. New York: McGraw-Hill, 1972.

ROSCOE, JOHN T., *The Funstat Package in Fortran IV*. New York: Holt, Rinehart and Winston, 1973.

SNEDECOR, GEORGE W., and WILLIAM G. COX, *Statistical Methods*, 6th ed. Ames: The Iowa State University Press, 1967.

WALBERG, HERBERT J., "Generalized Regression Models in Educational Research," *American Educational Research Journal*, **8**, 71–91 (January 1971).

WARD, JOE H., JR., and EARL JENNINGS, *Introduction to Linear Models*. Englewood Cliffs, N. J.: Prentice-Hall, 1973.

WILLIAMS, JOHN D., "Two way fixed effects analysis of variance with disproportionate cell frequencies," *Multivariate Behavioral Research*, **7**, 67–83 (January 1972).

46

THE MULTIPLE REGRESSION APPROACH TO COVARIANCE ANALYSIS

46.1 INTRODUCTION

The merits of the analysis of covariance in behavioral research have been discussed in Sections 17.6 and 41.1. The multiple regression approach to the analysis of covariance provides a convenient conceptual and computational framework for exploring the control of multiple covariates and/or covariates whose relationship to the criterion is curvilinear. When complex human behavior (such as classroom learning) is under investigation, the benefits to be derived from these options should be apparent to even the casual observer.

The statistical techniques discussed in this chapter are readily combined with those in the two preceding chapters. An experimental design involving two or more factors with or without nested variables and involving one or more covariates with or without curvilinear regression may be analyzed by the appropriate application of the principles presented in these chapters. Such research, of course, should only be undertaken with rather substantial numbers of subjects — the loss of degrees of freedom in multivariable research with small samples can appreciably affect the outcomes of the analysis.

The multiple regression approach to the analysis of covariance is surprisingly simple. Nevertheless, the computational strategies discussed in this chapter will ordinarily require the availability of an electronic computer and a multiple regression computer program such as that contained in *The Funstat Package*. However, the examples and exercises given in this chapter may be completed with paper-and-pencil computational strategies, and many practical research problems can be solved with an electronic desk calculator.

46.2   COMPUTATIONAL STRATEGY WITH EXAMPLE

Regression analysis especially lends itself to the sort of problem encountered in Chapter 41 — the analysis of covariance. The approach is surprisingly simple, even with many covariates. (1) One set of predictors is designated independent variables whose effects are to be studied; these will usually be binary variables organized in the fashion described in Section 44.2. (2) One set of predictors is designated independent variables whose effects are to be controlled; these may be either binary or continuous variables. (3) The full model includes all of the predictors. (4) The restricted model includes only those predictors whose effects are to be controlled. (5) An analysis of regression comparison of the two models yields the same results as the traditional analysis of covariance. When a computer is available, a wide variety of interesting hypotheses may be tested using this procedure.

The procedure is illustrated with the same problem used to illustrate the analysis of covariance in Chapter 41. The data have been arranged in Table 46.1 in a form convenient for the analysis of regression. The variable X_1 is the original predictor variable (X in Chapter 41) whose effects are to be controlled, and X_2 is a binary variable representing group membership (1 for Group I, 0 for Group II).

Table 46.1   Data for the analysis of covariance

Y	X_1	X_2	Y^2	X_1^2	X_2^2	X_1Y	X_2Y	X_1X_2
3	0	1	9	0	1	0	3	0
5	1	1	25	1	1	5	5	1
5	3	1	25	9	1	15	5	3
6	4	1	36	16	1	24	6	4
8	4	1	64	16	1	32	8	4
9	6	1	81	36	1	54	9	6
2	2	0	4	4	0	4	0	0
4	3	0	16	9	0	12	0	0
4	5	0	16	25	0	20	0	0
5	6	0	25	36	0	30	0	0
7	6	0	49	36	0	42	0	0
8	8	0	64	64	0	64	0	0
66	48	6	414	252	6	302	36	18

The series of calculations which follows will be used in developing both the full and restricted models:

$$\Sigma y^2 = 414 - 363 = 51$$
$$\Sigma x_1^2 = 252 - 192 = 60$$
$$\Sigma x_2^2 = 6 - 3 = 3$$
$$\Sigma x_1y = 302 - 264 = 38$$
$$\Sigma x_2y = 36 - 33 = 3$$
$$\Sigma x_1x_2 = 18 - 24 = -6$$

The full model is derived as follows:

$$b_1 = \frac{(38)(3) - (-6)(3)}{(60)(3) - (-6)^2} = \frac{132}{144} = 0.917$$

$$b_2 = \frac{(60)(3) - (-6)(38)}{(60)(3) - (-6)^2} = \frac{408}{144} = 2.833$$

$$c = \frac{(66) - 0.917(48) - 2.833(6)}{12} = \frac{4.986}{12} = 0.416$$

$$\hat{Y} = 0.917X_1 + 2.833X_2 + 0.416$$

$$SS_{reg(fm)} = 0.917(38) + 2.833(3) = 43.345 \quad \text{with } df = 2$$

$$SS_{res(fm)} = 51.000 - 43.345 = 7.655 \quad \text{with } df = 9$$

$$R^2_{fm} = \frac{43.345}{51} = 0.850$$

Now, the restricted model:

$$b = \frac{38}{60} = 0.633$$

$$c = 5.500 - 0.633(4) = 2.968$$

$$\hat{Y} = 0.633X_1 + 2.968$$

$$SS_{reg(rm)} = 0.633(38) = 24.054 \quad \text{with } df = 1$$

$$R^2_{rm} = \frac{24.054}{51} = 0.472$$

The two models may be compared using either the sums of squares-mean squares approach or the coefficients of determination approach outlined in Section 43.4. Using the sums of squares-mean squares:

$$F = \frac{\dfrac{43.345 - 24.054}{1}}{\dfrac{7.655}{9}} = \frac{19.29}{0.85} = 22.7 \quad \text{with } df = 1, 9$$

which is the same conclusion (within rounding error) which was arrived at using the traditional analysis of covariance.

46.3   EXAMPLE WITH THREE GROUPS AND THREE COVARIATES

The data recorded in Table 46.2 summarize the results from an imaginary experiment in classroom learning. The experiment involved the use of three instructional programs (the independent variable whose effects are to be studied) represented by X_1 and X_2. It also involved the use of three covariates (the independent variables whose effects are to be controlled) represented by X_3, X_4, and X_5.

Table 46.2   Data for analysis with multiple covariates

Y	X_1	X_2	X_3	X_4	X_5
26	1	0	1	82	10
37	1	0	1	99	9
27	1	0	2	94	9
51	1	0	1	112	7
46	1	0	2	105	8
58	1	0	1	120	6
49	1	0	2	116	6
60	1	0	2	140	1
40	0	1	1	84	8
58	0	1	1	119	6
44	0	1	1	100	8
50	0	1	1	108	7
55	0	1	2	114	7
59	0	1	2	123	2
38	0	1	2	93	9
41	0	1	2	101	8
50	0	0	1	102	5
54	0	0	2	101	7
60	0	0	2	125	3
45	0	0	2	94	9
59	0	0	1	117	2
43	0	0	1	98	7
55	0	0	2	102	6
42	0	0	1	92	8

where:

Y = score on a classroom achievement test
X_1 = 1 if group one, zero otherwise
X_2 = 1 if group two, zero otherwise
X_3 = 1 if male, 2 if female
X_4 = score on an IQ test
X_5 = score on a personal-social adjustment scale

The author used a computer to derive the full model:

$$\hat{Y} = -1.5612X_1 + 0.7037X_2 + 0.3808X_3 - 10.4332X_4$$
$$- 4.2234X_5 - 21.9956$$

This full model yielded an *R*-square, $R^2_{fm} = 0.8669$.

In similar fashion, a restricted model (from which the independent variable whose effects were to be studied was omitted) was derived:

$$\hat{Y} = -0.7565X_3 + 0.3721X_4 - 1.5215X_5 + 19.5485$$

This restricted model yielded an *R*-square, $R^2_{rm} = 0.7369$.

The *F*-ratio for testing the significance of differences among the adjusted means was calculated in the usual fashion for comparing two regression models:

$$F = \frac{18(0.8669 - 0.7369)}{2(1.0000 - 0.8669)} = 8.80 \quad \text{with } df = 2, 18$$

46.4   *R*-SQUARES AND SUMS OF SQUARES

Typically, the computer output from a multiple regression analysis reports *R*-squares rather than sums of squares. The adjusted sums of squares for the analysis of covariance may be calculated from a knowledge of the *R*-squares and the sum of squares for total (unadjusted) as follows:

$$SS'_{wy} = (1 - R^2_{fm})(SS_{ty})$$
$$SS'_{ty} = (1 - R^2_{rm})(SS_{ty})$$
$$SS'_{by} = SS'_{ty} - SS'_{wy}$$

For the example given in Sections 41.5 and 46.2:

$$SS'_{wy} = (1.000 - 0.850)(51.00) = 7.65$$
$$SS'_{ty} = (1.000 - 0.472)(51.00) = 29.93$$
$$SS'_{by} = 26.93 - 7.65 = 19.28$$

In the event that the computer output does not report the sum of squares for total but does report the standard deviation for the criterion, the sum of squares for total may be calculated from:

$$SS_t = N\sigma^2 = (N - 1)S^2$$

46.5 DETERMINATION OF ADJUSTED MEANS

The analysis of covariance is ordinarily interpreted with reference to adjusted criterion means (see Section 41.2 for a discussion of adjusted scores and adjusted means). However, adjusted means are not ordinarily a part of the output from a multiple regression solution to the analysis of covariance. The adjusted means are readily determined from a knowledge of sample means and regression coefficients from the full model.

The reader's attention is directed to the value of b_w calculated for the example in the traditional analysis of variance (Section 41.5, $b_w = 0.92$) and to the value of b_1 calculated for the full model in the analysis of regression (Section 46.2, $b_1 = 0.917$). In the situation in which a single covariate is being controlled, the adjusted means may be calculated in the traditional fashion (formula given in Section 41.2), using the regression coefficient for the covariate as derived for the full model as the value of b_w.

The procedure generalizes to the situation in which multiple covariates are controlled. For the situation in which there are three such covariates:

$$\bar{Y}'_i = \bar{Y}_j - b_1(\bar{X}_{1j} - M_{x1}) - b_2(\bar{X}_{2j} - M_{x2}) - b_3(\bar{X}_{3j} - M_{x3})$$

where:

$b_1, b_2,$ and b_3 are the regression coefficients for the three covariates (from the full model)

$\bar{X}_{1j}, \bar{X}_{2j},$ and $\bar{X}_{3j}$ are the means of the jth sample on the three covariates

$M_{x1}, M_{x2},$ and M_{x3} are the general means on the three covariates

For the example in Section 46.3, where the three covariates are identified as $X_3, X_4,$ and X_5:

$$\bar{Y}'_1 = 44.2500 - 0.3808(1.5000 - 1.5000) + 10.4332(108.5000 - 105.8750)$$
$$+ 4.2234(7.0000 - 6.5833) = 73.40$$

$$\bar{Y}'_2 = 48.1250 - 0.3808(1.5000 - 1.5000) + 10.4332(105.2500 - 105.8750)$$
$$+ 4.2234(6.8750 - 6.5833) = 42.84$$

$$\bar{Y}'_3 = 51.0000 - 0.3808(1.5000 - 1.5000) + 10.4332(103.8750 - 105.8750)$$
$$+ 4.2234(5.8750 - 6.5833) = 47.79$$

46.6 A TEST FOR HOMOGENEITY OF REGRESSION

A test for homogeneity of regression especially adapted to the multiple regression approach to the analysis of covariance may be developed as follows:

1. Generate a series of new variables by taking the product of each binary variable representing group membership and each covariate. For the example with two groups and one covariable (see Section 46.2), only one such variable need be generated, $X_3 = X_1X_2$. For the example with three groups and three covariates (see Section 46.3), six new variables need to be generated, $X_6 = X_1X_3$, $X_7 = X_1X_4$, $X_8 = X_1X_5$, $X_9 = X_2X_3$, $X_{10} = X_2X_4$, and $X_{11} = X_2X_5$.
2. Derive a full model which includes: (a) all of the binary variables representing group membership, (b) all of the covariates, and (c) all of the product variables generated in (1). The author derived such a model for the example with three groups and three covariates; the R-square was found to be 0.9526. Of course, the author used a computer to derive this regression equation with 11 predictors.
3. The restricted model is one in which the product variables generated above are removed from the full model. This will ordinarily be the same model used as a full model in the test of equal means.
4. Compare the two regression models in the usual fashion. For the example with three groups and three covariates:

$$F = \frac{12(0.9526 - 0.8669)}{6(1.0000 - 0.9526)} = 3.62 \quad \text{with } df = 6, 12$$

If the F-value is significant, the assumption of homogeneity of regression has been violated.

46.7 DISCUSSION

The computational procedures outlined in this chapter are mathematically equivalent to procedures treated in Chapter 41 (with the added option of multiple covariates). They have the same underlying mathematical assumptions.

The multiple regression approach to the analysis of covariance is a powerful and flexible tool for behavioral research. The option of using multiple covariates is likely to be an important asset in the study of complex human behavior. The reader is reminded that curvilinear regression is simply an adaptation of multiple regression (see Section 42.4 for a discussion of nonlinear regression and its relationship to multiple regression). The procedures outlined in this chapter are readily combined with those of the preceding chapters to analyze data from exceedingly complex designs. As the number of independent variables is increased, however, the number of subjects should be increased proportionately.

The computational strategies given here will ordinarily require the availability of an electronic computer and a multiple regression computer program of the sort contained in *The Funstat Package*. The examples and exercises may be completed with paper-and-pencil computational strategies and many practical research problems can be solved with an electronic desk calculator.

EXERCISES

46.1 Refer to the data of Exercise 41.1. Complete the analysis of covariance, using the multiple regression approach.
   (a) Derive the value of R-square for the full model.
   (b) Derive the value of R-square for the restricted model.
   (c) Calculate the F-ratio and its degrees of freedom for the test of equal means.

46.2 Refer to the data of Table 46.2. A scattergram of the data suggests a quadratic relationship between X_5 and the criterion. When the author derived a full model predicting from X_1, X_2, X_5, and the square of X_5, the model yielded an R-square of 0.8199.
   (a) Derive the R-square for a restricted model to be compared to the full model given above in the test of equal means with curvilinear control of X_5.
   (b) Calculate the F-ratio and its degrees of freedom for the test of equal means.

SELECTED REFERENCES

BOTTENBERG, ROBERT A., and JOE H. WARD, JR., *Applied Multiple Linear Regression*. (PRL-TDR-63-6) Lackland Air Force Base, Texas, 1963.

COHEN, JACOB, "Multiple regression as a general data-analytic system," *Psychological Bulletin*, **70**, 426–443 (1968).

FREESE, FRANK, *Linear Regression Methods for Forest Research*. U. S. Forest Service Research Paper, FPL 17, December 1964.

MENDENHALL, WILLIAM, *Introduction to Linear Models and the Design and Analysis of Experiments*. Belmont: Wadsworth, 1968.

OVERALL, JOHN E., "Multiple covariance analysis by the general least squares regression method," *Behavioral Science*, **17**, 313–320 (1972).

ROSCOE, JOHN T., *The Funstat Package in Fortran IV*. New York: Holt, Rinehart and Winston, 1973.

WARD, JOE H., JR., and EARL JENNINGS, *Introduction to Linear Models*. Englewood Cliffs, N. J.: Prentice-Hall, 1973.

WILSON, JAMES W., and L. RAY CARRY, "Homogeneity of regression — its rationale, computation and use," *American Educational Research Journal*, **6**, 80–90 (January 1969).

APPENDIX

Table 1   A Table of Random Numbers*

Row																	Column number															
	1	2	3	4	5	6	7	8	9	10	11	12	13	14	15	16	17	18	19	20	21	22	23	24	25	26	27	28	29	30	31	32
1	2	7	8	9	4	0	7	2	3	2	5	4	2	6	7	1	6	8	5	9	1	3	5	4	0	3	6	6	7	6	5	1
2	2	2	6	0	4	1	7	7	3	8	7	3	6	7	9	4	2	1	3	8	9	0	3	4	9	0	2	6	3	0	9	8
3	9	1	6	6	3	9	4	9	1	0	5	1	5	2	2	7	5	2	5	3	4	1	3	9	5	8	1	3	8	2	9	0
4	7	0	5	5	9	2	7	5	7	8	0	8	8	5	0	6	0	5	9	0	5	7	4	5	2	0	6	1	6	4	2	2
5	4	7	6	6	6	3	9	8	2	1	7	9	7	6	4	2	4	9	6	0	3	6	3	5	3	9	9	1	8	5	1	0
6	8	2	0	2	8	7	7	6	0	2	2	3	1	1	1	6	4	8	5	2	2	3	4	2	2	6	5	2	2	4	9	6
7	0	8	7	5	3	3	6	4	2	6	8	3	1	6	5	0	0	5	5	7	8	1	0	1	2	9	1	4	3	4	7	6
8	9	0	1	9	0	8	4	6	6	8	6	3	3	2	2	3	7	4	7	5	1	5	7	6	3	7	9	4	5	5	3	5
9	5	4	0	6	7	0	0	0	0	1	9	5	9	9	1	8	1	4	7	4	9	8	7	2	4	3	0	8	6	4	2	7
10	1	9	5	4	1	4	2	6	2	9	4	1	1	5	8	4	4	4	6	1	8	7	8	6	4	8	7	4	4	0	5	8
11	5	6	4	4	1	8	7	2	8	3	6	1	5	9	8	6	2	2	9	1	9	0	4	8	1	0	1	3	5	3	4	4
12	7	9	2	5	1	9	7	9	3	1	8	6	8	7	7	6	6	5	0	3	8	1	1	2	4	7	8	9	1	7	5	2
13	3	3	3	5	9	5	1	4	0	8	2	5	6	3	5	4	6	5	7	2	6	7	8	9	9	9	8	0	9	1	5	3
14	1	9	0	4	0	0	9	9	5	7	4	1	5	9	4	7	6	4	8	2	6	4	4	1	0	8	1	5	4	3	8	0
15	5	4	4	7	0	0	3	7	9	1	0	9	6	2	9	7	4	7	6	1	1	6	1	2	2	9	5	8	4	4	8	6
16	2	9	8	2	5	5	9	3	2	0	4	9	0	6	4	4	2	1	5	7	3	6	5	5	4	5	7	9	6	6	4	0
17	9	7	6	2	6	7	7	3	3	3	1	7	5	0	9	6	1	1	3	9	2	1	1	0	0	1	3	7	7	3	7	3
18	5	8	2	4	3	3	0	8	5	3	5	7	5	8	3	5	9	3	4	5	4	6	3	9	2	7	1	1	4	9	1	3
19	4	3	4	9	5	0	3	6	2	9	7	4	6	2	5	6	9	8	3	6	1	4	0	3	5	9	7	1	8	0	6	9
20	1	1	9	8	4	8	0	6	7	0	9	7	9	6	9	9	4	0	6	0	0	5	9	6	5	1	4	2	0	4	1	9
21	6	9	1	8	3	3	7	5	9	6	6	2	7	6	0	4	5	3	4	5	7	3	0	6	1	0	3	0	0	3	5	0
22	7	0	0	3	8	1	3	4	7	9	5	7	6	9	9	7	3	2	5	0	2	3	5	3	9	7	4	8	9	4	1	5
23	3	7	2	0	8	6	5	7	9	0	1	7	8	9	6	6	3	0	7	8	1	9	6	7	4	8	9	6	3	6	5	1

TABLE 1   411

24	2	1	5	5	4	2	3	4	0	9	4	8	2	2	7	1	3	9	2	3	0	6	5	6	0	6	6	0	0	2	2	**24**
25	2	1	3	6	4	8	9	0	7	7	1	8	3	4	0	6	9	4	7	4	2	8	2	4	9	5	6	7	0	7	3	**25**
26	8	6	2	1	7	6	1	7	1	4	4	4	5	7	4	2	2	4	6	4	3	2	2	5	7	7	3	3	9	6	6	**26**
27	3	0	7	0	9	5	6	6	4	3	3	1	6	7	9	8	1	2	7	4	6	9	0	7	5	8	1	1	2	2	5	**27**
28	2	7	2	0	2	7	7	8	2	3	2	9	5	2	5	5	9	0	1	2	2	6	3	0	8	0	4	2	7	4	0	**28**
29	9	0	1	3	1	3	5	8	8	0	9	8	0	7	0	1	3	7	8	3	9	3	2	7	9	5	7	9	6	3	3	**29**
30	6	6	3	8	6	2	5	9	4	3	1	2	7	5	8	8	4	4	0	2	8	2	8	9	6	9	2	9	8	6	2	**30**
31	9	5	7	5	1	7	1	6	5	5	5	1	2	1	6	6	5	3	8	1	7	9	3	9	5	6	2	9	2	3	8	**31**
32	0	4	4	5	1	4	5	0	4	1	8	7	3	9	3	1	3	5	9	0	5	5	4	1	5	9	9	8	7	5	8	**32**
33	1	0	9	6	1	8	2	2	9	5	6	8	3	1	3	4	9	1	9	0	7	7	3	4	4	4	8	8	7	9	8	**33**
34	2	3	0	8	6	4	5	0	9	6	0	9	1	6	6	5	4	3	9	6	8	5	8	2	3	6	6	8	8	3	7	**34**
35	1	7	8	8	7	2	9	8	5	3	0	3	1	3	2	6	1	1	4	0	5	7	5	9	8	8	6	0	5	9	8	**35**
36	0	6	8	6	2	6	2	2	7	7	4	2	3	8	5	0	5	2	4	6	7	6	1	6	7	8	8	3	8	9	0	**36**
37	9	9	8	3	4	4	7	3	8	6	2	6	2	1	0	3	7	0	3	2	6	1	8	8	6	9	3	3	8	9	0	**37**
38	8	9	3	4	8	4	4	8	3	3	7	4	0	8	4	2	0	8	7	1	9	7	6	0	8	8	6	9	4	6	2	**38**
39	4	3	5	6	8	2	1	0	2	3	0	1	4	7	0	3	5	7	1	8	1	2	7	2	9	9	5	0	9	8	0	**39**
40	2	9	3	7	0	7	6	1	2	4	1	1	8	7	0	4	6	0	2	8	9	6	9	4	0	5	5	6	0	0	4	**40**
41	6	0	4	0	5	4	5	3	8	7	8	6	1	5	0	4	8	7	4	3	8	8	9	3	8	9	4	0	2	3	9	**41**
42	9	3	5	8	7	0	1	3	5	8	9	1	2	9	2	6	2	8	4	0	5	1	8	2	2	2	3	3	7	4	0	**42**
43	3	6	3	9	1	7	2	8	7	3	1	5	8	6	8	7	1	0	7	0	4	5	5	2	8	5	2	2	4	8	0	**43**
44	4	1	8	9	1	9	0	1	4	6	4	6	0	8	6	5	4	7	8	1	4	7	2	3	6	5	0	9	5	2	4	**44**
45	7	8	4	2	9	7	4	9	0	0	6	4	3	6	1	1	7	9	8	0	5	6	5	8	0	5	5	6	8	5	3	**45**
46	1	0	1	0	6	4	1	9	5	6	7	3	1	0	9	2	5	2	6	2	4	2	8	3	3	5	5	3	2	3	9	**46**
47	6	8	7	7	7	3	0	1	8	3	3	8	5	6	4	9	7	7	5	2	4	0	1	2	1	5	0	0	7	4	0	**47**
48	8	5	7	8	2	3	6	3	1	5	7	9	5	3	2	0	6	1	4	8	0	8	8	2	7	0	4	3	4	8	0	**48**
49	7	7	1	6	1	9	0	0	1	1	6	1	2	2	0	3	5	9	7	2	5	1	0	2	5	4	8	4	5	2	4	**49**
50	0	2	5	4	6	3	3	3	5	9	8	5	2	4	9	0	9	9	3	6	1	7	9	5	5	7	7	5	8	6	3	**50**

*This table is reprinted from J. G. Peatman's and R. Schafer's "A Table of Random Numbers from Selective Service Numbers," copyright 1942 by *Jour. Psychol.*, 14, 296–297, and used by permission of the authors and editor.

Table 2 Table of Squares, Square Roots, and Reciprocals of Numbers from 1 to 1000*

N	N^2	$\sqrt{N}$	$1/N$	N	N^2	$\sqrt{N}$	$1/N$
1	1	1.0000	1.000000	41	1681	6.4031	.024390
2	4	1.4142	.500000	42	1764	6.4807	.023810
3	9	1.7321	.333333	43	1849	6.5574	.023256
4	16	2.0000	.250000	44	1936	6.6332	.022727
5	25	2.2361	.200000	45	2025	6.7082	.022222
6	36	2.4495	.166667	46	2116	6.7823	.021739
7	49	2.6458	.142857	47	2209	6.8557	.021277
8	64	2.8284	.125000	48	2304	6.9282	.020833
9	81	3.0000	.111111	49	2401	7.0000	.020408
10	100	3.1623	.100000	50	2500	7.0711	.020000
11	121	3.3166	.090909	51	2601	7.1414	.019608
12	144	3.4641	.083333	52	2704	7.2111	.019231
13	169	3.6056	.076923	53	2809	7.2801	.018868
14	196	3.7417	.071429	54	2916	7.3485	.018519
15	225	3.8730	.066667	55	3025	7.4162	.018182
16	256	4.0000	.062500	56	3136	7.4833	.017857
17	289	4.1231	.058824	57	3249	7.5498	.017544
18	324	4.2426	.055556	58	3364	7.6158	.017241
19	361	4.3589	.052632	59	3481	7.6811	.016949
20	400	4.4721	.050000	60	3600	7.7460	.016667
21	441	4.5826	.047619	61	3721	7.8102	.016393
22	484	4.6904	.045455	62	3844	7.8740	.016129
23	529	4.7958	.043478	63	3969	7.9373	.015873
24	576	4.8990	.041667	64	4096	8.0000	.015625
25	625	5.0000	.040000	65	4225	8.0623	.015385
26	676	5.0990	.038462	66	4356	8.1240	.015152
27	729	5.1962	.037037	67	4489	8.1854	.014925
28	784	5.2915	.035714	68	4624	8.2462	.014706
29	841	5.3852	.034483	69	4761	8.3066	.014493
30	900	5.4772	.033333	70	4900	8.3666	.014286
31	961	5.5678	.032258	71	5041	8.4261	.014085
32	1024	5.6569	.031250	72	5184	8.4853	.013889
33	1089	5.7446	.030303	73	5329	8.5440	.013699
34	1156	5.8310	.029412	74	5476	8.6023	.013514
35	1225	5.9161	.028571	75	5625	8.6603	.013333
36	1296	6.0000	.027778	76	5776	8.7178	.013158
37	1369	6.0828	.027027	77	5929	8.7750	.012987
38	1444	6.1644	.026316	78	6084	8.8318	.012821
39	1521	6.2450	.025641	79	6241	8.8882	.012658
40	1600	6.3246	.025000	80	6400	8.9443	.012500

*Portions of Table 2 have been reproduced from J. W. Dunlap and A. K. Kurtz. *Handbook of Statistical Nomographs, Tables, and Formulas*, World Book Company, New York (1932), by permission of the authors and publishers.

TABLE 2   413

Table 2   Table of Squares, Square Roots, and Reciprocals (*Continued*)

N	N^2	$\sqrt{N}$	$1/N$	N	N^2	$\sqrt{N}$	$1/N$
81	6561	9.0000	.012346	121	14641	11.0000	.00826446
82	6724	9.0554	.012195	122	14884	11.0454	.00819672
83	6889	9.1104	.012048	123	15129	11.0905	.00813008
84	7056	9.1652	.011905	124	15376	11.1355	.00806452
85	7225	9.2195	.011765	125	15625	11.1803	.00800000
86	7396	9.2736	.011628	126	15876	11.2250	.00793651
87	7569	9.3274	.011494	127	16129	11.2694	.00787402
88	7744	9.3808	.011364	128	16384	11.3137	.00781250
89	7921	9.4340	.011236	129	16641	11.3578	.00775194
90	8100	9.4868	.011111	130	16900	11.4018	.00769231
91	8281	9.5394	.010989	131	17161	11.4455	.00763359
92	8464	9.5917	.010870	132	17424	11.4891	.00757576
93	8649	9.6437	.010753	133	17689	11.5326	.00751880
94	8836	9.6954	.010638	134	17956	11.5758	.00746269
95	9025	9.7468	.010526	135	18225	11.6190	.00740741
96	9216	9.7980	.010417	136	18496	11.6619	.00735294
97	9409	9.8489	.010309	137	18769	11.7047	.00729927
98	9604	9.8995	.010204	138	19044	11.7473	.00724638
99	9801	9.9499	.010101	139	19321	11.7898	.00719424
100	10000	10.0000	.010000	140	19600	11.8322	.00714286
101	10201	10.0499	.00990099	141	19881	11.8743	.00709220
102	10404	10.0995	.00980392	142	20164	11.9164	.00704225
103	10609	10.1489	.00970874	143	20449	11.9583	.00699301
104	10816	10.1980	.00961538	144	20736	12.0000	.00694444
105	11025	10.2470	.00952381	145	21025	12.0416	.00689655
106	11236	10.2956	.00943396	146	21316	12.0830	.00684932
107	11449	10.3441	.00934579	147	21609	12.1244	.00680272
108	11664	10.3923	.00925926	148	21904	12.1655	.00675676
109	11881	10.4403	.00917431	149	22201	12.2066	.00671141
110	12100	10.4881	.00909091	150	22500	12.2474	.00666667
111	12321	10.5357	.00900901	151	22801	12.2882	.00662252
112	12544	10.5830	.00892857	152	23104	12.3288	.00657895
113	12769	10.6301	.00884956	153	23409	12.3693	.00653595
114	12996	10.6771	.00877193	154	23716	12.4097	.00649351
115	13225	10.7238	.00869565	155	24025	12.4499	.00645161
116	13456	10.7703	.00862069	156	24336	12.4900	.00641026
117	13689	10.8167	.00854701	157	24649	12.5300	.00636943
118	13924	10.8628	.00847458	158	24964	12.5698	.00632911
119	14161	10.9087	.00840336	159	25281	12.6095	.00628931
120	14400	10.9545	.00833333	160	25600	12.6491	.00625000

Table 2   Table of Squares, Square Roots, and Reciprocals (*Continued*)

N	N^2	$\sqrt{N}$	$1/N$	N	N^2	$\sqrt{N}$	$1/N$
161	25921	12.6886	.00621118	201	40401	14.1774	.00497512
162	26244	12.7279	.00617284	202	40804	14.2127	.00495050
163	26569	12.7671	.00613497	203	41209	14.2478	.00492611
164	26896	12.8062	.00609756	204	41616	14.2829	.00490196
165	27225	12.8452	.00606061	205	42025	14.3178	.00487805
166	27556	12.8841	.00602410	206	42436	14.3527	.00485437
167	27889	12.9228	.00598802	207	42849	14.3875	.00483092
168	28224	12.9615	.00595238	208	43264	14.4222	.00480769
169	28561	13.0000	.00591716	209	43681	14.4568	.00478469
170	28900	13.0384	.00588235	210	44100	14.4914	.00476190
171	29241	13.0767	.00584795	211	44521	14.5258	.00473934
172	29584	13.1149	.00581395	212	44944	14.5602	.00471698
173	29929	13.1529	.00578035	213	45369	14.5945	.00469484
174	30276	13.1909	.00574713	214	45796	14.6287	.00467290
175	30625	13.2288	.00571429	215	46225	14.6629	.00465116
176	30976	13.2665	.00568182	216	46656	14.6969	.00462963
177	31329	13.3041	.00564972	217	47089	14.7309	.00460829
178	31684	13.3417	.00561798	218	47524	14.7648	.00458716
179	32041	13.3791	.00558659	219	47961	14.7986	.00456621
180	32400	13.4164	.00555556	220	48400	14.8324	.00454545
181	32761	13.4536	.00552486	221	48841	14.8661	.00452489
182	33124	13.4907	.00549451	222	49284	14.8997	.00450450
183	33489	13.5277	.00546448	223	49729	14.9332	.00448430
184	33856	13.5647	.00543478	224	50176	14.9666	.00446429
185	34225	13.6015	.00540541	225	50625	15.0000	.00444444
186	34596	13.6382	.00537634	226	51076	15.0333	.00442478
187	34969	13.6748	.00534759	227	51529	15.0665	.00440529
188	35344	13.7113	.00531915	228	51984	15.0997	.00438596
189	35721	13.7477	.00529101	229	52441	15.1327	.00436681
190	36100	13.7840	.00526316	230	52900	15.1658	.00434783
191	36481	13.8203	.00523560	231	53361	15.1987	.00432900
192	36864	13.8564	.00520833	232	53824	15.2315	.00431034
193	37249	13.8924	.00518135	233	54289	15.2643	.00429185
194	37636	13.9284	.00515464	234	54756	15.2971	.00427350
195	38025	13.9642	.00512821	235	55225	15.3297	.00425532
196	38416	14.0000	.00510204	236	55696	15.3623	.00423729
197	38809	14.0357	.00507614	237	56169	15.3948	.00421941
198	39204	14.0712	.00505051	238	56644	15.4272	.00420168
199	39601	14.1067	.00502513	239	57121	15.4596	.00418410
200	40000	14.1421	.00500000	240	57600	15.4919	.00416667

TABLE 2   415

Table 2   Table of Squares, Square Roots, and Reciprocals (*Continued*)

N	N²	√N	1/N	N	N²	√N	1/N
241	58081	15.5242	.00414938	281	78961	16.7631	.00355872
242	58564	15.5563	.00413223	282	79524	16.7929	.00354610
243	59049	15.5885	.00411523	283	80089	16.8226	.00353357
244	59536	15.6205	.00409836	284	80656	16.8523	.00352113
245	60025	15.6525	.00408163	285	81225	16.8819	.00350877
246	60516	15.6844	.00406504	286	81796	16.9115	.00349650
247	61009	15.7162	.00404858	287	82369	16.9411	.00348432
248	61504	15.7480	.00403226	288	82944	16.9706	.00347222
249	62001	15.7797	.00401606	289	83521	17.0000	.00346021
250	62500	15.8114	.00400000	290	84100	17 0294	.00344828
251	63001	15.8430	.00398406	291	84681	17.0587	.00343643
252	63504	15.8745	.00396825	292	85264	17.0880	.00342466
253	64009	15.9060	.00395257	293	85849	17.1172	.00341297
254	64516	15.9374	.00393701	294	86436	17.1464	.00340136
255	65025	15.9687	.00392157	295	87025	17.1756	.00338983
256	65536	16.0000	.00390625	296	87616	17.2047	.00337838
257	66049	16.0312	.00389105	297	88209	17.2337	.00336700
258	66564	16.0624	.00387597	298	88804	17.2627	.00335570
259	67081	16.0935	.00386100	299	89401	17.2916	.00334448
260	67600	16.1245	.00384615	300	90000	17.3205	.00333333
261	68121	16.1555	.00383142	301	90601	17.3494	.00332226
262	68644	16.1864	.00381679	302	91204	17.3781	.00331126
263	69169	16.2173	.00380228	303	91809	17.4069	.00330033
264	69696	16.2481	.00378788	304	92416	17.4356	.00328947
265	70225	16.2788	.00377358	305	93025	17.4642	.00327869
266	70756	16.3095	.00375940	306	93636	17.4929	.00326797
267	71289	16.3401	.00374532	307	94249	17.5214	.00325733
268	71824	16.3707	.00373134	308	94864	17.5499	.00324675
269	72361	16.4012	.00371747	309	95481	17.5784	.00323625
270	72900	16.4317	.00370370	310	96100	17.6068	.00322581
271	73441	16.4621	.00369004	311	96721	17.6352	.00321543
272	73984	16.4924	.00367647	312	97344	17.6635	.00320513
273	74529	16.5227	.00366300	313	97969	17.6918	.00319489
274	75076	16.5529	.00364964	314	98596	17.7200	.00318471
275	75625	16.5831	.00363636	315	99225	17.7482	.00317460
276	76176	16.6132	.00362319	316	99856	17.7764	.00316456
277	76729	16.6433	.00361011	317	100489	17.8045	.00315457
278	77284	16.6733	.00359712	318	101124	17.8326	.00314465
279	77841	16.7033	.00358423	319	101761	17.8606	.00313480
280	78400	16.7332	.00357143	320	102400	17.8885	.00312500

Table 2   Table of Squares, Square Roots, and Reciprocals (*Continued*)

N	N^2	$\sqrt{N}$	1/N	N	N^2	$\sqrt{N}$	1/N
321	103041	17.9165	.00311526	361	130321	19.0000	.00277008
322	103684	17.9444	.00310559	362	131044	19.0263	.00276243
323	104329	17.9722	.00309598	363	131769	19.0526	.00275482
324	104976	18.0000	.00308642	364	132496	19.0788	.00274725
325	105625	18.0278	.00307692	365	133225	19.1050	.00273973
326	106276	18.0555	.00306748	366	133956	19.1311	.00273224
327	106929	18.0831	.00305810	367	134689	19.1572	.00272480
328	107584	18.1108	.00304878	368	135424	19.1833	.00271739
329	108241	18.1384	.00303951	369	136161	19.2094	.00271003
330	108900	18.1659	.00303030	370	136900	19.2354	.00270270
331	109561	18.1934	.00302115	371	137641	19.2614	.00269542
332	110224	18.2209	.00301205	372	138384	19.2873	.00268817
333	110889	18.2483	.00300300	373	139129	19.3132	.00268097
334	111556	18.2757	.00299401	374	139876	19.3391	.00267380
335	112225	18.3030	.00298507	375	140625	19.3649	.00266667
336	112896	18.3303	.00297619	376	141376	19.3907	.00265957
337	113569	18.3576	.00296736	377	142129	19.4165	.00265252
338	114244	18.3848	.00295858	378	142884	19.4422	.00264550
339	114921	18.4120	.00294985	379	143641	19.4679	.00263852
340	115600	18.4391	.00294118	380	144400	19.4936	.00263158
341	116281	18.4662	.00293255	381	145161	19.5192	.00262467
342	116964	18.4932	.00292398	382	145924	19.5448	.00261780
343	117649	18.5203	.00291545	383	146689	19.5704	.00261097
344	118336	18.5472	.00290698	384	147456	19.5959	.00260417
345	119025	18.5742	.00289855	385	148225	19.6214	.00259740
346	119716	18.6011	.00289017	386	148996	19.6469	.00259067
347	120409	18.6279	.00288184	387	149769	19.6723	.00258398
348	121104	18.6548	.00287356	388	150544	19.6977	.00257732
349	121801	18.6815	.00286533	389	151321	19.7231	.00257069
350	122500	18.7083	.00285714	390	152100	19.7484	.00256410
351	123201	18.7350	.00284900	391	152881	19.7737	.00255754
352	123904	18.7617	.00284091	392	153664	19.7990	.00255102
353	124609	18.7883	.00283286	393	154449	19.8242	.00254453
354	125316	18.8149	.00282486	394	155236	19.8494	.00253807
355	126025	18.8414	.00281690	395	156025	19.8746	.00253165
356	126736	18.8680	.00280899	396	156816	19.8997	.00252525
357	127449	18.8944	.00280112	397	157609	19.9249	.00251889
358	128164	18.9209	.00279330	398	158404	19.9499	.00251256
359	128881	18.9473	.00278552	399	159201	19.9750	.00250627
360	129600	18.9737	.00277778	400	160000	20.0000	.00250000

TABLE 2   417

Table 2   Table of Squares, Square Roots, and Reciprocals (*Continued*)

N	N²	√N	1/N	N	N²	√N	1/N
401	160801	20.0250	.00249377	441	194481	21.0000	.00226757
402	161604	20.0499	.00248756	442	195364	21.0238	.00226244
403	162409	20.0749	.00248139	443	196249	21.0476	.00225734
404	163216	20.0998	.00247525	444	197136	21.0713	.00225225
405	164025	20.1246	.00246914	445	198025	21.0950	.00224719
406	164836	20.1494	.00246305	446	198916	21.1187	.00224215
407	165649	20.1742	.00245700	447	199809	21.1424	.00223714
408	166464	20.1990	.00245098	448	200704	21.1660	.00223214
409	167281	20.2237	.00244499	449	201601	21.1896	.00222717
410	168100	20.2485	.00243902	450	202500	21.2132	.00222222
411	168921	20.2731	.00243309	451	203401	21.2368	.00221729
412	169744	20.2978	.00242718	452	204304	21.2603	.00221239
413	170569	20.3224	.00242131	453	205209	21.2838	.00220751
414	171396	20.3470	.00241546	454	206116	21.3073	.00220264
415	172225	20.3715	.00240964	455	207025	21.3307	.00219780
416	173056	20.3961	.00240385	456	207936	21.3542	.00219298
417	173889	20.4206	.00239808	457	208849	21.3776	.00218818
418	174724	20.4450	.00239234	458	209764	21.4009	.00218341
419	175561	20.4695	.00238663	459	210681	21.4243	.00217865
420	176400	20.4939	.00238095	460	211600	21.4476	.00217391
421	177241	20.5183	.00237530	461	212521	21.4709	.00216920
422	178084	20.5426	.00236967	462	213444	21.4942	.00216450
423	178929	20.5670	.00236407	463	214369	21.5174	.00215983
424	179776	20.5913	.00235849	464	215296	21.5407	.00215517
425	180625	20.6155	.00235294	465	216225	21.5639	.00215054
426	181476	20.6398	.00234742	466	217156	21.5870	.00214592
427	182329	20.6640	.00234192	467	218089	21.6102	.00214133
428	183184	20.6882	.00233645	468	219024	21.6333	.00213675
429	184041	20.7123	.00233100	469	219961	21.6564	.00213220
430	184900	20.7364	.00232558	470	220900	21.6795	.00212766
431	185761	20.7605	.00232019	471	221841	21.7025	.00212314
432	186624	20.7846	.00231481	472	222784	21.7256	.00211864
433	187489	20.8087	.00230947	473	223729	21.7486	.00211416
434	188356	20.8327	.00230415	474	224676	21.7715	.00210970
435	189225	20.8567	.00229885	475	225625	21.7945	.00210526
436	190096	20.8806	.00229358	476	226576	21.8174	.00210084
437	190969	20.9045	.00228833	477	227529	21.8403	.00209644
438	191844	20.9284	.00228311	478	228484	21.8632	.00209205
439	192721	20.9523	.00227790	479	229441	21.8861	.00208768
440	193600	20.9762	.00227273	480	230400	21.9089	.00208333

Table 2   Table of Squares, Square Roots, and Reciprocals (*Continued*)

N	N^2	$\sqrt{N}$	$1/N$	N	N^2	$\sqrt{N}$	$1/N$
481	231361	21.9317	.00207900	521	271441	22.8254	.00191939
482	232324	21.9545	.00207469	522	272484	22.8473	.00191571
483	233289	21.9773	.00207039	523	273529	22.8692	.00191205
484	234256	22.0000	.00206612	524	274576	22.8910	.00190840
485	235225	22.0227	.00206186	525	275625	22.9129	.00190476
486	236196	22.0454	.00205761	526	276676	22.9347	.00190114
487	237169	22.0681	.00205339	527	277729	22.9565	.00189753
488	238144	22.0907	.00204918	528	278784	22.9783	.00189394
489	239121	22.1133	.00204499	529	279841	23.0000	.00189036
490	240100	22.1359	.00204082	530	280900	23.0217	.00188679
491	241081	22.1585	.00203666	531	281961	23.0434	.00188324
492	242064	22.1811	.00203252	532	283024	23.0651	.00187970
493	243049	22.2036	.00202840	533	284089	23.0868	.00187617
494	244036	22.2261	.00202429	534	285156	23.1084	.00187266
495	245025	22.2486	.00202020	535	286225	23.1301	.00186916
496	246016	22.2711	.00201613	536	287296	23.1517	.00186567
497	247009	22.2935	.00201207	537	288369	23.1733	.00186220
498	248004	22.3159	.00200803	538	289444	23.1948	.00185874
499	249001	22.3383	.00200401	539	290521	23.2164	.00185529
500	250000	22.3607	.00200000	540	291600	23.2379	.00185185
501	251001	22.3830	.00199601	541	292681	23.2594	.00184843
502	252004	22.4054	.00199203	542	293764	23.2809	.00184502
503	253009	22.4277	.00198807	543	294849	23.3024	.00184162
504	254016	22.4499	.00198413	544	295936	23.3238	.00183824
505	255025	22.4722	.00198020	545	297025	23.3452	.00183486
506	256036	22.4944	.00197628	546	298116	23.3666	.00183150
507	257049	22.5167	.00197239	547	299209	23.3880	.00182815
508	258064	22.5389	.00196850	548	300304	23.4094	.00182482
509	259081	22.5610	.00196464	549	301401	23.4307	.00182149
510	260100	22.5832	.00196078	550	302500	23.4521	.00181818
511	261121	22.6053	.00195695	551	303601	23.4734	.00181488
512	262144	22.6274	.00195312	552	304704	23.4947	.00181159
513	263169	22.6495	.00194932	553	305809	23.5160	.00180832
514	264196	22.6716	.00194553	554	306916	23.5372	.00180505
515	265225	22.6936	.00194175	555	308025	23.5584	.00180180
516	266256	22.7156	.00193798	556	309136	23.5797	.00179856
517	267289	22.7376	.00193424	557	310249	23.6008	.00179533
518	268324	22.7596	.00193050	558	311364	23.6220	.00179211
519	269361	22.7816	.00192678	559	312481	23.6432	.00178891
520	270400	22.8035	.00192308	560	313600	23.6643	.00178571

TABLE 2   **419**

Table 2   Table of Squares, Square Roots, and Reciprocals (*Continued*)

N	N²	√N	1/N	N	N²	√N	1/N
561	314721	23.6854	.00178253	601	361201	24.5153	.00166389
562	315844	23.7065	.00177936	602	362404	24.5357	.00166113
563	316969	23.7276	.00177620	603	363609	24.5561	.00165837
564	318096	23.7487	.00177305	604	364816	24.5764	.00165563
565	319225	23.7697	.00176991	605	366025	24.5967	.00165289
566	320356	23.7908	.00176678	606	367236	24.6171	.00165017
567	321489	23.8118	.00176367	607	368449	24.6374	.00164745
568	322624	23.8328	.00176056	608	369664	24.6577	.00164474
569	323761	23.8537	.00175747	609	370881	24.6779	.00164204
570	324900	23.8747	.00175439	610	372100	24.6982	.00163934
571	326041	23.8956	.00175131	611	373321	24.7184	.00163666
572	327184	23.9165	.00174825	612	374544	24.7386	.00163399
573	328329	23.9374	.00174520	613	375769	24.7588	.00163132
574	329476	23.9583	.00174216	614	376996	24.7790	.00162866
575	330625	23.9792	.00173913	615	378225	24.7992	.00162602
576	331776	24.0000	.00173611	616	379456	24.8193	.00162338
577	332929	24.0208	.00173310	617	380689	24.8395	.00162075
578	334084	24.0416	.00173010	618	381924	24.8596	.00161812
579	335241	24.0624	.00172712	619	383161	24.8797	.00161551
580	336400	24.0832	.00172414	620	384400	24.8998	.00161290
581	337561	24.1039	.00172117	621	385641	24.9199	.00161031
582	338724	24.1247	.00171821	622	386884	24.9399	.00160772
583	339889	24.1454	.00171527	623	388129	24.9600	.00160514
584	341056	24.1661	.00171233	624	389376	24.9800	.00160256
585	342225	24.1868	.00170940	625	390625	25.0000	.00160000
586	343396	24.2074	.00170648	626	391876	25.0200	.00159744
587	344569	24.2281	.00170358	627	393129	25.0400	.00159490
588	345744	24.2487	.00170068	628	394384	25.0599	.00159236
589	346921	24.2693	.00169779	629	395641	25.0799	.00158983
590	348100	24.2899	.00169492	630	396900	25.0998	.00158730
591	349281	24.3105	.00169205	631	398161	25.1197	.00158479
592	350464	24.3311	.00168919	632	399424	25.1396	.00158228
593	351649	24.3516	.00168634	633	400689	25.1595	.00157978
594	352836	24.3721	.00168350	634	401956	25.1794	.00157729
595	354025	24.3926	.00168067	635	403225	25.1992	.00157480
596	355216	24.4131	.00167785	636	404496	25.2190	.00157233
597	356409	24.4336	.00167504	637	405769	25.2389	.00156986
598	357604	24.4540	.00167224	638	407044	25.2587	.00156740
599	358801	24.4745	.00166945	639	408321	25.2784	.00156495
600	360000	24.4949	.00166667	640	409600	25.2982	.00156250

Table 2   Table of Squares, Square Roots, and Reciprocals (*Continued*)

N	N^2	$\sqrt{N}$	$1/N$	N	N^2	$\sqrt{N}$	$1/N$
641	410881	25.3180	.00156006	681	463761	26.0960	.00146843
642	412164	25.3377	.00155763	682	465124	26.1151	.00146628
643	413449	25.3574	.00155521	683	466489	26.1343	.00146413
644	414736	25.3772	.00155280	684	467856	26.1534	.00146199
645	416025	25.3969	.00155039	685	469225	26.1725	.00145985
646	417316	25.4165	.00154799	686	470596	26.1916	.00145773
647	418609	25.4362	.00154560	687	471969	26.2107	.00145560
648	419904	25.4558	.00154321	688	473344	26.2298	.00145349
649	421201	25.4755	.00154083	689	474721	26.2488	.00145138
650	422500	25.4951	.00153846	690	476100	26.2679	.00144928
651	423801	25.5147	.00153610	691	477481	26.2869	.00144718
652	425104	25.5343	.00153374	692	478864	26.3059	.00144509
653	426409	25.5539	.00153139	693	480249	26.3249	.00144300
654	427716	25.5734	.00152905	694	481636	26.3439	.00144092
655	429025	25.5930	.00152672	695	483025	26.3629	.00143885
656	430336	25.6125	.00152439	696	484416	26.3818	.00143678
657	431649	25.6320	.00152207	697	485809	26.4008	.00143472
658	432964	25.6515	.00151976	698	487204	26.4197	.00143266
659	434281	25.6710	.00151745	699	488601	26.4386	.00143062
660	435600	25.6905	.00151515	700	490000	26.4575	.00142857
661	436921	25.7099	.00151286	701	491401	26.4764	.00142653
662	438244	25.7294	.00151057	702	492804	26.4953	.00142450
663	439569	25.7488	.00150830	703	494209	26.5141	.00142248
664	440896	25.7682	.00150602	704	495616	26.5330	.00142045
665	442225	25.7876	.00150376	705	497025	26.5518	.00141844
666	443556	25.8070	.00150150	706	498436	26.5707	.00141643
667	444889	25.8263	.00149925	707	499849	26.5895	.00141443
668	446224	25.8457	.00149701	708	501264	26.6083	.00141243
669	447561	25.8650	.00149477	709	502681	26.6271	.00141044
670	448900	25.8844	.00149254	710	504100	26.6458	.00140845
671	450241	25.9037	.00149031	711	505521	26.6646	.00140647
672	451584	25.9230	.00148810	712	506944	26.6833	.00140449
673	452929	25.9422	.00148588	713	508369	26.7021	.00140252
674	454276	25.9615	.00148368	714	509796	26.7208	.00140056
675	455625	25.9808	.00148148	715	511225	26.7395	.00139860
676	456976	26.0000	.00147929	716	512656	26.7582	.00139665
677	458329	26.0192	.00147710	717	514089	26.7769	.00139470
678	459684	26.0384	.00147493	718	515524	26.7955	.00139276
679	461041	26.0576	.00147275	719	516961	26.8142	.00139082
680	462400	26.0768	.00147059	720	518400	26.8328	.00138889

TABLE 2   421

Table 2   Table of Squares, Square Roots, and Reciprocals (*Continued*)

N	N²	√N	1/N	N	N²	√N	1/N
721	519841	26.8514	.00138696	761	579121	27.5862	.00131406
722	521284	26.8701	.00138504	762	580644	27.6043	.00131234
723	522729	26.8887	.00138313	763	582169	27.6225	.00131062
724	524176	26.9072	.00138122	764	583696	27.6405	.00130890
725	525625	26.9258	.00137931	765	585225	27.6586	.00130719
726	527076	26.9444	.00137741	766	586756	27.6767	.00130548
727	528529	26.9629	.00137552	767	588289	27.6948	.00130378
728	529984	26.9815	.00137363	768	589824	27.7128	.00130208
729	531441	27.0000	.00137174	769	591361	27.7308	.00130039
730	532900	27.0185	.00136986	770	592900	27.7489	.00129870
731	534361	27.0370	.00136799	771	594441	27.7669	.00129702
732	535824	27.0555	.00136612	772	595984	27.7849	.00129534
733	537289	27.0740	.00136426	773	597529	27.8029	.00129366
734	538756	27.0924	.00136240	774	599076	27.8209	.00129199
735	540225	27.1109	.00136054	775	600625	27.8388	.00129032
736	541696	27.1293	.00135870	776	602176	27.8568	.00128866
737	543169	27.1477	.00135685	777	603729	27.8747	.00128700
738	544644	27.1662	.00135501	778	605284	27.8927	.00128535
739	546121	27.1846	.00135318	779	606841	27.9106	.00128370
740	547600	27.2029	.00135135	780	608400	27.9285	.00128205
741	549081	27.2213	.00134953	781	609961	27.9464	.00128041
742	550564	27.2397	.00134771	782	611524	27.9643	.00127877
743	552049	27.2580	.00134590	783	613089	27.9821	.00127714
744	553536	27.2764	.00134409	784	614656	28.0000	.00127551
745	555025	27.2947	.00134228	785	616225	28.0179	.00127389
746	556516	27.3130	.00134048	786	617796	28.0357	.00127226
747	558009	27.3313	.00133869	787	619369	28.0535	.00127065
748	559504	27.3496	.00133690	788	620944	28.0713	.00126904
749	561001	27.3679	.00133511	789	622521	28.0891	.00126743
750	562500	27.3861	.00133333	790	624100	28.1069	.00126582
751	564001	27.4044	.00133156	791	625681	28.1247	.00126422
752	565504	27.4226	.00132979	792	627264	28.1425	.00126263
753	567009	27.4408	.00132802	793	628849	28.1603	.00126103
754	568516	27.4591	.00132626	794	630436	28.1780	.00125945
755	570025	27.4773	.00132450	795	632025	28.1957	.00125786
756	571536	27.4955	.00132275	796	633616	28.2135	.00125628
757	573049	27.5136	.00132100	797	635209	28.2312	.00125471
758	574564	27.5318	.00131926	798	636804	28.2489	.00125313
759	576081	27.5500	.00131752	799	638401	28.2666	.00125156
760	577600	27.5681	.00131579	800	640000	28.2843	.00125000

Table 2   Table of Squares, Square Roots, and Reciprocals (*Continued*)

N	N^2	$\sqrt{N}$	$1/N$	N	N^2	$\sqrt{N}$	$1/N$
801	641601	28.3019	.00124844	841	707281	29.0000	.00118906
802	643204	28.3196	.00124688	842	708964	29.0172	.00118765
803	644809	28.3373	.00124533	843	710649	29.0345	.00118624
804	646416	28.3549	.00124378	844	712336	29.0517	.00118483
805	648025	28.3725	.00124224	845	714025	29.0689	.00118343
806	649636	28.3901	.00124069	846	715716	29.0861	.00118203
807	651249	28.4077	.00123916	847	717409	29.1033	.00118064
808	652864	28.4253	.00123762	848	719104	29.1204	.00117925
809	654481	28.4429	.00123609	849	720801	29.1376	.00117786
810	656100	28.4605	.00123457	850	722500	29.1548	.00117647
811	657721	28.4781	.00123305	851	724201	29.1719	.00117509
812	659344	28.4956	.00123153	852	725904	29.1890	.00117371
813	660969	28.5132	.00123001	853	727609	29.2062	.00117233
814	662596	28.5307	.00122850	854	729316	29.2233	.00117096
815	664225	28.5482	.00122699	855	731025	29.2404	.00116959
816	665856	28.5657	.00122549	856	732736	29.2575	.00116822
817	667489	28.5832	.00122399	857	734449	29.2746	.00116686
818	669124	28.6007	.00122249	858	736164	29.2916	.00116550
819	670761	28.6182	.00122100	859	737881	29.3087	.00116414
820	672400	28.6356	.00121951	860	739600	29.3258	.00116279
821	674041	28.6531	.00121803	861	741321	29.3428	.00116144
822	675684	28.6705	.00121655	862	743044	29.3598	.00116009
823	677329	28.6880	.00121507	863	744769	29.3769	.00115875
824	678976	28.7054	.00121359	864	746496	29.3939	.00115741
825	680625	28.7228	.00121212	865	748225	29.4109	.00115607
826	682276	28.7402	.00121065	866	749956	29.4279	.00115473
827	683929	28.7576	.00120919	867	751689	29.4449	.00115340
828	685584	28.7750	.00120773	868	753424	29.4618	.00115207
829	687241	28.7924	.00120627	869	755161	29.4788	.00115075
830	688900	28.8097	.00120482	870	756900	29.4958	.00114943
831	690561	28.8271	.00120337	871	758641	29.5127	.00114811
832	692224	28.8444	.00120192	872	760384	29.5296	.00114679
833	693889	28.8617	.00120048	873	762129	29.5466	.00114548
834	695556	28.8791	.00119904	874	763876	29.5635	.00114416
835	697225	28.8964	.00119760	875	765625	29.5804	.00114286
836	698896	28.9137	.00119617	876	767376	29.5973	.00114155
837	700569	28.9310	.00119474	877	769129	29.6142	.00114025
838	702244	28.9482	.00119332	878	770884	29.6311	.00113895
839	703921	28.9655	.00119190	879	772641	29.6479	.00113766
840	705600	28.9828	.00119048	880	774400	29.6648	.00113636

TABLE 2   423

Table 2   Table of Squares, Square Roots, and Reciprocals (*Continued*)

N	N^2	$\sqrt{N}$	$1/N$	N	N^2	$\sqrt{N}$	$1/N$
881	776161	29.6816	.00113507	921	848241	30.3480	.00108578
882	777924	29.6985	.00113379	922	850084	30.3645	.00108460
883	779689	29.7153	.00113250	923	851929	30.3809	.00108342
884	781456	29.7321	.00113122	924	853776	30.3974	.00108225
885	783225	29.7489	.00112994	925	855625	30.4138	.00108108
886	784996	29.7658	.00112867	926	857476	30.4302	.00107991
887	786769	29.7825	.00112740	927	859329	30.4467	.00107875
888	788544	29.7993	.00112613	928	861184	30.4631	.00107759
889	790321	29.8161	.00112486	929	863041	30.4795	.00107643
890	792100	29.8329	.00112360	930	864900	30.4959	.00107527
891	793881	29.8496	.00112233	931	866761	30.5123	.00107411
892	795664	29.8664	.00112108	932	868624	30.5287	.00107296
893	797449	29.8831	.00111982	933	870489	30.5450	.00107181
894	799236	29.8998	.00111857	934	872356	30.5614	.00107066
895	801025	29.9166	.00111732	935	874225	30.5778	.00106952
896	802816	29.9333	.00111607	936	876096	30.5941	.00106838
897	804609	29.9500	.00111483	937	877969	30.6105	.00106724
898	806404	29.9666	.00111359	938	879844	30.6268	.00106610
899	808201	29.9833	.00111235	939	881721	30.6431	.00106496
900	810000	30.0000	.00111111	940	883600	30.6594	.00106383
901	811801	30.0167	.00110988	941	885481	30.6757	.00106270
902	813604	30.0333	.00110865	942	887364	30.6920	.00106157
903	815409	30.0500	.00110742	943	889249	30.7083	.00106045
904	817216	30.0666	.00110619	944	891136	30.7246	.00105932
905	819025	30.0832	.00110497	945	893025	30.7409	.00105820
906	820836	30.0998	.00110375	946	894916	30.7571	.00105708
907	822649	30.1164	.00110254	947	896809	30.7734	.00105597
908	824464	30.1330	.00110132	948	898704	30.7896	.00105485
909	826281	30.1496	.00110011	949	900601	30.8058	.00105374
910	828100	30.1662	.00109890	950	902500	30.8221	.00105263
911	829921	30.1828	.00109769	951	904401	30.8383	.00105152
912	831744	30.1993	.00109649	952	906304	30.8545	.00105042
913	833569	30.2159	.00109529	953	908209	30.8707	.00104932
914	835396	30.2324	.00109409	954	910116	30.8869	.00104822
915	837225	30.2490	.00109290	955	912025	30.9031	.00104712
916	839056	30.2655	.00109170	956	913936	30.9192	.00104603
917	840889	30.2820	.00109051	957	915849	30.9354	.00104493
918	842724	30.2985	.00108932	958	917764	30.9516	.00104384
919	844561	30.3150	.00108814	959	919681	30.9677	.00104275
920	846400	30.3315	.00108696	960	921600	30.9839	.00104167

Table 2   Table of Squares, Square Roots, and Reciprocals (*Continued*)

N	N^2	$\sqrt{N}$	$1/N$	N	N^2	$\sqrt{N}$	$1/N$
961	923521	31.0000	.00104058	981	962361	31.3209	.00101937
962	925444	31.0161	.00103950	982	964324	31.3369	.00101833
963	927369	31.0322	.00103842	983	966289	31.3528	.00101729
964	929296	31.0483	.00103734	984	968256	31.3688	.00101626
965	931225	31.0644	.00103627	985	970225	31.3847	.00101523
966	933156	31.0805	.00103520	986	972196	31.4006	.00101420
967	935089	31.0966	.00103413	987	974169	31.4166	.00101317
968	937024	31.1127	.00103306	988	976144	31.4325	.00101215
969	938961	31.1288	.00103199	989	978121	31.4484	.00101112
970	940900	31.1448	.00103093	990	980100	31.4643	.00101010
971	942841	31.1609	.00102987	991	982081	31.4802	.00100908
972	944784	31.1769	.00102881	992	984064	31.4960	.00100806
973	946729	31.1929	.00102775	993	986049	31.5119	.00100705
974	948676	31.2090	.00102669	994	988036	31.5278	.00100604
975	950625	31.2250	.00102564	995	990025	31.5436	.00100503
976	952576	31.2410	.00102459	996	992016	31.5595	.00100402
977	954529	31.2570	.00102354	997	994009	31.5753	.00100301
978	956484	31.2730	.00102249	998	996004	31.5911	.00100200
979	958441	31.2890	.00102145	999	998001	31.6070	.00100100
980	960400	31.3050	.00102041	1000	1000000	31.6228	.00100000

TABLE 3   425

Table 3   Areas under the Normal Curve

Proportion of Total Area Under the Normal Curve Between Mean Ordinate and Ordinate at Given z Distance from the Mean

$\frac{x}{\sigma}$ or z	Second decimal place in z									
	.00	.01	.02	.03	.04	.05	.06	.07	.08	.09
.0	.0000	.0040	.0080	.0120	.0160	.0199	.0239	.0279	.0319	.0359
.1	.0398	.0438	.0478	.0517	.0557	.0596	.0636	.0675	.0714	.0753
.2	.0793	.0832	.0871	.0910	.0948	.0987	.1026	.1064	.1103	.1141
.3	.1179	.1217	.1255	.1293	.1331	.1368	.1406	.1443	.1480	.1517
.4	.1554	.1591	.1628	.1664	.1700	.1736	.1772	.1808	.1844	.1879
.5	.1915	.1950	.1985	.2019	.2054	.2088	.2123	.2157	.2190	.2224
.6	.2257	.2291	.2324	.2357	.2389	.2422	.2454	.2486	.2517	.2549
.7	.2580	.2611	.2642	.2673	.2704	.2734	.2764	.2794	.2823	.2852
.8	.2881	.2910	.2939	.2967	.2995	.3023	.3051	.3078	.3106	.3133
.9	.3159	.3186	.3212	.3238	.3264	.3289	.3315	.3340	.3365	.3389
1.0	.3413	.3438	.3461	.3485	.3508	.3531	.3554	.3577	.3599	.3621
1.1	.3643	.3665	.3686	.3708	.3729	.3749	.3770	.3790	.3810	.3830
1.2	.3849	.3869	.3888	.3907	.3925	.3944	.3962	.3980	.3997	.4015
1.3	.4032	.4049	.4066	.4082	.4099	.4115	.4131	.4147	.4162	.4177
1.4	.4192	.4207	.4222	.4236	.4251	.4265	.4279	.4292	.4306	.4319
1.5	.4332	.4345	.4357	.4370	.4382	.4394	.4406	.4418	.4429	.4441
1.6	.4452	.4463	.4474	.4484	.4495	.4505	.4515	.4525	.4535	.4545
1.7	.4554	.4564	.4573	.4582	.4591	.4599	.4608	.4616	.4625	.4633
1.8	.4641	.4649	.4656	.4664	.4671	.4678	.4686	.4693	.4699	.4706
1.9	.4713	.4719	.4726	.4732	.4738	.4744	.4750	.4756	.4761	.4767
2.0	.4772	.4778	.4783	.4788	.4793	.4798	.4803	.4808	.4812	.4817
2.1	.4821	.4826	.4830	.4834	.4838	.4842	.4846	.4850	.4854	.4857
2.2	.4861	.4864	.4868	.4871	.4875	.4878	.4881	.4884	.4887	.4890
2.3	.4893	.4896	.4898	.4901	.4904	.4906	.4909	.4911	.4913	.4916
2.4	.4918	.4920	.4922	.4925	.4927	.4929	.4931	.4932	.4934	.4936
2.5	.4938	.4940	.4941	.4943	.4945	.4946	.4948	.4949	.4951	.4952
2.6	.4953	.4955	.4956	.4957	.4959	.4960	.4961	.4962	.4963	.4964
2.7	.4965	.4966	.4967	.4968	.4969	.4970	.4971	.4972	.4973	.4974
2.8	.4974	.4975	.4976	.4977	.4977	.4978	.4979	.4979	.4980	.4981
2.9	.4981	.4982	.4982	.4983	.4984	.4984	.4985	.4985	.4986	.4986
3.0	.4987	.4987	.4987	.4988	.4988	.4989	.4989	.4989	.4990	.4990
3.1	.4990	.4991	.4991	.4991	.4992	.4992	.4992	.4992	.4993	.4993
3.2	.4993	.4993	.4994	.4994	.4994	.4994	.4994	.4995	.4995	.4995
3.3	.4995	.4995	.4995	.4996	.4996	.4996	.4996	.4996	.4996	.4997
3.4	.4997	.4997	.4997	.4997	.4997	.4997	.4997	.4997	.4997	.4998
3.5	.4998									
4.0	.49997									
4.5	.499997									
5.0	.4999997									

Table 4   Ordinates of the Normal Curve

p	p	y	pq/y	p	p	y	pq/y
.01	.99	.027	.372	.26	.74	.324	.593
.02	.98	.048	.405	.27	.73	.331	.596
.03	.97	.068	.428	.28	.72	.337	.599
.04	.96	.086	.446	.29	.71	.342	.602
.05	.95	.103	.461	.30	.70	.348	.604
.06	.94	.119	.474	.31	.69	.353	.606
.07	.93	.134	.485	.32	.68	.358	.609
.08	.92	.149	.495	.33	.67	.362	.612
.09	.91	.162	.504	.34	.66	.366	.612
.10	.90	.176	.513	.35	.65	.370	.614
.11	.89	.188	.521	.36	.64	.374	.616
.12	.88	.200	.528	.37	.63	.378	.617
.13	.87	.212	.535	.38	.62	.381	.619
.14	.86	.223	.541	.39	.61	.384	.620
.15	.85	.233	.547	.40	.60	.386	.621
.16	.84	.243	.552	.41	.59	.389	.622
.17	.83	.253	.558	.42	.58	.391	.623
.18	.82	.262	.563	.43	.57	.393	.624
.19	.81	.271	.567	.44	.56	.394	.625
.20	.80	.280	.572	.45	.55	.396	.625
.21	.79	.288	.576	.46	.54	.397	.626
.22	.78	.296	.580	.47	.53	.398	.626
.23	.77	.304	.583	.48	.52	.398	.626
.24	.76	.311	.587	.49	.51	.399	.627
.25	.75	.318	.590	.50	.50	.399	.627

This table prepared by the author for this volume.

TABLE 5   427

Table 5   Tetrachoric Correlation Coefficient from Phi Coefficient

Phi	r_{tet}	Phi	r_{tet}	Phi	r_{tet}	Phi	r_{tet}	Phi	r_{tet}
.000	.000	.200	.309	.400	.588	.600	.809	.800	.951
.005	.008	.205	.317	.405	.594	.605	.814	.805	.953
.010	.016	.210	.324	.410	.600	.610	.818	.810	.956
.015	.024	.215	.331	.415	.607	.615	.823	.815	.958
.020	.031	.220	.339	.420	.613	.620	.827	.820	.960
.025	.039	.225	.346	.425	.619	.625	.832	.825	.963
.030	.047	.230	.354	.430	.625	.630	.836	.830	.965
.035	.055	.235	.361	.435	.631	.635	.840	.835	.967
.040	.063	.240	.368	.440	.637	.640	.844	.840	.969
.045	.071	.245	.375	.445	.643	.645	.849	.845	.971
.050	.079	.250	.383	.450	.649	.650	.853	.850	.972
.055	.086	.255	.390	.455	.655	.655	.857	.855	.974
.060	.094	.260	.397	.460	.661	.660	.861	.860	.976
.065	.102	.265	.404	.465	.667	.665	.865	.865	.978
.070	.110	.270	.412	.470	.673	.670	.869	.870	.979
.075	.118	.275	.419	.475	.679	.675	.873	.875	.981
.080	.125	.280	.426	.480	.685	.680	.876	.880	.982
.085	.133	.285	.433	.485	.690	.685	.880	.885	.984
.090	.141	.290	.440	.490	.696	.690	.884	.890	.985
.095	.149	.295	.447	.495	.702	.695	.887	.895	.986
.100	.156	.300	.454	.500	.707	.700	.891	.900	.988
.105	.164	.305	.461	.505	.713	.705	.895	.905	.989
.110	.172	.310	.468	.510	.718	.710	.898	.910	.990
.115	.180	.315	.475	.515	.724	.715	.902	.915	.991
.120	.187	.320	.482	.520	.729	.720	.905	.920	.992
.125	.195	.325	.489	.525	.734	.725	.908	.925	.993
.130	.203	.330	.496	.530	.740	.730	.911	.930	.994
.135	.211	.335	.502	.535	.745	.735	.915	.935	.995
.140	.218	.340	.509	.540	.750	.740	.918	.940	.996
.145	.226	.345	.516	.545	.755	.745	.921	.945	.996
.150	.234	.350	.523	.550	.760	.750	.924	.950	.997
.155	.241	.355	.529	.555	.766	.755	.927	.955	.998
.160	.249	.360	.536	.560	.771	.760	.930	.960	.998
.165	.256	.365	.542	.565	.776	.765	.933	.965	.999
.170	.264	.370	.549	.570	.780	.770	.935	.970	.999
.175	.271	.375	.556	.575	.785	.775	.938	.975	.999
.180	.279	.380	.562	.580	.790	.780	.941	.980	.999
.185	.287	.385	.569	.585	.795	.785	.944	.985	1.00
.190	.294	.390	.575	.590	.800	.790	.946	.990	1.00
.195	.302	.395	.581	.595	.804	.795	.949	.995	1.00

This table was compiled by the author.

Table 6 Coefficients of the Binomial Distribution

N	$\binom{N}{0}$	$\binom{N}{1}$	$\binom{N}{2}$	$\binom{N}{3}$	$\binom{N}{4}$	$\binom{N}{5}$	$\binom{N}{6}$	$\binom{N}{7}$	$\binom{N}{8}$	$\binom{N}{9}$	$\binom{N}{10}$
0	1										
1	1	1									
2	1	2	1								
3	1	3	3	1							
4	1	4	6	4	1						
5	1	5	10	10	5	1					
6	1	6	15	20	15	6	1				
7	1	7	21	35	35	21	7	1			
8	1	8	28	56	70	56	28	8	1		
9	1	9	36	84	126	126	84	36	9	1	
10	1	10	45	120	210	252	210	120	45	10	1
11	1	11	55	165	330	462	462	330	165	55	11
12	1	12	66	220	495	792	924	792	495	220	66
13	1	13	78	286	715	1287	1716	1716	1287	715	286
14	1	14	91	364	1001	2002	3003	3432	3003	2002	1001
15	1	15	105	455	1365	3003	5005	6435	6435	5005	3003
16	1	16	120	560	1820	4368	8008	11440	12870	11440	8008
17	1	17	136	680	2380	6188	12376	19448	24310	24310	19448
18	1	18	153	816	3060	8568	18564	31824	43758	48620	43758
19	1	19	171	969	3876	11628	27132	50388	75582	92378	92378
20	1	20	190	1140	4845	15504	38760	77520	125970	167960	184756

This table is reprinted from Sydney Siegel, *Nonparametric Statistics for the Behavioral Sciences*, (New York: McGraw-Hill Book Company, 1956), p. 288.

TABLE 7   429

Table 7   Distribution of *t* for Given Probability Levels

df	Level of significance for one-tailed test					
	.10	.05	.025	.01	.005	.0005
	Level of significance for two-tailed test					
	.20	.10	.05	.02	.01	.001
1	3.078	6.314	12.706	31.821	63.657	636.619
2	1.886	2.920	4.303	6.965	9.925	31.598
3	1.638	2.353	3.182	4.541	5.841	12.941
4	1.533	2.132	2.776	3.747	4.604	8.610
5	1.476	2.015	2.571	3.365	4.032	6.859
6	1.440	1.943	2.447	3.143	3.707	5.959
7	1.415	1.895	2.365	2.998	3.499	5.405
8	1.397	1.860	2.306	2.896	3.355	5.041
9	1.383	1.833	2.262	2.821	3.250	4.781
10	1.372	1.812	2.228	2.764	3.169	4.587
11	1.363	1.796	2.201	2.718	3.106	4.437
12	1.356	1.782	2.179	2.681	3.055	4.318
13	1.350	1.771	2.160	2.650	3.012	4.221
14	1.345	1.761	2.145	2.624	2.977	4.140
15	1.341	1.753	2.131	2.602	2.947	4.073
16	1.337	1.746	2.120	2.583	2.921	4.015
17	1.333	1.740	2.110	2.567	2.898	3.965
18	1.330	1.734	2.101	2.552	2.878	3.992
19	1.328	1.729	2.093	2.539	2.861	3.883
20	1.325	1.725	2.086	2.528	2.845	3.850
21	1.323	1.721	2.080	2.518	2.831	3.819
22	1.321	1.717	2.074	2.508	2.819	3.792
23	1.319	1.714	2.069	2.500	2.807	3.767
24	1.318	1.711	2.064	2.492	2.797	3.745
25	1.316	1.708	2.060	2.485	2.787	3.725
26	1.315	1.706	2.056	2.479	2.779	3.707
27	1.314	1.703	2.052	2.473	2.771	3.690
28	1.313	1.701	2.048	2.467	2.763	3.674
29	1.311	1.699	2.045	2.462	2.756	3.659
30	1.310	1.697	2.042	2.457	2.750	3.646
40	1.303	1.684	2.021	2.423	2.704	3.551
60	1.296	1.671	2.000	2.390	2.660	3.460
120	1.289	1.658	1.980	2.358	2.617	3.373
∞	1.282	1.645	1.960	2.326	2.576	3.291

* This table is abridged from Table III of R. A. Fisher and F. Yates: *Statistical Tables for Biological, Agricultural, and Medical Research*, published by Oliver and Boyd, Ltd., Edinburgh, by permission of the authors and publishers.

Table 8   Mann-Whitney U
One-Tailed Test at .05 Level; Two-Tailed Test at .10 Level

m	1	2	3	4	5	6	7	8	9	10	11	12	13	14	15	16	17	18	19	20
1	—																			
2	—	—																		
3	—	—	0																	
4	—	—	0	1																
5	—	0	1	2	4															
6	—	0	2	3	5	7														
7	—	0	2	4	6	8	11													
8	—	1	3	5	8	10	13	15												
9	—	1	4	6	9	12	15	18	21											
10	—	1	4	7	11	14	17	20	24	27										
11	—	1	5	8	12	16	19	23	27	31	34									
12	—	2	5	9	13	17	21	26	30	34	38	42								
13	—	2	6	10	15	19	24	28	33	37	42	47	51							
14	—	3	7	11	16	21	26	31	36	41	46	51	56	61						
15	—	3	7	12	18	23	28	33	39	44	50	55	61	66	72					
16	—	3	8	14	19	25	30	36	42	48	54	60	65	71	77	83				
17	—	3	9	15	20	26	33	39	45	51	57	64	70	77	83	89	96			
18	—	4	9	16	22	28	35	41	48	55	61	68	75	82	88	95	102	109		
19	0	4	10	17	23	30	37	44	51	58	65	72	80	87	94	101	109	116	123	
20	0	4	11	18	25	32	39	47	54	62	69	77	84	92	100	107	115	123	130	138
21	0	5	11	19	26	34	41	49	57	65	73	81	89	97	105	113	121	130	138	146
22	0	5	12	20	28	36	44	52	60	68	77	85	94	102	111	119	128	136	145	154
23	0	5	13	21	29	37	46	54	63	72	81	90	98	107	116	125	134	143	152	161
24	0	6	13	22	30	39	48	57	66	75	85	94	103	113	122	131	141	150	160	162
25	0	6	14	23	32	41	50	60	69	79	89	98	108	118	128	137	147	157	167	177
26	0	6	15	24	33	43	53	62	72	82	92	103	113	123	133	143	154	164	174	185
27	0	7	15	25	35	45	55	65	75	86	96	107	117	128	139	149	160	171	182	192
28	0	7	16	26	36	46	57	68	78	89	100	111	122	133	144	156	167	178	189	200
29	0	7	17	27	38	48	59	70	82	93	104	116	127	138	150	162	173	185	196	208
30	0	7	17	28	39	50	61	73	85	96	108	120	132	144	156	168	180	192	204	216
31	0	8	18	29	40	52	64	76	88	100	112	124	136	149	161	174	186	199	211	224
32	0	8	19	30	42	54	66	78	91	103	116	128	141	154	167	180	193	206	218	231
33	0	8	19	31	43	56	68	81	94	107	120	133	146	159	172	186	199	212	226	239
34	0	9	20	32	45	57	70	84	97	110	124	137	151	164	178	192	206	219	233	247
35	0	9	21	33	46	59	73	86	100	114	128	141	156	170	184	198	212	226	241	255
36	0	9	21	34	48	61	75	89	103	117	131	146	160	175	189	204	219	233	248	263
37	0	10	22	35	49	63	77	91	106	121	135	150	165	180	195	210	225	240	255	271
38	0	10	23	36	50	65	79	94	109	124	139	154	170	185	201	216	232	247	263	278
39	1	10	23	38	52	67	82	97	112	128	143	159	175	190	206	222	238	254	270	286
40	1	11	24	39	53	68	84	99	115	131	147	163	179	196	212	228	245	261	278	294

This table is reprinted from *American Statistical Association Journal* (September 1964), pp. 927–932.

TABLE 8   431

Table 8   Mann-Whitney U (*Continued*)
One-Tailed Test at .025 Level; Two-Tailed Test at .05 Level

m	1	2	3	4	5	6	7	8	9	10	11	12	13	14	15	16	17	18	19	20
1	—																			
2	—	—																		
3	—	—	—																	
4	—	—	—	0																
5	—	—	0	1	2															
6	—	—	1	2	3	5														
7	—	—	1	3	5	6	8													
8	—	0	2	4	6	8	10	13												
9	—	0	2	4	7	10	12	15	17											
10	—	0	3	5	8	11	14	17	20	23										
11	—	0	3	6	9	13	16	19	23	26	30									
12	—	1	4	7	11	14	18	22	26	29	33	37								
13	—	1	4	8	12	16	20	24	28	33	37	41	45							
14	—	1	5	9	13	17	22	26	31	36	40	45	50	55						
15	—	1	5	10	14	19	24	29	34	39	44	49	54	59	64					
16	—	1	6	11	15	21	26	31	37	42	47	53	59	64	70	75				
17	—	2	6	11	17	22	28	34	39	45	51	57	63	69	75	81	87			
18	—	2	7	12	18	24	30	36	42	48	55	61	67	74	80	86	93	99		
19	—	2	7	13	19	25	32	38	45	52	58	65	72	78	85	92	99	106	113	
20	—	2	8	14	20	27	34	41	48	55	62	69	76	83	90	98	105	112	119	127
21	—	3	8	15	22	29	36	43	50	58	65	73	80	88	96	103	111	119	126	134
22	—	3	9	16	23	30	38	45	53	61	69	77	85	93	101	109	117	125	133	141
23	—	3	9	17	24	32	40	48	56	64	73	81	89	98	106	115	123	132	140	149
24	—	3	10	17	25	33	42	50	59	67	76	85	94	102	111	120	129	138	147	156
25	—	3	10	18	27	35	44	53	62	71	80	89	98	107	117	126	135	145	154	163
26	—	4	11	19	28	37	46	55	64	74	83	93	102	112	122	132	141	151	161	171
27	—	4	11	20	29	38	48	57	67	77	87	97	107	117	127	137	147	158	168	178
28	—	4	12	21	30	40	50	60	70	80	90	101	111	122	132	143	154	164	175	186
29	—	4	13	22	32	42	52	62	73	83	94	105	116	127	138	149	160	171	182	193
30	—	5	13	23	33	43	54	65	76	87	98	109	120	131	143	154	166	177	189	200
31	—	5	14	24	34	45	56	67	78	90	101	113	125	136	148	160	172	184	196	208
32	—	5	14	24	35	46	58	69	81	93	105	117	129	141	153	166	178	190	203	215
33	—	5	15	25	37	48	60	72	84	96	108	121	133	146	159	171	184	197	210	222
34	—	5	15	26	38	50	62	74	87	99	112	125	138	151	164	177	190	203	217	230
35	—	6	16	27	39	51	64	77	89	103	116	129	142	156	169	183	196	210	224	237
36	—	6	16	28	40	53	66	79	92	106	119	133	147	161	174	188	202	216	231	245
37	—	6	17	29	41	55	68	81	95	109	123	137	151	165	180	194	209	223	238	252
38	—	6	17	30	43	56	70	84	98	112	127	141	156	170	185	200	215	230	245	259
39	0	7	18	31	44	58	72	86	101	115	130	145	160	175	190	206	321	236	252	267
40	0	7	18	31	45	59	74	89	103	119	134	149	165	180	196	211	227	243	258	274

Table 8   Mann-Whitney *U* (*Continued*)
One-Tailed Test at .01 Level; Two-Tailed Test at .02 Level

m	1	2	3	4	5	6	7	8	9	10	11	12	13	14	15	16	17	18	19	20
1	—																			
2	—	—																		
3	—	—	—																	
4	—	—	—	—																
5	—	—	—	0	1															
6	—	—	—	1	2	3														
7	—	—	0	1	3	4	6													
8	—	—	0	2	4	6	7	9												
9	—	—	1	3	5	7	9	11	14											
10	—	—	1	3	6	8	11	13	16	19										
11	—	—	1	4	7	9	12	15	18	22	25									
12	—	—	2	5	8	11	14	17	21	24	28	31								
13	—	0	2	5	9	12	16	20	23	27	31	35	39							
14	—	0	2	6	10	13	17	22	26	30	34	38	43	47						
15	—	0	3	7	11	15	19	24	28	33	37	42	47	51	56					
16	—	0	3	7	12	16	21	26	31	36	41	46	51	56	61	66				
17	—	0	4	8	13	18	23	28	33	38	44	49	55	60	66	71	77			
18	—	0	4	9	14	19	24	30	36	41	47	53	59	65	70	76	82	88		
19	—	1	4	9	15	20	26	32	38	44	50	56	63	69	75	82	88	94	101	
20	—	1	5	10	16	22	28	34	40	47	53	60	67	73	80	87	93	100	107	114
21	—	1	5	11	17	23	30	36	43	50	57	64	71	78	85	92	99	106	113	121
22	—	1	6	11	18	24	31	38	45	53	60	67	75	82	90	97	105	112	120	127
23	—	1	6	12	19	26	33	40	48	55	63	71	79	87	94	102	110	118	126	134
24	—	1	6	13	20	27	35	42	50	58	66	75	83	91	99	108	116	124	133	141
25	—	1	7	13	21	29	36	45	53	61	70	78	87	95	104	113	122	130	139	148
26	—	1	7	14	22	30	38	47	55	64	73	82	91	100	109	118	127	136	146	155
27	—	2	7	15	23	31	40	49	58	67	76	85	95	104	114	123	133	142	152	162
28	—	2	8	16	24	33	42	51	60	70	79	89	99	109	119	129	139	149	159	169
29	—	2	8	16	25	34	43	53	63	73	83	93	103	113	123	134	144	155	165	176
30	—	2	9	17	26	35	45	55	65	76	86	96	107	118	128	139	150	161	172	182
31	—	2	9	18	27	37	47	57	68	78	89	100	111	122	133	144	156	167	178	189
32	—	2	9	18	28	38	49	59	70	81	92	104	115	127	138	150	161	173	185	196
33	—	2	10	19	29	40	50	61	73	84	96	107	119	131	143	155	167	179	191	203
34	—	3	10	20	30	41	52	64	75	87	99	111	123	135	148	160	173	185	198	210
35	—	3	11	20	31	42	54	66	78	90	102	115	127	140	153	165	178	191	204	217
36	—	3	11	21	32	44	56	68	80	93	106	118	131	144	158	171	184	197	211	224
37	—	3	11	22	33	45	57	70	83	96	109	122	135	149	162	176	190	203	217	231
38	—	3	12	22	34	46	59	72	85	99	112	126	139	153	167	181	195	209	224	238
39	—	3	12	23	35	48	61	74	88	101	115	129	144	158	172	187	201	216	230	245
40	—	3	13	24	36	49	63	76	90	104	119	133	148	162	177	192	207	222	237	252

TABLE 8   433

Table 8   Mann-Whitney U (*Continued*)
One-Tailed Test at .005 Level; Two-Tailed Test at .01 Level

m	n																			
	1	2	3	4	5	6	7	8	9	10	11	12	13	14	15	16	17	18	19	20
1	—																			
2	—	—																		
3	—	—	—																	
4	—	—	—	—																
5	—	—	—	—	0															
6	—	—	—	0	1	2														
7	—	—	—	0	1	3	4													
8	—	—	—	1	2	4	6	7												
9	—	—	0	1	3	5	7	9	11											
10	—	—	0	2	4	6	9	11	13	16										
11	—	—	0	2	5	7	10	13	16	18	21									
12	—	—	1	3	6	9	12	15	18	21	24	27								
13	—	—	1	3	7	10	13	17	20	24	27	31	34							
14	—	—	1	4	7	11	15	18	22	26	30	34	38	42						
15	—	—	2	5	8	12	16	20	24	29	33	37	42	46	51					
16	—	—	2	5	9	13	18	22	27	31	36	41	45	50	55	60				
17	—	—	2	6	10	15	19	24	29	34	39	44	49	54	60	65	70			
18	—	—	2	6	11	16	21	26	31	37	42	47	53	58	64	70	75	81		
19	—	0	3	7	12	17	22	28	33	39	45	51	57	63	69	74	81	87	93	
20	—	0	3	8	13	18	24	30	36	42	48	54	60	67	73	79	86	92	99	105
21	—	0	3	8	14	19	25	32	38	44	51	58	64	71	78	84	91	98	105	112
22	—	0	4	9	14	21	27	34	40	47	54	61	68	75	82	89	96	104	111	118
23	—	0	4	9	15	22	29	35	43	50	57	64	72	79	87	94	102	109	117	125
24	—	0	4	10	16	23	30	37	45	52	60	68	75	83	91	99	107	115	123	131
25	—	0	5	10	17	24	32	39	47	55	63	71	79	87	96	104	112	121	129	138
26	—	0	5	11	18	25	33	41	49	58	66	74	83	92	100	109	118	127	135	144
27	—	1	5	12	19	27	35	43	52	60	69	78	87	96	105	114	123	132	142	151
28	—	1	5	12	20	28	36	45	54	63	72	81	91	100	109	119	128	138	148	157
29	—	1	6	13	21	29	38	47	56	66	75	85	94	104	114	124	134	144	154	164
30	—	1	6	13	22	30	40	49	58	68	78	88	98	108	119	129	139	150	160	170
31	—	1	6	14	22	32	41	51	61	71	81	92	102	113	123	134	145	155	166	177
32	—	1	7	14	23	33	43	53	63	74	84	95	106	117	128	139	150	161	172	184
33	—	1	7	15	24	34	44	55	65	76	87	98	110	121	132	144	155	167	179	190
34	—	1	7	16	25	35	46	57	68	79	90	102	113	125	137	149	161	173	185	197
35	—	1	8	16	26	37	47	59	70	82	93	105	117	129	142	154	166	179	191	203
36	—	1	8	17	27	38	49	60	72	84	96	109	121	134	146	159	172	184	197	210
37	—	1	8	17	28	39	51	62	75	87	99	112	125	138	151	164	177	190	203	217
38	—	1	9	18	29	40	52	64	77	90	102	116	129	142	155	169	182	196	210	223
39	—	2	9	19	30	41	54	66	79	92	106	119	133	146	160	174	188	202	216	230
40	—	2	9	19	31	43	55	68	81	95	109	122	136	150	165	179	193	208	222	237

Table 9   Wilcoxon Matched-Pairs Signed-Ranks Test*

N	Level of significance for one-tailed test		
	.025	.01	.005
	Level of significance for two-tailed test		
	.05	.02	.01
6	0	—	—
7	2	0	—
8	4	2	0
9	6	3	2
10	8	5	3
11	11	7	5
12	14	10	7
13	17	13	10
14	21	16	13
15	25	20	16
16	30	24	20
17	35	28	23
18	40	33	28
19	46	38	32
20	52	43	38
21	59	49	43
22	66	56	49
23	73	62	55
24	81	69	61
25	89	77	68

* Adapted from Table I of F. Wilcoxon, *Some Rapid Approximate Statistical Procedures*, p. 13, American Cyanamid Company, New York, 1949, with their kind permission.

TABLE 10   435

Table 10   Distribution of Chi-Square for Given Probability Levels

df	.99	.98	.95	.90	.80	.70	Probability .50	.30	.20	.10	.05	.02	.01	.001
1	.00016	.00663	.00393	.0158	.0642	.148	.455	1.074	1.642	2.706	3.841	5.412	6.635	10.827
2	.0201	.0404	.103	.211	.446	.713	1.386	2.408	3.219	4.605	5.991	7.824	9.210	13.815
3	.115	.185	.352	.584	1.005	1.424	2.366	3.665	4.642	6.251	7.815	9.837	11.345	16.266
4	.297	.429	.711	1.064	1.649	2.195	3.357	4.878	5.989	7.779	9.488	11.668	13.277	18.467
5	.554	.752	1.145	1.610	2.343	3.000	4.351	6.064	7.289	9.236	11.070	13.388	15.086	20.515
6	.872	1.134	1.635	2.204	3.070	3.828	5.348	7.231	8.558	10.645	12.592	15.033	16.812	22.457
7	1.239	1.564	2.167	2.833	3.822	4.671	6.346	8.383	9.803	12.017	14.067	16.622	18.475	24.322
8	1.646	2.032	2.733	3.490	4.594	5.527	7.344	9.524	11.030	13.362	15.507	18.168	20.090	26.125
9	2.088	2.532	3.325	4.168	5.380	6.393	8.343	10.656	12.242	14.684	16.919	19.679	21.666	27.877
10	2.558	3.059	3.940	4.865	6.179	7.267	9.342	11.781	13.442	15.987	18.307	21.161	23.209	29.588
11	3.053	3.609	4.575	5.578	6.989	8.148	10.341	12.899	14.631	17.275	19.675	22.618	24.725	31.264
12	3.571	4.178	5.226	6.304	7.807	9.034	11.340	14.011	15.812	18.549	21.026	24.054	26.217	32.909
13	4.107	4.765	5.892	7.042	8.634	9.926	12.340	15.119	16.985	19.812	22.362	25.472	27.688	34.528
14	4.660	5.368	6.571	7.790	9.467	10.821	13.339	16.222	18.151	21.064	23.685	26.873	29.141	36.123
15	5.229	5.985	7.261	8.547	10.307	11.721	14.339	17.322	19.311	22.307	24.996	28.259	30.578	37.697
16	5.812	6.614	7.962	9.312	11.152	12.624	15.338	18.418	20.465	23.542	26.296	29.633	32.000	39.252
17	6.408	7.255	8.672	10.085	12.002	13.531	16.338	19.511	21.615	24.769	27.587	30.995	33.409	40.790
18	7.015	7.906	9.390	10.865	12.857	14.440	17.338	20.601	22.760	25.989	28.869	32.346	34.805	42.312
19	7.633	8.567	10.117	11.651	13.716	15.352	18.338	21.689	23.900	27.204	30.144	33.687	36.191	43.820
20	8.260	9.237	10.851	12.443	14.578	16.266	19.337	22.775	25.038	28.412	31.410	35.020	37.566	45.315
21	8.897	9.915	11.591	13.240	15.445	17.182	20.337	23.858	26.171	29.615	32.671	36.343	38.932	46.797
22	9.542	10.600	12.338	14.041	16.314	18.101	21.337	24.939	27.301	30.813	33.924	37.659	40.289	48.268

For larger values of df, the expression $\sqrt{\chi^2 2} - \sqrt{2df - 1}$ may be used as a normal deviate with unit variance, remembering that the probability for χ^2 corresponds with that of a single tail of the normal curve.

This table is adapted from R. A. Fisher and F. Yates, *Statistical Tables for Biological, Agricultural and Medical Research*, Oliver and Boyd, Ltd., Edinburgh, by permission of the authors and publishers.

(Continued)

Table 10 Distribution of Chi-Square for Given Probability Levels (*Continued*)

df	.99	.98	.95	.90	.80	.70	.50	.30	.20	.10	.05	.02	.01	.001
23	10.196	11.293	13.091	14.848	17.187	19.021	22.337	26.018	28.429	32.007	35.172	38.968	41.638	49.728
24	10.856	11.992	13.848	15.659	18.062	19.943	23.337	27.096	29.553	33.196	36.415	40.270	42.980	51.179
25	11.524	12.697	14.611	16.473	18.940	20.867	24.337	28.172	30.675	34.382	37.652	41.566	44.314	52.620
26	12.198	13.409	15.379	17.292	19.820	21.792	25.336	29.246	31.795	35.563	38.885	42.856	45.642	54.052
27	12.879	14.125	16.151	18.114	20.703	22.719	26.336	30.319	32.912	36.741	40.113	44.140	46.963	55.476
28	13.565	14.847	16.928	18.939	21.588	23.647	27.336	31.391	34.027	37.916	41.337	45.419	48.278	56.893
29	14.256	15.574	17.708	19.768	22.475	24.577	28.336	32.461	35.139	39.087	42.557	46.693	49.588	58.302
30	14.953	16.306	18.493	20.599	23.364	25.508	29.336	33.530	36.250	40.256	43.773	47.962	50.892	59.703
32	16.362	17.783	20.072	22.271	25.148	27.373	31.336	35.665	38.466	42.585	46.194	50.487	53.486	62.487
34	17.789	19.275	21.664	23.952	26.938	29.242	33.336	37.795	40.676	44.903	48.602	52.995	56.061	65.247
36	19.233	20.783	23.269	25.643	28.735	31.115	35.336	39.922	42.879	47.212	50.999	55.489	58.619	67.985
38	20.691	22.304	24.884	27.343	30.537	32.992	37.335	42.045	45.076	49.513	53.384	57.969	61.162	70.703
40	22.164	23.838	26.509	29.051	32.345	34.872	39.335	44.165	47.269	51.805	55.759	60.436	63.691	73.402
42	23.650	25.383	28.144	30.765	34.157	36.755	41.335	46.282	49.456	54.090	58.124	62.892	66.206	76.084
44	25.148	26.939	29.787	32.487	35.794	38.641	43.335	48.396	51.639	56.369	60.481	65.337	68.710	78.750
46	26.657	28.504	31.439	34.215	37.795	40.529	45.335	50.507	53.818	58.641	62.830	67.771	71.201	81.400
48	28.177	30.080	33.098	35.949	39.621	42.420	47.335	52.616	55.993	60.907	65.171	70.197	73.683	84.037
50	29.707	31.664	34.764	37.689	41.449	44.313	49.335	54.723	58.164	63.167	67.505	72.613	76.154	86.661
52	31.246	33.256	36.437	39.433	43.281	46.209	51.335	56.827	60.332	65.422	69.832	75.021	78.616	89.272
54	32.793	34.856	38.116	41.183	45.117	48.106	53.335	58.930	62.496	67.673	72.153	77.422	81.069	91.872
56	34.350	36.464	39.801	42.937	46.955	50.005	55.335	61.031	64.658	69.919	74.468	79.815	83.513	94.461
58	35.913	38.078	41.492	44.696	48.797	51.906	57.335	63.129	66.816	72.160	76.778	82.201	85.950	97.039
60	37.485	39.699	43.188	46.459	50.641	53.809	59.335	65.227	68.972	74.397	79.082	84.580	88.379	99.607
62	39.063	41.327	44.889	48.226	52.487	55.714	61.335	67.322	71.125	76.630	81.381	86.953	90.802	102.166
64	40.649	42.960	46.595	49.996	54.336	57.620	63.335	69.416	73.276	78.860	83.675	89.320	93.217	104.716
66	42.240	44.599	48.305	51.770	56.188	59.527	65.335	71.508	75.424	81.085	85.965	91.681	95.626	107.258
68	43.838	46.244	50.020	53.548	58.042	61.436	67.335	73.600	77.571	83.308	88.250	94.037	98.028	109.791
	45.442	47.893	51.739	55.329	59.898	63.346	69.334	75.689	79.715	85.527	90.531	96.388	100.425	112.317

TABLE 11   437

Table 11   Transformation of r to Z

r	z	r	z	r	z	r	z	r	z
.000	.000	.200	.203	.400	.424	.600	.693	.800	1.099
.005	.005	.205	.208	.405	.430	.605	.701	.805	1.113
.010	.010	.210	.213	.410	.436	.610	.709	.810	1.127
.015	.015	.215	.218	.415	.442	.615	.717	.815	1.142
.020	.020	.220	.224	.420	.448	.620	.725	.820	1.157
.025	.025	.225	.229	.425	.454	.625	.733	.825	1.172
.030	.030	.230	.234	.430	.460	.630	.741	.830	1.188
.035	.035	.235	.239	.435	.466	.635	.750	.835	1.204
.040	.040	.240	.245	.440	.472	.640	.758	.840	1.221
.045	.045	.245	.250	.445	.478	.645	.767	.845	1.238
.050	.050	.250	.255	.450	.485	.650	.775	.850	1.256
.055	.055	.255	.261	.455	.491	.655	.784	.855	1.274
.060	.060	.260	.266	.460	.497	.660	.793	.860	1.293
.065	.065	.265	.271	.465	.504	.665	.802	.865	1.313
.070	.070	.270	.277	.470	.510	.670	.811	.870	1.333
.075	.075	.275	.282	.475	.517	.675	.820	.875	1.354
.080	.080	.280	.288	.480	.523	.680	.829	.880	1.376
.085	.085	.285	.293	.485	.530	.685	.838	.885	1.398
.090	.090	.290	.299	.490	.536	.690	.848	.890	1.422
.095	.095	.295	.304	.495	.543	.695	.858	.895	1.447
.100	.100	.300	.310	.500	.549	.700	.867	.900	1.472
.105	.105	.305	.315	.505	.556	.705	.877	.905	1.499
.110	.110	.310	.321	.510	.563	.710	.887	.910	1.528
.115	.116	.315	.326	.515	.570	.715	.897	.915	1.557
.120	.121	.320	.332	.520	.576	.720	.908	.920	1.589
.125	.126	.325	.337	.525	.583	.725	.918	.925	1.623
.130	.131	.330	.343	.530	.590	.730	.929	.930	1.658
.135	.136	.335	.348	.535	.597	.735	.940	.935	1.697
.140	.141	.340	.354	.540	.604	.740	.950	.940	1.738
.145	.146	.345	.360	.545	.611	.745	.962	.945	1.783
.150	.151	.350	.365	.550	.618	.750	.973	.950	1.832
.155	.156	.355	.371	.555	.626	.755	.984	.955	1.886
.160	.161	.360	.377	.560	.633	.760	.996	.960	1.946
.165	.167	.365	.383	.565	.640	.765	1.008	.965	2.014
.170	.172	.370	.388	.570	.648	.770	1.020	.970	2.092
.175	.177	.375	.394	.575	.655	.775	1.033	.975	2.185
.180	.182	.380	.400	.580	.662	.780	1.045	.980	2.298
.185	.187	.385	.406	.585	.670	.785	1.058	.985	2.443
.190	.192	.390	.412	.590	.678	.790	1.071	.990	2.647
.195	.198	.395	.418	.595	.685	.795	1.085	.995	2.994

This table was constructed by F. P. Kilpatrick and D. A. Buchanan.

This table is reprinted from Allen L. Edwards, *Statistical Methods*, 2nd ed. New York: Holt Rinehart and Winston, 1967, p. 427.

Table 12   Critical Values of the Pearson Correlation Coefficient*

df	Level of significance for one-tailed test			
	.05	.025	.01	.005
	Level of significance for two-tailed test			
	.10	.05	.02	.01
1	.988	.997	.9995	.9999
2	.900	.950	.980	.990
3	.805	.878	.934	.959
4	.729	.811	.882	.917
5	.669	.754	.833	.874
6	.622	.707	.789	.834
7	.582	.666	.750	.798
8	.549	.632	.716	.765
9	.521	.602	.685	.735
10	.497	.576	.658	.708
11	.576	.553	.634	.684
12	.458	.532	.612	.661
13	.441	.514	.592	.641
14	.426	.497	.574	.623
15	.412	.482	.558	.606
16	.400	.468	.542	.590
17	.389	.456	.528	.575
18	.378	.444	.516	.561
19	.369	.433	.503	.549
20	.360	.423	.492	.537
21	.352	.413	.482	.526
22	.344	.404	.472	.515
23	.337	.396	.462	.505
24	.330	.388	.453	.496
25	.323	.381	.445	.487
26	.317	.374	.437	.479
27	.311	.367	.430	.471
28	.306	.361	.423	.463
29	.301	.355	.416	.486
30	.296	.349	.409	.449
35	.275	.325	.381	.418
40	.257	.304	.358	.393
45	.243	.288	.338	.372
50	.231	.273	.322	.354
60	.211	.250	.295	.325
70	.195	.232	.274	.303
80	.183	.217	.256	.283
90	.173	.205	.242	.267
100	.164	.195	.230	.254

* Abridged from R. A. Fisher and F. Yates, *Statistical Tables for Biological, Agricultural, and Medical Research*, Oliver & Boyd, Ltd., Edinburgh, by permission of the authors and publishers.

TABLE 13   439

Table 13   Critical Values of the Spearman Correlation Coefficient

	Level of significance for a one-tailed test			
	.05	.025	.01	.005
	Level of significance for a two-tailed test			
N	.10	.05	.02	.01
4	1.000			
5	.900	1.000	1.000	
6	.829	.886	.943	1.000
7	.714	.786	.893	.929
8	.643	.738	.833	.881
9	.600	.683	.783	.833
10	.564	.648	.746	.794
12	.506	.591	.712	.777
14	.456	.544	.645	.715
16	.425	.506	.601	.665
18	.399	.475	.564	.625
20	.377	.450	.534	.591
22	.359	.428	.508	.562
24	.343	.409	.485	.537
26	.329	.392	.465	.515
28	.317	.377	.448	.496
30	.306	.364	.432	.478

This table was compiled by author.

Table 14   Critical Values of the Kendall Tau Correlation Coefficient

	Level of significance for a one-tailed test			
	.05	.025	.01	.005
	Level of significance for a two-tailed test			
N	.10	.05	.02	.01
4	1.000			
5	.800	1.000	1.000	
6	.600	.733	.867	.867
7	.619	.714	.810	.810
8	.500	.571	.714	.714
9	.444	.555	.611	.667
10	.422	.511	.600	.644
11	.382	.455	.564	.600
12	.364	.424	.515	.576
13	.333	.410	.487	.538
14	.319	.385	.473	.516
15	.314	.371	.448	.505
16	.300	.367	.433	.483
17	.294	.353	.412	.456
18	.281	.333	.399	.451
19	.275	.333	.392	.427
20	.263	.316	.379	.421
25	.233	.280	.327	.360
30	.214	.251	.297	.329
35	.193	.230	.274	.301
40	.182	.215	.254	.282

This table was compiled by the author.

TABLE 15   **441**

Table 15   Table of Critical Values of D in the Kolmogorov One-Sample Test*

Sample size (N)	Level of significance for $D = maximum \ |F_0(X) - S_N(X)|$				
	.20	.15	.10	.05	.01
1	.900	.925	.950	.975	.995
2	.684	.726	.776	.842	.929
3	.565	.597	.642	.708	.828
4	.494	.525	.564	.624	.733
5	.446	.474	.510	.565	.669
6	.410	.436	.470	.521	.618
7	.381	.405	.438	.486	.577
8	.358	.381	.411	.457	.543
9	.339	.360	.388	.432	.514
10	.322	.342	.368	.410	.490
11	.307	.326	.352	.391	.468
12	.295	.313	.338	.375	.450
13	.284	.302	.325	.361	.433
14	.274	.292	.314	.349	.418
15	.266	.283	.304	.338	.404
16	.258	.274	.295	.328	.392
17	.250	.266	.286	.318	.381
18	.244	.259	.278	.309	.371
19	.237	.252	.272	.301	.363
20	.231	.246	.264	.294	.356
25	.21	.22	.24	.27	.32
30	.19	.20	.22	.24	.29
35	.18	.19	.21	.23	.27
Over 35	$\dfrac{1.07}{\sqrt{N}}$	$\dfrac{1.14}{\sqrt{N}}$	$\dfrac{1.22}{\sqrt{N}}$	$\dfrac{1.36}{\sqrt{N}}$	$\dfrac{1.63}{\sqrt{N}}$

* Adapted from F. J. Massey, Jr., "The Kolmogorov-Smirnov Test for Goodness of Fit," *J. Amer. Statist. Ass.*, **46**, 70, 1951, with the kind permission of the author and publisher.

Table 16 Table of Critical Values of K_D in the Kolmogorov-Smirnov Test for Two Samples of Equal Size (N is the Size of Each Sample)

N	One-tailed test*		Two-tailed test†	
	$\alpha = .05$	$\alpha = .01$	$\alpha = .05$	$\alpha = .01$
3	3	—	—	—
4	4	—	4	—
5	4	5	5	5
6	5	6	5	6
7	5	6	6	6
8	5	6	6	7
9	6	7	6	7
10	6	7	7	8
11	6	8	7	8
12	6	8	7	8
13	7	8	7	9
14	7	8	8	9
15	7	9	8	9
16	7	9	8	10
17	8	9	8	10
18	8	10	9	10
19	8	10	9	10
20	8	10	9	11
21	8	10	9	11
22	9	11	9	11
23	9	11	10	11
24	9	11	10	12
25	9	11	10	12
26	9	11	10	12
27	9	12	10	12
28	10	12	11	13
29	10	12	11	13
30	10	12	11	13
35	11	13	12	
40	11	14	13	

* Abridged from L. A. Goodman, "Kolomo-gorov-Smirnov tests for psychological research," *Psychol. Bull.*, **51**, 167, 1954, with the kind permission of the author and the American Psychological Association.

† Derived from Table 1 of F. J. Massey, Jr., "The distribution of the maximum deviation between two sample cumulative step functions," *Ann. Math. Statist.*, **22**, 126–127, 1951, with the kind permission of the author and the publisher.

TABLE 17   443

Table 17   Table of Critical Values of D in the Kolmogorov-Smirnov Test for Two Large Samples*

Level of significance	Value of D so large as to call for rejection of H_0 at the indicated level of significance, where $D = maximum \ |S_{n_1}(X) - S_{n_2}(X)|$
.10	$1.22 \sqrt{\dfrac{n_1 + n_2}{n_1 n_2}}$
.05	$1.36 \sqrt{\dfrac{n_1 + n_2}{n_1 n_2}}$
.025	$1.48 \sqrt{\dfrac{n_1 + n_2}{n_1 n_2}}$
.01	$1.63 \sqrt{\dfrac{n_1 + n_2}{n_1 n_2}}$
.005	$1.73 \sqrt{\dfrac{n_1 + n_2}{n_1 n_2}}$
.001	$1.95 \sqrt{\dfrac{n_1 + n_2}{n_1 n_2}}$

* Adapted from N. Smirnov, "Tables for estimating the goodness of fit of empirical distributions," *Ann. Math. Statist.*, **19**, 280–281, 1948, with the kind permission of the publisher.

Table 18 Table of Critical Values for the Fisher Test

Totals in right margin		B (or A)	Level of significance			
			.05	.025	.01	.005
A + B = 3	C + D = 3	3	0	—	—	—
A + B = 4	C + D = 4	4	0	0	—	—
	C + D = 3	4	0	—	—	—
A + B = 5	C + D = 5	5	1	1	0	0
		4	0	0	—	—
	C + D = 4	5	1	0	0	—
		4	0	—	—	—
	C + D = 3	5	0	0	—	—
	C + D = 2	5	0	—	—	—
A + B = 6	C + D = 6	6	2	1	1	0
		5	1	0	0	—
		4	0	—	—	—
	C + D = 5	6	1	0	0	0
		5	0	0	—	—
		4	0	—	—	—
	C + D = 4	6	1	0	0	0
		5	0	0	—	—
	C + D = 3	6	0	0	—	—
		5	0	—	—	—
	C + D = 2	6	0	—	—	—
A + B = 7	C + D = 7	7	3	2	1	1
		6	1	1	0	0
		5	0	0	—	—
		4	0	—	—	—
	C + D = 6	7	2	2	1	1
		6	1	0	0	0
		5	0	0	—	—
		4	0	—	—	—
	C + D = 5	7	2	1	0	0
		6	1	0	0	—
		5	0	—	—	—

NOTE: Critical values are maximum values of D (or C).

* Adapted from D. J. Finney, "The Fisher-Yates test of significance in 2 × 2 contingency tables," *Biometrika*, **35**, 149–154, 1948, with the kind permission of the author and the publisher.

TABLE 18   **445**

Table 18   Fisher Test (*Continued*)

Totals in right margin		B (or A)	Level of significance			
			.05	.025	.01	.005
A + B = 7	C + D = 4	7	1	1	0	0
		6	0	0	—	—
		5	0	—	—	—
	C + D = 3	7	0	0	0	—
		6	0	—	—	—
	C + D = 2	7	0	—	—	—
A + B = 8	C + D = 8	8	4	3	2	2
		7	2	2	1	0
		6	1	1	0	0
		5	0	0	—	—
		4	0	—	—	—
	C + D = 7	8	3	2	2	1
		7	2	1	1	0
		6	1	0	0	—
		5	0	0	—	—
	C + D = 6	8	2	2	1	1
		7	1	1	0	0
		6	0	0	0	—
		5	0	—	—	—
	C + D = 5	8	2	1	1	0
		7	1	0	0	0
		6	0	0	—	—
		5	0	—	—	—
	C + D = 4	8	1	1	0	0
		7	0	0	—	—
		6	0	—	—	—
	C + D = 3	8	0	0	0	—
		7	0	0	—	—
	C + D = 2	8	0	0	—	—
A + B = 9	C + D = 9	9	5	4	3	3
		8	3	3	2	1
		7	2	1	1	0
		6	1	1	0	0
		5	0	0	—	—
		4	0	—	—	—

NOTE: Critical values are maximum values of D (or C).

* When B is entered in the middle column, the significance levels are for D. When A is used in place of B, the significance levels are for C.

Table 18 **Fisher Test** (*Continued*)

Totals in right margin		B (or A)	Level of significance			
			.05	.025	.01	.005
A + B = 9	C + D = 8	9	4	3	3	2
		8	3	2	1	1
		7	2	1	0	0
		6	1	0	0	—
		5	0	0	—	—
	C + D = 7	9	3	3	2	2
		8	2	2	1	0
		7	1	1	0	0
		6	0	0	—	—
		5	0	—	—	—
	C + D = 6	9	3	2	1	1
		8	2	1	0	0
		7	1	0	0	—
		6	0	0	—	—
		5	0	—	—	—
	C + D = 5	9	2	1	1	1
		8	1	1	0	0
		7	0	0	—	—
		6	0	—	—	—
	C + D = 4	9	1	1	0	0
		8	0	0	0	—
		7	0	0	—	—
		6	0	—	—	—
	C + D = 3	9	1	0	0	0
		8	0	0	—	—
		7	0	—	—	—
	C + D = 2	9	0	0	—	—
A + B = 10	C + D = 10	10	6	5	4	3
		9	4	3	3	2
		8	3	2	1	1
		7	2	1	1	0
		6	1	0	0	—
		5	0	0	—	—
		4	0	—	—	—
	C + D = 9	10	5	4	3	3
		9	4	3	2	2

NOTE: Critical values are maximum values of D (or C).

TABLE 18   447

Table 18   Fisher Test (*Continued*)

Totals in right margin		B (or A)	Level of significance			
			.05	.025	.01	.005
A + B = 10	C + D = 9	8	2	2	1	1
		7	1	1	0	0
		6	1	0	0	—
		5	0	0	—	—
	C + D = 8	10	4	4	3	2
		9	3	2	2	1
		8	2	1	1	0
		7	1	1	0	0
		6	0	0	—	—
		5	0	—	—	—
	C + D = 7	10	3	3	2	2
		9	2	2	1	1
		8	1	1	0	0
		7	1	0	0	—
		6	0	0	—	—
		5	0	—	—	—
	C + D = 6	10	3	2	2	1
		9	2	1	1	0
		8	1	1	0	0
		7	0	0	—	—
		6	0	—	—	—
	C + D = 5	10	2	2	1	1
		9	1	1	0	0
		8	1	0	0	—
		7	0	0	—	—
		6	0	—	—	—
	C + D = 4	10	1	1	0	0
		9	1	0	0	0
		8	0	0	—	—
		7	0	—	—	—
	C + D = 3	10	1	0	0	0
		9	0	0	—	—
		8	0	—	—	—
	C + D = 2	10	0	0	—	—
		9	0	—	—	—
A + B = 11	C + D = 11	11	7	6	5	4
		10	5	4	3	3

NOTE: Critical values are maximum values of D (or C).

Table 18 Fisher Test (*Continued*)

Totals in right margin		B (or A)	Level of significance			
			.05	.025	.01	.005
A + B = 11	C + D = 11	9	4	3	2	2
		8	3	2	1	1
		7	2	1	0	0
		6	1	0	0	—
		5	0	0	—	—
		4	0	—	—	—
	C + D = 10	11	6	5	4	4
		10	4	4	3	2
		9	3	3	2	1
		8	2	2	1	0
		7	1	1	0	0
		6	1	0	0	—
		5	0	—	—	—
	C + D = 9	11	5	4	4	3
		10	4	3	2	2
		9	3	2	1	1
		8	2	1	1	0
		7	1	1	0	0
		6	0	0	—	—
		5	0	—	—	—
	C + D = 8	11	4	4	3	3
		10	3	3	2	1
		9	2	2	1	1
		8	1	1	0	0
		7	1	0	0	—
		6	0	0	—	—
		5	0	—	—	—
	C + D = 7	11	4	3	2	2
		10	3	2	1	1
		9	2	1	1	0
		8	1	1	0	0
		7	0	0	—	—
		6	0	0	—	—
	C + D = 6	11	3	2	2	1
		10	2	1	1	0
		9	1	1	0	0
		8	1	0	0	—
		7	0	0	—	—
		6	0	—	—	—

NOTE: Critical values are maximum values of D (or C).

TABLE 18   449

Table 18   Fisher Test (*Continued*)

Totals in right margin		B (or A)	Level of significance			
			.05	.025	.01	.005
A + B = 11	C + D = 5	11	2	2	1	1
		10	1	1	0	0
		9	1	0	0	0
		8	0	0	—	—
		7	0	—	—	—
	C + D = 4	11	1	1	1	0
		10	1	0	0	0
		9	0	0	—	—
		8	0	—	—	—
	C + D = 3	11	1	0	0	0
		10	0	0	—	—
		9	0	—	—	—
	C + D = 2	11	0	0	—	—
		10	0	—	—	—
A + B = 12	C + D = 12	12	8	7	6	5
		11	6	5	4	4
		10	5	4	3	2
		9	4	3	2	1
		8	3	2	1	1
		7	2	1	0	0
		6	1	0	0	—
		5	0	0	—	—
		4	0	—	—	—
	C + D = 11	12	7	6	5	5
		11	5	5	4	3
		10	4	3	2	2
		9	3	2	2	1
		8	2	1	1	0
		7	1	1	0	0
		6	1	0	0	—
		5	0	0	—	—
	C + D = 10	12	6	5	5	4
		11	5	4	3	3
		10	4	3	2	2
		9	3	2	1	1
		8	2	1	0	0
		7	1	0	0	0
		6	0	0	—	—
		5	0	—	—	—

NOTE: Critical values are maximum values of D (or C).

Table 18 Fisher Test (*Continued*)

Totals in right margin		B (or A)	Level of significance			
			.05	.025	.01	.005
A + B = 12	C + D = 9	12	5	5	4	3
		11	4	3	3	2
		10	3	2	2	1
		9	2	2	1	0
		8	1	1	0	0
		7	1	0	0	—
		6	0	0	—	—
		5	0	—	—	—
	C + D = 8	12	5	4	3	3
		11	3	3	2	2
		10	2	2	1	1
		9	2	1	1	0
		8	1	1	0	0
		7	0	0	—	—
		6	0	0	—	—
	C + D = 7	12	4	3	3	2
		11	3	2	2	1
		10	2	1	1	0
		9	1	1	0	0
		8	1	0	0	—
		7	0	0	—	—
		6	0	—	—	—
	C + D = 6	12	3	3	2	2
		11	2	2	1	1
		10	1	1	0	0
		9	1	0	0	0
		8	0	0	—	—
		7	0	0	—	—
		6	0	—	—	—
	C + D = 5	12	2	2	1	1
		11	1	1	1	0
		10	1	0	0	0
		9	0	0	0	—
		8	0	0	—	—
		7	0	—	—	—
	C + D = 4	12	2	1	1	0
		11	1	0	0	0
		10	0	0	0	—
		9	0	0	—	—
		8	0	—	—	—

NOTE: Critical values are maximum values of D (or C).

TABLE 18   451

Table 18   Fisher Test (*Continued*)

Totals in right margin		B (or A)	Level of significance			
			.05	.025	.01	.005
A + B = 12	C + D = 3	12	1	0	0	0
		11	0	0	0	—
		10	0	0	—	—
		9	0	—	—	—
	C + D = 2	12	0	0	—	—
		11	0	—	—	—
A + B = 13	C + D = 13	13	9	8	7	6
		12	7	6	5	4
		11	6	5	4	3
		10	4	4	3	2
		9	3	3	2	1
		8	2	2	1	0
		7	2	1	0	0
		6	1	0	0	—
		5	0	0	—	—
		4	0	—	—	—
	C + D = 12	13	8	7	6	5
		12	6	5	5	4
		11	5	4	3	3
		10	4	3	2	2
		9	3	2	1	1
		8	2	1	1	0
		7	1	1	0	0
		6	1	0	0	—
		5	0	0	—	—
	C + D = 11	13	7	6	5	5
		12	6	5	4	3
		11	4	4	3	2
		10	3	3	2	1
		9	3	2	1	1
		8	2	1	0	0
		7	1	0	0	0
		6	0	0	—	—
		5	0	—	—	—
	C + D = 10	13	6	6	5	4
		12	5	4	3	3
		11	4	3	2	2
		10	3	2	1	1

NOTE: Critical values are maximum values of *D* (or *C*).

Table 18   Fisher Test (*Continued*)

Totals in right margin		B (or A)	Level of significance			
			.05	.025	.01	.005
$A + B = 13$	$C + D = 10$	9	2	1	1	0
		8	1	1	0	0
		7	1	0	0	—
		6	0	0	—	—
		5	0	—	—	—
	$C + D = 9$	13	5	5	4	4
		12	4	4	3	2
		11	3	3	2	1
		10	2	2	1	1
		9	2	1	0	0
		8	1	1	0	0
		7	0	0	—	—
		6	0	0	—	—
		5	0	—	—	—
	$C + D = 8$	13	5	4	3	3
		12	4	3	2	2
		11	3	2	1	1
		10	2	1	1	0
		9	1	1	0	0
		8	1	0	0	—
		7	0	0	—	—
		6	0	—	—	—
	$C + D = 7$	13	4	3	3	2
		12	3	2	2	1
		11	2	2	1	1
		10	1	1	0	0
		9	1	0	0	0
		8	0	0	—	—
		7	0	0	—	—
		6	0	—	—	—
	$C + D = 6$	13	3	3	2	2
		12	2	2	1	1
		11	2	1	1	0
		10	1	1	0	0
		9	1	0	0	—
		8	0	0	—	—
		7	0	—	—	—
	$C + D = 5$	13	2	2	1	1
		12	2	1	1	0

NOTE: Critical values are maximum values of D (or C).

TABLE 18   453

Table 18   Fisher Test (*Continued*)

Totals in right margin		B (or A)	Level of significance			
			.05	.025	.01	.005
A + B = 13	C + D = 5	11	1	1	0	0
		10	1	0	0	—
		9	0	0	—	—
		8	0	—	—	—
	C + D = 4	13	2	1	1	0
		12	1	1	0	0
		11	0	0	0	—
		10	0	0	—	—
		9	0	—	—	—
	C + D = 3	13	1	1	0	0
		12	0	0	0	—
		11	0	0	—	—
		10	0	—	—	—
	C + D = 2	13	0	0	0	—
		12	0	—	—	—
A + B = 14	C + D = 14	14	10	9	8	7
		13	8	7	6	5
		12	6	6	5	4
		11	5	4	3	3
		10	4	3	2	2
		9	3	2	2	1
		8	2	2	1	0
		7	1	1	0	0
		6	1	0	0	—
		5	0	0	—	—
		4	0	—	—	—
	C + D = 13	14	9	·8	7	6
		13	7	6	5	5
		12	6	5	4	3
		11	5	4	3	2
		10	4	3	2	2
		9	3	2	1	1
		8	2	1	1	0
		7	1	1	0	0
		6	1	0	—	—
		5	0	0	—	—
	C + D = 12	14	8	7	6	6
		13	6	6	5	4

NOTE: Critical values are maximum values of *D* (or *C*).

Table 18   Fisher Test (*Continued*)

Totals in right margin		B (or A)	Level of significance			
			.05	.025	.01	.005
$A + B = 14$	$C + D = 12$	12	5	4	4	3
		11	4	3	3	2
		10	3	3	2	1
		9	2	2	1	1
		8	2	1	0	0
		7	1	0	0	—
		6	0	0	—	—
		5	0	—	—	—
	$C + D = 11$	14	7	6	6	5
		13	6	5	4	4
		12	5	4	3	3
		11	4	3	2	2
		10	3	2	1	1
		9	2	1	1	0
		8	1	1	0	0
		7	1	0	0	—
		6	0	0	—	—
		5	0	—	—	—
	$C + D = 10$	14	6	6	5	4
		13	5	4	4	3
		12	4	3	3	2
		11	3	3	2	1
		10	2	2	1	1
		9	2	1	0	0
		8	1	1	0	0
		7	0	0	0	—
		6	0	0	—	—
		5	0	—	—	—
	$C + D = 9$	14	6	5	4	4
		13	4	4	3	3
		12	3	3	2	2
		11	3	2	1	1
		10	2	1	1	0
		9	1	1	0	0
		8	1	0	0	—
		7	0	0	—	—
		6	0	—	—	—
	$C + D = 8$	14	5	4	4	3
		13	4	3	2	2

NOTE: Critical values are maximum values of D (or C).

TABLE 18   455

Table 18   Fisher Test (*Continued*)

Totals in right margin		B (or A)	Level of significance			
			.05	.025	.01	.005
A + B = 14	C + D = 8	12	3	2	2	1
		11	2	2	1	1
		10	2	1	0	0
		9	1	0	0	0
		8	0	0	0	—
		7	0	0	—	—
		6	0	—	—	—
	C + D = 7	14	4	3	3	2
		13	3	2	2	1
		12	2	2	1	1
		11	2	1	1	0
		10	1	1	0	0
		9	1	0	0	—
		8	0	0	—	—
		7	0	—	—	—
	C + D = 6	14	3	3	2	2
		13	2	2	1	1
		12	2	1	1	0
		11	1	1	0	0
		10	1	0	0	—
		9	0	0	—	—
		8	0	0	—	—
		7	0	—	—	—
	C + D = 5	14	2	2	1	1
		13	2	1	1	0
		12	1	1	0	0
		11	1	0	0	0
		10	0	0	—	—
		9	0	0	—	—
		8	0	—	—	—
	C + D = 4	14	2	1	1	1
		13	1	1	0	0
		12	1	0	0	0
		11	0	0	—	—
		10	0	0	—	—
		9	0	—	—	—
	C + D = 3	14	1	1	0	0
		13	0	0	0	—
		12	0	0	—	—
		11	0	—	—	—

NOTE: Critical values are maximum values of D (or C).

Table 18 Fisher Test (*Continued*)

Totals in right margin	B (or A)	Level of significance			
		.05	.025	.01	.005
A + B = 14 C + D = 2	14	0	0	0	—
	13	0	0	—	—
	12	0	—	—	—
A + B = 15 C + D = 15	15	11	10	9	8
	14	9	8	7	6
	13	7	6	5	5
	12	6	5	4	4
	11	5	4	3	3
	10	4	3	2	2
	9	3	2	1	1
	8	2	1	1	0
	7	1	1	0	0
	6	1	0	0	—
	5	0	0	—	—
	4	0	—	—	—
C + D = 14	15	10	9	8	7
	14	8	7	6	6
	13	7	6	5	4
	12	6	5	4	3
	11	5	4	3	2
	10	4	3	2	1
	9	3	2	1	1
	8	2	1	1	0
	7	1	1	0	0
	6	1	0	—	—
	5	0	—	—	—
C + D = 13	15	9	8	7	7
	14	7	7	6	5
	13	6	5	4	4
	12	5	4	3	3
	11	4	3	2	2
	10	3	2	2	1
	9	2	2	1	0
	8	2	1	0	0
	7	1	0	0	—
	6	0	0	—	—
	5	0	—	—	—

NOTE: Critical values are maximum values of D (or C).

TABLE 18   457

Table 18   Fisher Test (*Continued*)

Totals in right margin		B (or A)	Level of significance			
			.05	.025	.01	.005
A + B = 15	C + D = 12	15	8	7	7	6
		14	7	6	5	4
		13	6	5	4	3
		12	5	4	3	2
		11	4	3	2	2
		10	3	2	1	1
		9	2	1	1	0
		8	1	1	0	0
		7	1	0	0	—
		6	0	0	—	—
		5	0	—	—	—
	C + D = 11	15	7	7	6	5
		14	6	5	4	4
		13	5	4	3	3
		12	4	3	2	2
		11	3	2	2	1
		10	2	2	1	1
		9	2	1	0	0
		8	1	1	0	0
		7	1	0	0	—
		6	0	0	—	—
		5	0	—	—	—
	C + D = 10	15	6	6	5	5
		14	5	5	4	3
		13	4	4	3	2
		12	3	3	2	2
		11	3	2	1	1
		10	2	1	1	0
		9	1	1	0	0
		8	1	0	0	—
		7	0	0	—	—
		6	0	—	—	—
	C + D = 9	15	6	5	4	4
		14	5	4	3	3
		13	4	3	2	2
		12	3	2	2	1
		11	2	2	1	1
		10	2	1	0	0
		9	1	1	0	0

NOTE: Critical values are maximum values of D (or C).

Table 18   Fisher Test (*Continued*)

Totals in right margin		B (or A)	Level of significance			
			.05	.025	.01	.005
A + B = 15	C + D = 9	8	1	0	0	—
		7	0	0	—	—
		6	0	—	—	—
	C + D = 8	15	5	4	4	3
		14	4	3	3	2
		13	3	2	2	1
		12	2	2	1	1
		11	2	1	1	0
		10	1	1	0	0
		9	1	0	0	—
		8	0	0	—	—
		7	0	—	—	—
		6	0	—	—	—
	C + D = 7	15	4	4	3	3
		14	3	3	2	2
		13	2	2	1	1
		12	2	1	1	0
		11	1	1	0	0
		10	1	0	0	0
		9	0	0	—	—
		8	0	0	—	—
		7	0	—	—	—
	C + D = 6	15	3	3	2	2
		14	2	2	1	1
		13	2	1	1	0
		12	1	1	0	0
		11	1	0	0	0
		10	0	0	0	—
		9	0	0	—	—
		8	0	—	—	—
	C + D = 5	15	2	2	2	1
		14	2	1	1	1
		13	1	1	0	0
		12	1	0	0	0
		11	0	0	0	—
		10	0	0	—	—
		9	0	—	—	—
	C + D = 4	15	2	1	1	1
		14	1	1	0	0

NOTE: Critical values are maximum values of *D* (or *C*).

TABLE 18   459

Table 18   Fisher Test (*Continued*)

Totals in right margin		B (or A)	Level of significance			
			.05	.025	.01	.005
A + B = 15	C + D = 4	13	1	0	0	0
		12	0	0	0	—
		11	0	0	—	—
		10	0	—	—	—
	C + D = 3	15	1	1	0	0
		14	0	0	0	0
		13	0	0	—	—
		12	0	0	—	—
		11	0	—	—	—
	C + D = 2	15	0	0	0	—
		14	0	0	—	—
		13	0	—	—	—

NOTE: Critical values are maximum values of D (or C).

Table 19 The F-Distribution* (.05 Level)

df_2 \ df_1	1	2	3	4	5	6	7	8	9	10	12	15	20	24	30	40	60	120	∞
1	161.4	199.5	215.7	224.6	230.2	234.0	236.8	238.9	240.5	241.9	243.9	245.9	248.0	249.1	250.1	251.1	252.2	253.3	254.3
2	18.51	19.00	19.16	19.25	19.30	19.33	19.35	19.37	19.38	19.40	19.41	19.43	19.45	19.45	19.46	19.47	19.48	19.49	19.50
3	10.13	9.55	9.28	9.12	9.01	8.94	8.89	8.85	8.81	8.79	8.74	8.70	8.66	8.64	8.62	8.59	8.57	8.55	8.53
4	7.71	6.94	6.59	6.39	6.26	6.16	6.09	6.04	6.00	5.96	5.91	5.86	5.80	5.77	5.75	5.72	5.69	5.66	5.63
5	6.61	5.79	5.41	5.19	5.05	4.95	4.88	4.82	4.77	4.74	4.68	4.62	4.56	4.53	4.50	4.46	4.43	4.40	4.36
6	5.99	5.14	4.76	4.53	4.39	4.28	4.21	4.15	4.10	4.06	4.00	3.94	3.87	3.84	3.81	3.77	3.74	3.70	3.67
7	5.59	4.74	4.35	4.12	3.97	3.87	3.79	3.73	3.68	3.64	3.57	3.51	3.44	3.41	3.38	3.34	3.30	3.27	3.23
8	5.32	4.46	4.07	3.84	3.69	3.58	3.50	3.44	3.39	3.35	3.28	3.22	3.15	3.12	3.08	3.04	3.01	2.97	2.93
9	5.12	4.26	3.86	3.63	3.48	3.37	3.29	3.23	3.18	3.14	3.07	3.01	2.94	2.90	2.86	2.83	2.79	2.75	2.71
10	4.96	4.10	3.71	3.48	3.33	3.22	3.14	3.07	3.02	2.98	2.91	2.85	2.77	2.74	2.70	2.66	2.62	2.58	2.54
11	4.84	3.98	3.59	3.36	3.20	3.09	3.01	2.95	2.90	2.85	2.79	2.72	2.65	2.61	2.57	2.53	2.49	2.45	2.40
12	4.75	3.89	3.49	3.26	3.11	3.00	2.91	2.85	2.80	2.75	2.69	2.62	2.54	2.51	2.47	2.43	2.38	2.34	2.30
13	4.67	3.81	3.41	3.18	3.03	2.92	2.83	2.77	2.71	2.67	2.60	2.53	2.46	2.42	2.38	2.34	2.30	2.25	2.21
14	4.60	3.74	3.34	3.11	2.96	2.85	2.76	2.70	2.65	2.60	2.53	2.46	2.39	2.35	2.31	2.27	2.22	2.18	2.13
15	4.54	3.68	3.29	3.06	2.90	2.79	2.71	2.64	2.59	2.54	2.48	2.40	2.33	2.29	2.25	2.20	2.16	2.11	2.07
16	4.49	3.63	3.24	3.01	2.85	2.74	2.66	2.59	2.54	2.49	2.42	2.35	2.28	2.24	2.19	2.15	2.11	2.06	2.01
17	4.45	3.59	3.20	2.96	2.81	2.70	2.61	2.55	2.49	2.45	2.38	2.31	2.23	2.19	2.15	2.10	2.06	2.01	1.96
18	4.41	3.55	3.16	2.93	2.77	2.66	2.58	2.51	2.46	2.41	2.34	2.27	2.19	2.15	2.11	2.06	2.02	1.97	1.92
19	4.38	3.52	3.13	2.90	2.74	2.63	2.54	2.48	2.42	2.38	2.31	2.23	2.16	2.11	2.07	2.03	1.98	1.93	1.88
20	4.35	3.49	3.10	2.87	2.71	2.60	2.51	2.45	2.39	2.35	2.28	2.20	2.12	2.08	2.04	1.99	1.95	1.90	1.84
21	4.32	3.47	3.07	2.84	2.68	2.57	2.49	2.42	2.37	2.32	2.25	2.18	2.10	2.05	2.01	1.96	1.92	1.87	1.81
22	4.30	3.44	3.05	2.82	2.66	2.55	2.46	2.40	2.34	2.30	2.23	2.15	2.07	2.03	1.98	1.94	1.89	1.84	1.78
23	4.28	3.42	3.03	2.80	2.64	2.53	2.44	2.37	2.32	2.27	2.20	2.13	2.05	2.01	1.96	1.91	1.86	1.81	1.76
24	4.26	3.40	3.01	2.78	2.62	2.51	2.42	2.36	2.30	2.25	2.18	2.11	2.03	1.98	1.94	1.89	1.84	1.79	1.73
25	4.24	3.39	2.99	2.76	2.60	2.49	2.40	2.34	2.28	2.24	2.16	2.09	2.01	1.96	1.92	1.87	1.82	1.77	1.71
26	4.23	3.37	2.98	2.74	2.59	2.47	2.39	2.32	2.27	2.22	2.15	2.07	1.99	1.95	1.90	1.85	1.80	1.75	1.69
27	4.21	3.35	2.96	2.73	2.57	2.46	2.37	2.31	2.25	2.20	2.13	2.06	1.97	1.93	1.88	1.84	1.79	1.73	1.67
28	4.20	3.34	2.95	2.71	2.56	2.45	2.36	2.29	2.24	2.19	2.12	2.04	1.96	1.91	1.87	1.82	1.77	1.71	1.65
29	4.18	3.33	2.93	2.70	2.55	2.43	2.35	2.28	2.22	2.18	2.10	2.03	1.94	1.90	1.85	1.81	1.75	1.70	1.64
30	4.17	3.32	2.92	2.69	2.53	2.42	2.33	2.27	2.21	2.16	2.09	2.01	1.93	1.89	1.84	1.79	1.74	1.68	1.62
40	4.08	3.23	2.84	2.61	2.45	2.34	2.25	2.18	2.12	2.08	2.00	1.92	1.84	1.79	1.74	1.69	1.64	1.58	1.51
60	4.00	3.15	2.76	2.53	2.37	2.25	2.17	2.10	2.04	1.99	1.92	1.84	1.75	1.70	1.65	1.59	1.53	1.47	1.39
120	3.92	3.07	2.68	2.45	2.29	2.17	2.09	2.02	1.96	1.91	1.83	1.75	1.66	1.61	1.55	1.50	1.43	1.35	1.25
∞	3.84	3.00	2.60	2.37	2.21	2.10	2.01	1.94	1.88	1.83	1.75	1.67	1.57	1.52	1.46	1.39	1.32	1.22	1.00

This table is abridged from Table 18 of the *Biometrika Tables for Statisticians*, Vol. 1 (ed. 1), edited by E. S. Pearson and H. O. Hartley. Reproduced by the

TABLE 19   461

Table 19 The F-Distribution* (.01 Level) (Continued)

df_2 \ df_1	1	2	3	4	5	6	7	8	9	10	12	15	20	24	30	40	60	120	∞
1	4052	4999.5	5403	5625	5764	5859	5928	5982	6022	6056	6106	6157	6209	6235	6261	6287	6313	6339	6366
2	98.5	99.00	99.17	99.25	99.30	99.33	99.36	99.37	99.39	99.40	99.42	99.43	99.45	99.46	99.47	99.47	99.48	99.49	99.50
3	34.12	30.82	29.46	28.71	28.24	27.91	27.67	27.49	27.35	27.23	27.05	26.87	26.69	26.60	26.50	26.41	26.32	26.22	26.13
4	21.20	18.00	16.69	15.98	15.52	15.21	14.98	14.80	14.66	14.55	14.37	14.20	14.02	13.93	13.84	13.75	13.65	13.56	13.46
5	16.26	13.27	12.06	11.39	10.97	10.67	10.46	10.29	10.16	10.05	9.89	9.72	9.55	9.47	9.38	9.29	9.20	9.11	9.02
6	13.75	10.92	9.78	9.15	8.75	8.47	8.26	8.10	7.98	7.87	7.72	7.56	7.40	7.31	7.23	7.14	7.06	6.97	6.88
7	12.25	9.55	8.45	7.85	7.46	7.19	6.99	6.84	6.72	6.62	6.47	6.31	6.16	6.07	5.99	5.91	5.82	5.74	5.65
8	11.26	8.65	7.59	7.01	6.63	6.37	6.18	6.03	5.91	5.81	5.67	5.52	5.36	5.28	5.20	5.12	5.03	4.95	4.86
9	10.56	8.02	6.99	6.42	6.06	5.80	5.61	5.47	5.35	5.26	5.11	4.96	4.81	4.73	4.65	4.57	4.48	4.40	4.31
10	10.04	7.56	6.55	5.99	5.64	5.39	5.20	5.06	4.94	4.85	4.71	4.56	4.41	4.33	4.25	4.17	4.08	4.00	3.91
11	9.65	7.21	6.22	5.67	5.32	5.07	4.89	4.74	4.63	4.54	4.40	4.25	4.10	4.02	3.94	3.86	3.78	3.69	3.60
12	9.33	6.93	5.95	5.41	5.06	4.82	4.64	4.50	4.39	4.30	4.16	4.01	3.86	3.78	3.70	3.62	3.54	3.45	3.36
13	9.07	6.70	5.74	5.21	4.86	4.62	4.44	4.30	4.19	4.10	3.96	3.82	3.66	3.59	3.51	3.43	3.34	3.25	3.17
14	8.86	6.51	5.56	5.04	4.69	4.46	4.28	4.14	4.03	3.94	3.80	3.66	3.51	3.43	3.35	3.27	3.18	3.09	3.00
15	8.68	6.36	5.42	4.89	4.56	4.32	4.14	4.00	3.89	3.80	3.67	3.52	3.37	3.29	3.21	3.13	3.05	2.96	2.87
16	8.53	6.23	5.29	4.77	4.44	4.20	4.03	3.89	3.78	3.69	3.55	3.41	3.26	3.18	3.10	3.02	2.93	2.84	2.75
17	8.40	6.11	5.18	4.67	4.34	4.10	3.93	3.79	3.68	3.59	3.46	3.31	3.16	3.08	3.00	2.92	2.83	2.75	2.65
18	8.29	6.01	5.09	4.58	4.25	4.01	3.84	3.71	3.60	3.51	3.37	3.23	3.08	3.00	2.92	2.84	2.75	2.66	2.57
19	8.18	5.93	5.01	4.50	4.17	3.94	3.77	3.63	3.52	3.43	3.30	3.15	3.00	2.92	2.84	2.76	2.67	2.58	2.49
20	8.10	5.85	4.94	4.43	4.10	3.87	3.70	3.56	3.46	3.37	3.23	3.09	2.94	2.86	2.78	2.69	2.61	2.52	2.42
21	8.02	5.78	4.87	4.37	4.04	3.81	3.64	3.51	3.40	3.31	3.17	3.03	2.88	2.80	2.72	2.64	2.55	2.46	2.36
22	7.95	5.72	4.82	4.31	3.99	3.76	3.59	3.45	3.35	3.26	3.12	2.98	2.83	2.75	2.67	2.58	2.50	2.40	2.31
23	7.88	5.66	4.76	4.26	3.94	3.71	3.54	3.41	3.30	3.21	3.07	2.93	2.78	2.70	2.62	2.54	2.45	2.35	2.26
24	7.82	5.61	4.72	4.22	3.90	3.67	3.50	3.36	3.26	3.17	3.03	2.89	2.74	2.66	2.58	2.49	2.40	2.31	2.21
25	7.77	5.57	4.68	4.18	3.85	3.63	3.46	3.32	3.22	3.13	2.99	2.85	2.70	2.62	2.54	2.45	2.36	2.27	2.17
26	7.72	5.53	4.64	4.14	3.82	3.59	3.42	3.29	3.18	3.09	2.96	2.81	2.66	2.58	2.50	2.42	2.33	2.23	2.13
27	7.68	5.49	4.60	4.11	3.78	3.56	3.39	3.26	3.15	3.06	2.93	2.78	2.63	2.55	2.47	2.38	2.29	2.20	2.10
28	7.64	5.45	4.57	4.07	3.75	3.53	3.36	3.23	3.12	3.03	2.90	2.75	2.60	2.52	2.44	2.35	2.26	2.17	2.06
29	7.60	5.42	4.54	4.04	3.73	3.50	3.33	3.20	3.09	3.00	2.87	2.73	2.57	2.49	2.41	2.33	2.23	2.14	2.03
30	7.56	5.39	4.51	4.02	3.70	3.47	3.30	3.17	3.07	2.98	2.84	2.70	2.55	2.47	2.39	2.30	2.21	2.11	2.01
40	7.31	5.18	4.31	3.83	3.51	3.29	3.12	2.99	2.89	2.80	2.66	2.52	2.37	2.29	2.20	2.11	2.02	1.92	1.80
60	7.08	4.98	4.13	3.65	3.34	3.12	2.95	2.82	2.72	2.63	2.50	2.35	2.20	2.12	2.03	1.94	1.84	1.73	1.60
120	6.85	4.79	3.95	3.48	3.17	2.96	2.79	2.66	2.56	2.47	2.34	2.19	2.03	1.95	1.86	1.76	1.66	1.53	1.38
∞	6.63	4.61	3.78	3.32	3.02	2.80	2.64	2.51	2.41	2.32	2.18	2.04	1.88	1.79	1.70	1.59	1.47	1.32	1.00

Table 20   The Hartley F_{max} Test for Homogeneity of Variances

$df =$ $n - 1$	α	$k = number\ of\ variances$										
		2	3	4	5	6	7	8	9	10	11	12
4	.05	9.60	15.5	20.6	25.2	29.5	33.6	37.5	41.4	44.6	48.0	51.4
	.01	23.2	37.	49.	59.	69.	79.	89.	97.	106.	113.	120.
5	.05	7.15	10.8	13.7	16.3	18.7	20.8	22.9	24.7	26.5	28.2	29.9
	.01	14.9	22.	28.	33.	38.	42.	46.	50.	54.	57.	60.
6	.05	5.82	8.38	10.4	12.1	13.7	15.0	16.3	17.5	18.6	19.7	20.7
	.01	11.1	15.5	19.1	22.	25.	27.	30.	32.	34.	36.	37.
7	.05	4.99	6.94	8.44	9.70	10.8	11.8	12.7	13.5	14.3	15.1	15.8
	.01	8.89	12.1	14.5	16.5	18.4	20.	22.	23.	24.	26.	27.
8	.05	4.43	6.00	7.18	8.12	9.03	9.78	10.5	11.1	11.7	12.2	12.7
	.01	7.50	9.9	11.7	13.2	14.5	15.8	16.9	17.9	18.9	19.8	21.
9	.05	4.03	5.34	6.31	7.11	7.80	8.41	8.95	9.45	9.91	10.3	10.7
	.01	6.54	8.5	9.9	11.1	12.1	13.1	13.9	14.7	15.3	16.0	16.6
10	.05	3.72	4.85	5.67	6.34	6.92	7.42	7.87	8.28	8.66	9.01	9.34
	.01	5.85	7.4	8.6	9.6	10.4	11.1	11.8	12.4	12.9	13.4	13.9
12	.05	3.28	4.16	4.79	5.30	5.72	6.09	6.42	6.72	7.00	7.25	7.48
	.01	4.91	6.1	6.9	7.6	8.2	8.7	9.1	9.5	9.9	10.2	10.6
15	.05	2.86	3.54	4.01	4.37	4.68	4.95	5.19	5.40	5.59	5.77	5.93
	.01	4.07	4.9	5.5	6.0	6.4	6.7	7.1	7.3	7.5	7.8	8.0
20	.05	2.46	2.95	3.29	3.54	3.76	3.94	4.10	4.24	4.37	4.49	4.59
	.01	3.32	3.8	4.3	4.6	4.9	5.1	5.3	5.5	5.6	5.8	5.9
30	.05	2.07	2.40	2.61	2.78	2.91	3.02	3.12	3.21	3.29	3.36	3.39
	.01	2.63	3.0	3.3	3.4	3.6	3.7	3.8	3.9	4.0	4.1	4.2
60	.05	1.67	1.85	1.96	2.04	2.11	2.17	2.22	2.26	2.30	2.33	2.36
	.01	1.96	2.2	2.3	2.4	2.4	2.5	2.5	2.6	2.6	2.7	2.7
∞	.05	1.00	1.00	1.00	1.00	1.00	1.00	1.00	1.00	1.00	1.00	1.00
	.01	1.00	1.00	1.00	1.00	1.00	1.00	1.00	1.00	1.00	1.00	1.00

This table is abridged from Table 31 in *Biometrika Tables for Statisticians*, vol. 1, 2nd ed. New York: Cambridge, 1958. Edited by E. S. Pearson and H. O. Hartley. Reproduced with the kind permission of the editors and the trustees of *Biometrika*.

TABLE 21   463

Table 21   The Cochran Test for Homogeneity of Variances (.01 Level)

| n | \multicolumn{10}{c}{k = number of variances} |
	2	3	4	5	6	7	8	9	10	12	15
5	.959	.834	.721	.633	.564	.508	.463	.425	.393	.343	.288
6	.937	.793	.676	.588	.520	.466	.423	.387	.357	.310	.259
7	.917	.761	.641	.553	.487	.435	.393	.359	.331	.286	.239
8	.899	.734	.613	.526	.461	.411	.370	.338	.311	.268	.223
9	.882	.711	.590	.504	.440	.391	.352	.321	.295	.254	.210
10	.867	.691	.570	.485	.423	.375	.337	.307	.281	.242	.200
11	.855	.675	.554	.470	.408	.361	.325	.295	.270	.232	.191
12	.843	.660	.539	.456	.396	.350	.314	.285	.261	.223	.184
13	.831	.647	.527	.445	.385	.340	.305	.276	.253	.216	.178
14	.821	.635	.515	.434	.376	.331	.297	.269	.246	.210	.173
15	.812	.624	.505	.425	.367	.323	.289	.262	.240	.205	.168
16	.803	.615	.496	.417	.360	.317	.283	.256	.234	.200	.164
17	.795	.606	.488	.409	.353	.311	.278	.251	.230	.196	.161
18	.788	.598	.481	.403	.347	.305	.272	.246	.225	.191	.157
19	.781	.591	.474	.396	.341	.300	.267	.242	.220	.188	.154
20	.775	.584	.468	.391	.336	.295	.263	.238	.217	.185	.151
21	.769	.578	.462	.385	.331	.291	.259	.234	.213	.182	.149
22	.763	.572	.457	.381	.327	.287	.255	.231	.210	.179	.147
23	.758	.566	.452	.376	.323	.283	.252	.227	.207	.176	.144
24	.753	.561	.447	.372	.319	.279	.249	.224	.205	.174	.142
25	.748	.556	.443	.368	.315	.276	.246	.222	.202	.172	.140
26	.744	.552	.439	.364	.312	.273	.243	.219	.200	.170	.139
27	.739	.548	.435	.361	.309	.270	.241	.217	.197	.168	.137
28	.735	.544	.431	.358	.306	.268	.238	.215	.195	.166	.136
29	.732	.540	.428	.355	.303	.265	.236	.212	.193	.164	.134
30	.728	.536	.424	.352	.301	.263	.234	.210	.192	.163	.133
32	.721	.530	.418	.346	.296	.258	.230	.207	.188	.160	.130
34	.715	.523	.413	.341	.291	.255	.226	.203	.185	.157	.128
36	.709	.518	.408	.337	.287	.251	.223	.200	.182	.154	.126
38	.704	.513	.403	.333	.284	.248	.220	.198	.180	.152	.124
40	.699	.508	.399	.329	.281	.245	.217	.195	.177	.150	.122
42	.694	.504	.395	.326	.277	.242	.215	.193	.175	.148	.121
44	.690	.500	.392	.323	.275	.239	.212	.191	.173	.147	.119
46	.686	.496	.388	.320	.272	.237	.210	.189	.171	.145	.118
48	.682	.492	.385	.317	.270	.235	.208	.187	.170	.143	.117
50	.679	.489	.382	.314	.267	.234	.206	.185	.168	.142	.116
145	.606	.423	.325	.264	.223	.193	.170	.152	.138	.116	.093
∞	.500	.333	.250	.200	.143	.125	.111	.100	.083	.067	.050

Table 21   The Cochran Test for Homogeneity of Variances (.05 Level) (*Continued*)

					$k = $ number of variances						
n	2	3	4	5	6	7	8	9	10	12	15
5	.906	.746	.629	.544	.480	.431	.391	.358	.331	.288	.242
6	.877	.707	.590	.507	.445	.397	.360	.329	.303	.262	.220
7	.853	.677	.560	.478	.418	.373	.336	.307	.282	.244	.203
8	.833	.653	.537	.456	.398	.354	.319	.290	.267	.230	.191
9	.816	.633	.518	.439	.382	.338	:304	.277	.254	.219	.182
10	.801	.617	.502	.424	.368	.326	.293	.266	.244	.210	.174
11	.789	.603	.489	.412	.357	.315	.283	.257	.235	.202	.167
12	.777	.590	.477	.401	.347	.306	.275	.249	.228	.195	.161
13	.767	.580	.467	.392	.339	.299	.267	.242	.222	.190	.157
14	.757	.570	.458	.384	.331	.292	.261	.237	.216	.185	.153
15	.749	.561	.450	.377	.325	.286	.256	.231	.211	.181	.149
16	.741	.554	.443	:370	.319	.280	.251	.227	.207	.177	.146
17	.734	.547	.437	.365	.314	.276	.246	.223	.203	.174	.143
18	.728	.540	.431	.359	.309	.271	.242	.219	.200	.170	.140
19	.722	.534	.425	.354	.304	.267	.238	.215	.197	.168	.138
20	.717	.529	.421	.350	.300	.264	.235	.212	.194	.165	.135
21	.712	.524	.416	.346	.297	.260	.232	.209	.191	.163	.133
22	.707	.519	.412	.342	.293	.257	.229	.207	.188	.160	.132
23	.702	.515	.408	.339	.290	.254	.226	.204	.186	.158	.130
24	.698	.511	.404	.335	.287	.251	.224	.202	.184	.157	.128
25	.694	.507	.401	.332	.284	.249	.222	.200	.182	.155	.127
26	.691	.504	.398	.330	.282	.247	.219	.199	.180	.153	.125
27	.687	.500	.395	.327	.279	.244	.217	.196	.178	.152	.124
28	.684	.497	.392	.324	.277	.242	.215	.194	.177	.150	.123
29	.681	.494	.389	.322	.275	.240	.214	.193	.175	.149	.122
30	.678	.491	.387	.320	.273	.238	.212	.191	.174	.148	.121
32	.672	.486	.382	.315	.269	.235	.209	.188	.171	.145	.119
34	.667	.481	.378	.312	.266	.232	.206	.185	.169	.143	.117
36	.662	.477	.374	.308	.263	.229	.203	.183	.166	.141	.115
38	.658	.473	.370	.305	.260	.227	.201	.181	.164	.139	.114
40	.654	.469	.367	.302	.257	.224	.199	.179	.163	.138	.112
42	.651	.466	.364	.299	.255	.222	.197	.177	.161	.136	.111
44	.647	.463	.361	.297	.253	.220	.195	.175	.159	.135	.110
46	.644	.460	.359	.295	.251	.218	.193	.174	.158	.134	.109
48	.641	.457	.357	.293	.249	.216	.192	.172	.156	.132	.108
50	.638	.454	.354	.290	.247	.215	.190	.171	.155	.131	.107
145	.581	.403	.309	.251	.212	.183	.162	.145	.131	.110	.089
∞	.500	.333	.250	.167	.143	.125	.111	.100	.083	.067	.050

This table was derived by John T. Roscoe, William R. Veitch, and Donaldson G. Woods.

TABLE 22   465

Table 22   Critical Values of the Studentized Range

| Error df | α | \multicolumn{10}{c}{$k = number\ of\ means$} |
		2	3	4	5	6	7	8	9	10	11
5	.05	.364	4.60	5.22	5.67	6.03	6.33	6.58	6.80	6.99	7.17
	.01	5.70	6.98	7.80	8.42	8.91	9.32	9.67	9.97	10.24	10.48
6	.05	3.46	4.34	4.90	5.30	5.63	5.90	6.12	6.32	6.49	6.65
	.01	5.24	6.33	7.03	7.56	7.97	8.32	8.61	8.87	9.10	9.30
7	.05	3.34	4.16	4.68	5.06	5.36	5.61	5.82	6.00	6.16	6.30
	.01	4.95	5.92	6.54	7.01	7.37	7.68	7.94	8.17	8.37	8.55
8	.05	3.26	4.04	4.53	4.89	5.17	5.40	5.60	5.77	5.92	6.05
	.01	4.75	5.64	6.20	6.62	6.96	7.24	7.47	7.68	7.86	8.03
9	.05	3.20	3.95	4.41	4.76	5.02	5.24	5.43	5.59	5.74	5.87
	.01	4.60	5.43	5.96	6.35	6.66	6.91	7.13	7.33	7.49	7.65
10	.05	3.15	3.88	4.33	4.65	4.91	5.12	5.30	5.46	5.60	5.72
	.01	4.48	5.27	5.77	6.14	6.43	6.67	6.87	7.05	7.21	7.36
11	.05	3.11	3.82	4.26	4.57	4.82	5.03	5.20	5.35	5.49	5.61
	.01	4.39	5.15	5.62	5.97	6.25	6.48	6.67	6.84	6.99	7.13
12	.05	3.08	3.77	4.20	4.51	4.75	4.95	5.12	5.27	5.39	5.51
	.01	4.32	5.05	5.50	5.84	6.10	6.32	6.51	6.67	6.81	6.94
13	.05	3.06	3.73	4.15	4.45	4.69	4.88	5.05	5.19	5.32	5.43
	.01	4.26	4.96	5.40	5.73	5.98	6.19	6.37	6.53	6.67	6.79
14	.05	3.03	3.70	4.11	4.41	4.64	4.83	4.99	5.13	5.25	5.36
	.01	4.21	4.89	5.32	5.63	5.88	6.08	6.26	6.41	6.54	6.66
15	.05	3.01	3.67	4.08	4.37	4.59	4.78	4.94	5.08	5.20	5.31
	.01	4.17	4.84	5.25	5.56	5.80	5.99	6.16	6.31	6.44	6.55
16	.05	3.00	3.65	4.05	4.33	4.56	4.74	4.90	5.03	5.15	5.26
	.01	4.13	4.79	5.19	5.49	5.72	5.92	6.08	6.22	6.35	6.46
17	.05	2.98	3.63	4.02	4.30	4.52	4.70	4.86	4.99	5.11	5.21
	.01	4.10	4.74	5.14	5.43	5.66	5.85	6.01	6.15	6.27	6.38
18	.05	2.97	3.61	4.00	4.28	4.49	4.67	4.82	4.96	5.07	5.17
	.01	4.07	4.70	5.09	5.38	5.60	5.79	5.94	6.08	6.20	6.31
19	.05	2.96	3.59	3.98	4.25	4.47	4.65	4.79	4.92	5.04	5.14
	.01	4.05	4.67	5.05	5.33	5.55	5.73	5.89	6.02	6.14	6.25
20	.05	2.95	3.58	3.96	4.23	4.45	4.62	4.77	4.90	5.01	5.11
	.01	4.02	4.64	5.02	5.29	5.51	5.69	5.84	5.97	6.09	6.19

Table 22   (*Continued*)

24	.05	2.92	3.53	3.90	4.17	4.37	4.54	4.68	4.81	4.92	5.01
	.01	3.96	4.55	4.91	5.17	5.37	5.54	5.69	5.81	5.92	6.02
30	.05	2.89	3.49	3.85	4.10	4.30	4.46	4.60	4.72	4.82	4.92
	.01	3.89	4.45	4.80	5.05	5.24	5.40	5.54	5.65	5.76	5.85
40	.05	2.86	3.44	3.79	4.04	4.23	4.39	4.52	4.63	4.73	4.82
	.01	3.82	4.37	4.70	4.93	5.11	5.26	5.39	5.50	5.60	5.69
60	.05	2.83	3.40	3.74	3.98	4.16	4.31	4.44	4.55	4.65	4.73
	.01	3.76	4.28	4.59	4.82	4.99	5.13	5.25	5.36	5.45	5.53
120	.05	2.80	3.36	3.68	3.92	4.10	4.24	4.36	4.47	4.56	4.64
	.01	3.70	4.20	4.50	4.71	4.87	5.01	5.12	5.21	5.30	5.37
∞	.05	2.77	3.31	3.63	3.86	4.03	4.17	4.29	4.39	4.47	4.55
	.01	3.64	4.12	4.40	4.60	4.76	4.88	4.99	5.08	5.16	5.23

This table is abridged from Table 29 in *Biometrika Tables for Statisticians*, vol. 1, 2nd ed. New York: Cambridge, 1958. Edited by E. S. Pearson and H. O. Hartley. Reproduced with the kind permission of the editors and the trustees of *Biometrika*.

TABLE 23   467

Table 23   The Dunnett Test* (.05 Level, One-Tailed) (*Continued*)

df	K = Number of treatment means (including the control)								
	2	3	4	5	6	7	8	9	10
5	2.02	2.44	2.68	2.85	2.98	3.08	3.16	3.24	3.30
6	1.94	2.34	2.56	2.71	2.83	2.92	3.00	3.07	3.12
7	1.89	2.27	2.48	2.62	2.73	2.82	2.89	2.95	3.01
8	1.86	2.22	2.42	2.55	2.66	2.74	2.81	2.87	2.92
9	1.83	2.18	2.37	2.50	2.60	2.68	2.75	2.81	2.86
10	1.81	2.15	2.34	2.47	2.56	2.64	2.70	2.76	2.81
11	1.80	2.13	2.31	2.44	2.53	2.60	2.67	2.72	2.77
12	1.78	2.11	2.29	2.41	2.50	2.58	2.64	2.69	2.74
13	1.77	2.09	2.27	2.39	2.48	2.55	2.61	2.66	2.71
14	1.76	2.08	2.25	2.37	2.46	2.53	2.59	2.64	2.69
15	1.75	2.07	2.24	2.36	2.44	2.51	2.57	2.62	2.67
16	1.75	2.06	2.23	2.34	2.43	2.50	2.56	2.61	2.65
17	1.74	2.05	2.22	2.33	2.42	2.49	2.54	2.59	2.64
18	1.73	2.04	2.21	2.32	2.41	2.48	2.53	2.58	2.62
19	1.73	2.03	2.20	2.31	2.40	2.47	2.52	2.57	2.61
20	1.72	2.03	2.19	2.30	2.39	2.46	2.51	2.56	2.60
24	1.71	2.01	2.17	2.28	2.36	2.43	2.48	2.53	2.57
30	1.70	1.99	2.15	2.25	2.33	2.40	2.45	2.50	2.54
40	1.68	1.97	2.13	2.23	2.31	2.37	2.42	2.47	2.51
60	1.67	1.95	2.10	2.21	2.28	2.35	2.39	2.44	2.48
120	1.66	1.93	2.08	2.18	2.26	2.32	2.37	2.41	2.45
inf.	1.64	1.92	2.06	2.16	2.23	2.29	2.34	2.38	2.42

* This table is reprinted from C. W. Dunnett, "A multiple comparison procedure for comparing several treatments with a control," *J. Amer. Statist. Ass.*, **50**, 1096–1121, 1955, by permission of the author and the editors of the *Journal of the American Statistical Association*.

Table 23   The Dunnett Test (.05 Level, Two-Tailed) (*Continued*)

K = Number of treatment means (including the control)

df	2	3	4	5	6	7	8	9	10
5	2.57	3.03	3.39	3.66	3.88	4.06	4.22	4.36	4.49
6	2.45	2.86	3.18	3.41	3.60	3.75	3.88	4.00	4.11
7	2.36	2.75	3.04	3.24	3.41	3.54	3.66	3.76	3.86
8	2.31	2.67	2.94	3.13	3.28	3.40	3.51	3.60	3.68
9	2.26	2.61	2.86	3.04	3.18	3.29	3.39	3.48	3.55
10	2.23	2.57	2.81	2.97	3.11	3.21	3.31	3.39	3.46
11	2.20	2.53	2.76	2.92	3.05	3.15	3.24	3.31	3.38
12	2.18	2.50	2.72	2.88	3.00	3.10	3.18	3.25	3.32
13	2.16	2.48	2.69	2.84	2.96	3.06	3.14	3.21	3.27
14	2.14	2.46	2.67	2.81	2.93	3.02	3.10	3.17	3.23
15	2.13	2.44	2.64	2.79	2.90	2.99	3.07	3.13	3.19
16	2.12	2.42	2.63	2.77	2.88	2.96	3.04	3.10	3.16
17	2.11	2.41	2.61	2.75	2.85	2.94	3.01	3.08	3.13
17	2.10	2.40	2.59	2.73	2.84	2.92	2.99	3.05	3.11
19	2.09	2.39	2.58	2.72	2.82	2.90	2.97	3.04	3.09
20	2.09	2.38	2.57	2.70	2.81	2.89	2.96	3.02	3.07
24	2.06	2.35	2.53	2.66	2.76	2.84	2.91	2.96	3.01
30	2.04	2.32	2.50	2.62	2.72	2.79	2.86	2.91	2.96
40	2.02	2.29	2.47	2.58	2.67	2.75	2.81	2.86	2.90
60	2.00	2.27	2.43	2.55	2.63	2.70	2.76	2.81	2.85
120	1.98	2.24	2.40	2.51	2.59	2.66	2.71	2.76	2.80
inf.	1.96	2.21	2.37	2.47	2.55	2.62	2.67	2.71	2.75

TABLE 23   469

Table 23   The Dunnett Test (.01 Level, one-Tailed) (*Continued*)

K = *Number of treatment means (including the control)*

df	2	3	4	5	6	7	8	9	10
5	3.37	3.90	4.21	4.43	4.60	4.73	4.85	4.94	5.03
6	3.14	3.61	3.88	4.07	4.21	4.33	4.43	4.51	4.59
7	3.00	3.42	3.66	3.83	3.96	4.07	4.15	4.23	4.30
8	2.90	3.29	3.51	3.67	3.79	3.88	3.96	4.03	4.09
9	2.82	3.19	3.40	3.55	3.66	3.75	3.82	3.89	3.94
10	2.76	3.11	3.31	3.45	3.56	3.64	3.71	3.78	3.83
11	2.72	3.06	3.25	3.38	3.48	3.56	3.63	3.69	3.74
12	2.68	3.01	3.19	3.32	3.42	3.50	3.56	3.62	3.67
13	2.65	2.97	3.15	3.27	3.37	3.44	3.51	3.56	3.61
14	2.62	2.94	3.11	3.23	3.32	3.40	3.46	3.51	3.56
15	2.60	2.91	3.08	3.20	3.29	3.36	3.42	3.47	3.52
16	2.58	2.88	3.05	3.17	3.26	3.33	3.39	3.44	3.48
17	2.57	2.86	3.03	3.14	3.23	3.30	3.36	3.41	3.45
18	2.55	2.84	3.01	3.12	3.21	3.27	3.33	3.38	3.42
19	2.54	2.83	2.99	3.10	3.18	3.25	3.31	3.36	3.40
20	2.53	2.81	2.97	3.08	3.17	3.23	3.29	3.34	3.38
24	2.49	2.77	2.92	3.03	3.11	3.17	3.22	3.27	3.31
30	2.46	2.72	2.87	2.97	3.05	3.11	3.16	3.21	3.24
40	2.42	2.68	2.82	2.92	2.99	3.05	3.10	3.14	3.18
60	2.39	2.64	2.78	2.87	2.94	3.00	3.04	3.08	3.12
120	2.36	2.60	2.73	2.82	2.89	2.94	2.99	3.03	3.06
inf.	2.33	2.56	2.68	2.77	2.84	2.89	2.93	2.97	3.00

Table 23   The Dunnett Test (.01 Level, Two-tailed) (*Continued*)

$K =$ *Number of treatment means (including the control)*

df	2	3	4	5	6	7	8	9	10
5	4.03	4.63	5.09	5.44	5.73	5.97	6.18	6.36	6.53
6	3.71	4.22	4.60	4.88	5.11	5.30	5.47	5.61	5.74
7	3.50	3.95	4.28	4.52	4.71	4.87	5.01	5.13	5.24
8	3.36	3.77	4.06	4.27	4.44	4.58	4.70	4.81	4.90
9	3.25	3.63	3.90	4.09	4.24	4.37	4.48	4.57	4.65
10	3.17	3.53	3.78	3.95	4.10	4.21	4.31	4.40	4.47
11	3.11	3.45	3.68	3.85	3.98	4.09	4.18	4.26	4.33
12	3.05	3.39	3.61	3.76	3.89	3.99	4.08	4.15	4.22
13	3.01	3.33	3.54	3.69	3.81	3.91	3.99	4.06	4.13
14	2.98	3.29	3.49	3.64	3.75	3.84	3.92	3.99	4.05
15	2.95	3.25	3.45	3.59	3.70	3.79	3.86	3.93	3.99
16	2.92	3.22	3.41	3.55	3.65	3.74	3.82	3.88	3.93
17	2.90	3.19	3.38	3.51	3.62	3.70	3.77	3.83	3.89
18	2.88	3.17	3.35	3.48	3.58	3.67	3.74	3.80	3.85
19	2.86	3.15	3.33	3.46	3.55	3.64	3.70	3.76	3.81
20	2.85	3.13	3.31	3.43	3.53	3.61	3.67	3.73	3.78
24	2.80	3.07	3.24	3.36	3.45	3.52	3.58	3.64	3.69
30	2.75	3.01	3.17	3.28	3.37	3.44	3.50	3.55	3.59
40	2.70	2.95	3.10	3.21	3.29	3.36	3.41	3.46	3.50
60	2.66	2.90	3.04	3.14	3.22	3.28	3.33	3.38	3.42
120	2.62	2.84	2.98	3.08	3.15	3.21	3.25	3.30	3.33
inf.	2.58	2.79	2.92	3.01	3.08	3.14	3.18	3.22	3.25

ANSWERS TO SELECTED EXERCISES

2.1 (a) Quantitative; (b) quantitative; (c) quantitative; (d) quantitative; (e) qualitative; (f) quantitative; (g) quantitative; (h) quantitative; (i) quantitative; (j) qualitative; (k) quantitative; (l) qualitative.

2.2 (a) Ratio; (b) ratio; (c) interval; (d) ordinal; (e) nominal; (f) ordinal; (g) ratio; (h) ordinal; (i) ordinal; (j) nominal; (k) ordinal; (l) nominal.

2.3 (a) Discrete; (b) probably continuous; (c) discrete, possibly continuous if distributed throughout the range k through six; (d) discrete, possibly continuous if distributed throughout the range A through F; (e) discrete; (f) probably continous; (g) probably continuous; (h) probably continuous; (i) probably continuous, especially if distributed throughout the range one through nine; (j) discrete; (k) discrete; (l) discrete.

2.4 For most practical purposes, the measures should probably be regarded as (a) quantitative; (b) continuous; (c) ordinal or interval.

2.5 (a) Discrete, possibly continuous if distributed throughout the range one through five on all items; (b) continuous; (c) ordinal; (d) interval.

2.6 (a) Discrete, possibly continuous if distributed throughout the range zero through four; (b) probably continuous.

3.2 (a)

X	f	fX	cf	rf	rcf
10	1	10	20	.05	1.00
9	2	18	19	.10	.95
8	3	24	17	.15	.85
7	1	7	14	.05	.70
6	4	24	13	.20	.65
5	6	30	9	.30	.45
4	2	8	3	.10	.15
3	1	3	1	.05	.05
	20	124		1.00	

(b) $N = 20$, $T = 124$. For answers to (c), (d), and (e), see (a) above.

3.3

X	f	fX	cf	rf	rcf	PR
8	4	32	50	.08	1.00	96
7	4	28	46	.08	.92	88
6	11	66	42	.22	.84	73
5	10	50	31	.20	.62	52
4	10	40	21	.20	.42	32
3	4	12	11	.08	.22	18
2	2	4	7	.04	.14	12
1	3	3	5	.06	.10	7
0	2	0	2	.04	.04	2

3.4

	Agree	Disagree	No opinion
Boys	22	44	34
Girls	62	27	11

The independent variable is sex; the dependent variable is response. Girls are more likely to agree with the policy; they are also more likely to have an opinion.

4.1 $Q_1 = 90.5$, $Q_2 = 99.5$, $Q_3 = 108.5$.

4.2 $D_2 = 4.67$, $D_4 = 5.33$, $D_6 = 6.25$, $D_8 = 8.17$.

4.3   See answers to 3.3.

5.1   See answers to 4.1.

5.2   See answers to 4.2.

6.1   (a) 50; (b) 340; (c) 2500; (d) 90.

6.2   (a) 647; (b) 16.1; (c) 1792.8; (d) 10,416.7; (e) 56.7.

6.3   (a) 30; (b) 16; (c) 75; (d) 285; (e) 5625.

6.4   (a) 340; (b) 400; (c) 357; (d) 2500; (e) 3600; (f) 3000; (g) 90; (h) 40; (i) 57.

7.1   (a) 6; (b) 5.5; (c) 5; (d) 30; (e) 16; (f) 0.

7.2   (a) 12.03; (b) 11.3; (c) 10.

7.3   (a) 4.70; (b) 4.90; (c) 6.

7.4   (a) 3.00, 2.50, 2.29, 2.38; (b) 2.50.

7.5   43.422.

8.1   (a) 6; (b) 24; (c) 4; (d) 2; (e) 2; (f) 20; (g) 16.

8.2   (a) 9; (b) 2.875; (c) 2.60; (d) 90; (e) 9; (f) 3.

8.3   (a) 98.917; (b) 8.243; (c) 2.871.

8.4   (a) 40; (b) 4; (c) 2; (d) 4.44.

8.5   (a) 1912.5917; (b) 15.9383; (c) 3.9923.

9.1   (a) 5, 2; (b) -2.0, $-.5$, $-.5$, 0, 0, .5, 1.0, 1.5; (c) 30, 45, 45, 50, 50, 55, 60, 65.

9.2   (a) 10, 2; (b) -1.5, -1.5, -1.0, $-.5$, $-.5$, 0, 0, .5, .5, 1.0, 1.5, 1.5; (c) 77.5, 77.5, 85, 92.5, 92.5, 100, 100, 107.5, 107.5, 115, 122.5, 122.5.

9.3   (a) 11.06, 3.99; (b) -2.52, -2.27, -2.02, -1.77, -1.52, -1.27, -1.02, $-.77$, $-.52$, $-.27$, $-.02$, .24, .49, .74, .99, 1.24, 1.49, 1.74, 1.99, 2.24.

10.1 (a) .4429; (b) .9429; (c) .0571; (d) 94.
10.2 (a) .4429; (b) .0571; (c) .9429; (d) 6.
10.3 (a) .4975; (b) .9975; (c) .0025; (d) 99.
10.4 (a) .4975; (b) .0025; (c) .9975; (d) 0+.
10.5 −.77.
10.6 1.34.
10.7 0.
10.8 (a) .0013; (b) .9987.
10.9 1.645.
10.10 −.67 to +.67.

11.1

PR	z-score	T-score	Stanine
90	+1.28	62.8	8
80	+0.84	58.4	7
70	+0.52	55.2	6
60	+0.25	52.5	6
50	0.00	50.0	5
40	−0.25	47.5	4
30	−0.52	44.8	4
20	−0.84	41.6	3
10	−1.28	37.2	2

11.2 −2.65, −2.41, −2.12, −1.79, −1.44, −1.19, −.97, −.74, −.49, −.27, −.03, .23, .48, .70, .90, 1.13, 1.44, 1.77, 2.12, 2.65.
11.3 (a) For IQ of 80, $z = -1.37$; for 90, $z = -.69$; for 100, $z = .09$; for 110, $z = .71$; for 120, $z = 1.41$; (b) For IQ of 80, $T = 36$, for 90, $T = 43$, for 100, $T = 51$; for 110, $T = 57$; for 120, $T = 64$; (c) for IQ of 80, 2; for 90, 4; for 100, 5; for 110, 6; for 120, 8.
12.1 (b) .517.
12.2 (b) −.792.
12.3 (a) −.535; (b) Girls scored higher than boys; (c) .287.
12.4 (a) +0.80; (b) +0.80; (c) +1.00; (d) +0.80; (e) +1.00; (f) +0.80.
13.1 0.70.
13.2 0.612.
13.3 0.271 without tie correction, 0.176 with tie correction.
13.4 +0.60.
13.5 0.488.
13.6 0.71.
13.7 −0.535.
13.8 −0.671.
13.9 0.351.
13.10 0.53.
14.1 (b) .90; (c) .09; (d) −7; (e) $\hat{Y} = .09$, $X − 7$; (g) 3.8, 2.9, 2.0, 1.1, .2; (h) 1.90; (i) 2; (k) .796; (l) 1.55.
14.2 (a) −.85; (b) 9.25; (c) 6.70; (d) 2.45; (e) 1.18.
16.1 (a) 15/67 or .224; (b) 27/67 or .403.
16.2 (a) 1/2 or .500; (b) 1/256 or .004; (c) 1/256 or .004.

16.3 (a) 1/3 or .333; (b) 2/3 or .667; (c) 1/243 or .004.

16.4 (a) 24; (b) 1/24 or 0.04.

16.5 35.

16.6 (a) 336; (b) 1/8 or .125; 1/336 or .003.

17.1 (a) 1/64 or .016; (b) 1/64 or .016; (c) 6/64 or .094; (d) 7/64 or .109.

17.2 (a) 1/729 or .0014; (b) 64/729 or .0878; (c) 12/729 or .0165; (d) 13/729 or .0178.

17.3 201/59,049 or .0034.

17.4 19/4,096 or .0046.

17.5 (a) Less than .0001; (b) .0287; (c) .5398.

17.6 (a) Less than .0001; (b) less than .0001; (c) less than .0001.

18.1 (a) .0228; (b) .9987; (c) .0000003; (d) .00003; (e) .999997.

18.2 (a) 1.25; (b) 47.944–52–056.

18.3 (a) 3.0; (b) .548; (c) 20.926–23.074.

18.4 .032–.437.

19.1 (a) $z \leq -1.960$ or $z \geq +1.960$; (b) $z = 1.80$. (c) Retain the null hypothesis. There is no significant difference between the two means.

19.2 (a) $z \geq +2.326$; (b) $z = 2.50$. (c) Reject the null hypothesis. The students scored significantly higher than the national average.

19.3 (a) $z \leq -2.326$; (b) $z = -1.80$. (c) Retain the null hypothesis. The sample mean does not differ significantly from that of population used to norm the test.

19.4 (a) $z \geq +3.291$; (b) $z = 3.50$. (c) Reject the null hypothesis. The subject has significantly more right answers than could be expected if he were only guessing.

19.5 (a) $r \geq 9$. (b) Retain. The subject's responses do not differ significantly from what could be expected if he were only guessing.

23.1 $t = -1.50$, $df = 9$. Retain the null hypothesis.

23.2 17.46–32.54.

23.3 (a) $t \geq +1.711$; (b) $S_M = 0.5$; (c) $t = 1.60$. (d) Retain the null hypothesis. The students do not differ significantly from the national norm.

14.1 $t = 1.99$, $df = 16$. Retain the null hypothesis (two-tailed test). There is no significant difference in the mean scores of the two groups.

24.2 $t = 2.69$, $df = 18$. Reject the null hypothesis. The girls scored significantly higher than the boys.

24.3 $t = 2.11$, $df = 57$. Reject the null hypothesis. Sample two scored significantly higher than sample one.

24.4 (a) $t \leq -2.763$ or $t \geq +2.763$; (b) $S_{M_1-M_2} = 3.00$; (c) $t = 2.00$. (d) Retain the null hypothesis. There is no significant difference in the mean performance of the two groups.

25.1 $t = 2.00$, $df = 7$. Reject the null hypothesis (one-tailed test). There was a significant gain in the IQ scores.

25.2 $t = 2.31$, $df = 11$. Reject the null hypothesis (one-tailed test). There was a significant downward shift in the scores.

26.1 $U = 18.5$. Retain the null hypothesis (two-tailed test).

26.2 $U = 20.5$. Reject the null hypothesis.

26.3 $U = 63$. Reject the null hypothesis.

27.1 $T = 3.5$. Retain at the .025 level, N reduced to 7.

27.2 $T = 12$. Retain the null hypothesis, N reduced to 11.

28.1 Chi-square $= 21.74$, $df = 19$. Retain the null hypothesis. The sample does not deviate significantly from the hypothesized variance.

28.2 Chi-square $= 24.15$, $df = 29$. Retain the null hypothesis. The sample does not deviate significantly from the hypothesized variance.

28.3 Chi-square $= 43.84$, $df = 39$. Retain the null hypothesis. The sample does not deviate significantly from the hypothesized variance.

29.1 Chi-square $= 26$, $df = 19$. Retain the null hypothesis. The sample does not deviate significantly from the hypothesized normal distribution with mean of zero and standard deviation of one.

29.2 Chi-square $= 110.0$, $df = 19$. Reject the null hypothesis. The sample deviates significantly from the hypothesized normal distribution with mean of 50 and standard deviation of 10.

29.3 Chi-square $= 2.96$, $df = 3$. Retain the null hypothesis. The sample does not deviate significantly from the hypothesized distribution.

30.1 Chi-square $= 38.1$, $df = 4$. Reject the null hypothesis. There is a significant difference in the distribution of the three groups. Unfortunately, the precise nature of this difference is left undefined. A comparison of the expected and observed frequencies in the various cells might aid in the interpretation of the data.

30.2 Chi-square $= 6.10$, $df = 4$. Retain the null hypothesis. There is no significant difference in the two groups.

30.3 Chi-square $= 3.33$, $df = 1$. Retain the null hypothesis. There is no significant difference in the performance of the two groups.

30.4 Chi-square $= 87.1$, $df = 9$. Reject the null hypothesis. There is a significant relationship between the two sets of ratings. $C = .68$.

31.1 Retain the null hypothesis. There is no significant relationship between the two variables.

31.2 Reject the null hypothesis. There is a significant negative relationship between the two variables.

31.3 Reject the null hypothesis. There is a significant negative relationship between the two variables (girls tend to score higher than boys).

31.4 Retain the null hypothesis. There is no significant relationship between the two sets of ratings.

31.5 Retain the null hypothesis. There is no significant relationship between the two sets of ratings.

32.1 $D = .30$. Reject the null hypothesis. The data deviate significantly from the hypothesized normal distribution with mean of zero and standard deviation of one (20 intervals used).

32.2 $D = .19$. Retain the null hypothesis. The sample does not deviate significantly from the hypothesized normal distribution with mean of 50 and standard deviation of 10.

32.3 $D = .25$ (using 10 intervals). Retain the null hypothesis. $D = .30$ (using 20 intervals). Reject the null hypothesis.

33.1 $K_D = 7$ (using unit intervals). Retain the null hypothesis.

33.2 $D = .15$. Retain the null hypothesis.

34.1 Reject. The sexes differ on the issue with the businesswomen having a more favorable view of the issue than the businessmen.

34.2 Chi-square $= 6.90$, $df = 1$. Reject the null hypothesis.

35.1 (a) $F' \leq 0.357$ or $F \geq 2.74$, $df = 21$, 24; (b) $F = 0.342$ or $F = 2.92$. (c) Reject the null hypothesis. Population two has a larger variance than population one.

35.2 (a) $F_{max} \geq 5.34$; (b) $F_{max} = 4.97$. (c) Retain the hypothesis of equal variances.

35.3 (a) $C \geq .617$; (b) $C = .691$. (c) Reject the null hypothesis. The samples were drawn from populations having different variances.

36.1 (a) $F \geq 4.26$, $df = 2$, 9; (b) $MS_b = 7.000$, $MS_w = .4444$, $F = 15.77$. (c) Reject the null hypothesis. There is a significant difference in the means.

36.2 (a) $F \geq 3.24$, $df = 3$, 16; (b) $MS_b = 6.53$, $MS_w = 7.04$, $F = 0.93$. (c) Retain the null hypothesis. There is no significant difference in the means.

36.3 (a) $F \geq 4.49$, $df = 1$, 16; (b) $MS_b = 17.8$, $MS_w = 4.5$, $F = 3.95$. (c) Retain the null hypothesis. There is no significant difference in the means.

36.4 (a) $F \geq 4.64$, $df = 3$, 26; (b) $MS_b = 30.57$, $MS_w = 6.14$, $F = 4.98$. (c) Reject the null hypothesis. There is a significant difference in the means.

36.5 (a) $F \geq 5.57$, $df = 2$, 25; (b) $MS_b = 10.50$, $MS_w = 0.80$, $F = 13.125$. (c) Reject the null hypothesis. There is a significant difference in the means.

36.6 (a) $F \geq 4.31$, $df = 3$, 44; (b) $MS_b = 24.00$, $MS_w = 3.00$, $F = 8.00$. (c) Reject the null hypothesis. There is a significant difference in the means.

37.1 (a) Chi-square equal to or greater than 9.210, $df \doteq 2$. (b) Chi-square $= 11.94$ (13.42 corrected for ties). (c) Reject the null hypothesis. There is a significant difference in the ranks of the three samples.

37.2 (a) Chi-square equal to or greater than 11.345, $df = 3$. (b) Chi-square $= 12.24$ (13.81 corrected for ties). (c) Reject the null hypothesis. There is a significant difference in the ranks of the four samples.

38.1 $t = 3.5$, $df = 25$. Sample 3 scored significantly higher than sample 2. (b) $t = 3.74$, $df = 25$. Sample 1 scored significantly lower than the other two samples combined.

38.2 $F = 6.12$, $df = 2$, 25. Sample 3 scored significantly higher than sample 2; (b) $F = 7.00$, $df = 2$, 25. Sample 1 scored significantly lower than the other two samples combined.

38.3 (a) $d = 2.54$. (b) Sample 1 scored significantly higher than sample 3 and also significantly higher than sample 4. There were no other significant differences.

38.4 (a) $d = 2.35$. (b) Sample 1 scored significantly higher than sample 3 and sample 4. There were no other significant differences.

38.5 (a) $t = 1.65$, $df = 25$. Retain. There is no significant difference in the two means; (b) $t = 4.95$, $df = 25$. Reject. Sample 3 scored significantly higher than sample 1.

39.1 (a) $F \geq 3.55$, $df = 2$, 18; (b) $SS_r = 14.80$, $SS_c = 10.40$, $SS_{rc} = 11.60$, $SS_t = 36.80$; (c) $F = 5.200/0.644 = 8.07$. Reject the null hypothesis. One or both of the drugs is effective in reducing the mean activity of the rats.

39.2 (a) $F \geq 10.92$, $df = 2$, 6; (b) $F \geq 9.78$, $df = 3$, 6; (c) $SS_r = 19.50$, $SS_c = 6.00$, $SS_{rc} = 2.50$, $SS_t = 28.00$; (d) $F = 23.40$. Reject the null hypothesis. There is a significant difference in the row means; (e) $F = 4.80$. Retain the null hypothesis. There is no significant difference in the column means.

40.1 (a) $F = .20/1.30 = 0.15$, $df = 1$, 16. Retain the null hypothesis. There is no significant interaction; (b) $F = 1.80/1.30 = 1.38$, $df = 1$, 16. Retain the null hypothesis. There is no significant difference in the mean performance of the urban and rural students; (c) $F = 5.00/1.30 = 3.85$, $df = 1$, 16. Retain the null hypothesis. There is no significant difference in the mean performance of the male and female students.

40.2 (a) $F = 10.00/6.00 = 1.67$, $df = 2$, 24. Retain; (b) $F = 30.00/10.00 = 3.00$, $df = 2$, 2. Retain; (c) $F = 70.00/6.00 = 11.67$, $df = 2$, 24. Reject.

40.3 (a) $SS_r = 1.50$, $SS_c = 4.00$, $SS_{rc} = 2.00$, $SS_w = 13.00$, $SS_t = 20.50$; (b) $F = 0.33/1.08 = 0.31$, $df = 6$, 12. Retain; (c) $F = 0.50/1.08 = 0.46$, $df = 3$, 12. Retain; (d) $F = 2.00/1.08 = 1.85$, $df = 2$, 12. Retain.

40.4 (a) $SS_c = 4.00$, $SS_{nf} = 3.50$, $SS_w = 13.00$, $SS_t = 20.50$; (b) $F = 2.00/1.08 = 1.85$, $df = 2$, 12. Retain.

41.1 (a) $F = 5.00/.278 = 18.0$, $df = 1$, 6. Reject the null hypothesis. There is a significant difference in the adjusted means; (b) $\bar{Y}_1' = 4.0$, $\bar{Y}_2' = 5.5$.

41.2 (a) $F = 1.80/1.24 = 1.45$, $df = 1$, 17. Retain the null hypothesis. There is no significant difference in the adjusted means; (b) $\bar{Y}_1' = 2.2$, $\bar{Y}_2' = 1.6$.

41.3 (a) $SS_{by}' = 17.89$, $SS_{wy}' = 129.00$, $SS_{ty}' = 146.89$; (b) $F = 8.95/1.50 = 5.97$, $df = 2$, 86. Reject the null hypothesis. There is a significant difference in the adjusted means; (c) $\bar{Y}_1' = 24.16$, $\bar{Y}_2' = 21.48$, $\bar{Y}_3' = 20.36$.

42.1 (a) $\hat{Y} = .714 X_1 + .054 X_2 - 3.857$; (b) .846; (c) .878.

42.2 (a) $\hat{Y} = 1.56X_1 + 1.06$; (b) .781; (c) .854; (d) $\hat{Y} = 2.5\ X - .5X^2 + 1$; (e) .80; (f) 1.00.

42.3 (a) $\hat{Y} = -.4286X_1 - .2857X_2 + 8.5715$; (b) $R^2 = .7286$. (c) Standard error = 1.245.

43.1 (a) $F = 21.91$, $df = 2$, 27. Reject the null hypothesis. The two-predictor model accounts for a significant proportion of the variance of the dependent variable; (b) $F = 9.29$, $df = 1$, 27. Reject the null hypothesis. The two-predictor model is significantly better than the one-predictor model using high school percentile rank. The admissions test score makes a significant contribution to the two-predictor model. (c) $F = 18.73$, $df = 1$, 27. Reject the null hypothesis. The two-predictor model is significantly better than the one-predictor model using admissions test score. High school percentile rank makes a significant contribution to the two-predictor model.

43.2 (a) $F = 37.03$, $df = 2$, 17. Reject the null hypothesis. The quadratic model accounts for a significant proportion of the variance of the dependent variables; (b) $F = 11.73$, $df = 1$, 17. Reject the null hypothesis. The quadratic model is significantly better than the linear model; (c) The evidence considered indicates that the quadratic model is superior to the linear model. There is the possibility that a higher order (cubic or quartic) relationship should be explored. However, the number of subjects is hardly sufficient to explore these higher order relationships.

43.3 (a) $R_{rm}^2 = .5124$, $F = 79.51$, $df = 1$, 96. Reject the null hypothesis. IQ contributes significantly to the three-predictor model; (b) $R_{rm}^2 = .4102$, $F = 116.30$, $df = 1$, 96. Reject the null hypothesis. Sex contributes significantly to the three-predictor model; (c) $R_{rm}^2 = .5987$, $F = 48.45$, $df = 1$, 96. Reject the null hypothesis. Socioeconomic rating contributes significantly to the three-predictor model.

44.1 $R^2 = .778$, $F = 15.77$, $df = 2, 9$. Reject the null hypothesis.

44.2 (a) $R_{rm}^2 = .3795$, $F = 6.12$, $df = 2, 20$; (b) $R_{rm}^2 = .2084$, $F = 5.51$, $df = 1, 20$; (c) $R_{rm}^2 = .3396$, $F = 1.29$, $df = 1, 20$; (d) $R_{rm}^2 = .0201$, $F = 11.58$, $df = 1, 20$.

44.3 (a) $F = 5.515/2 = 2.76$, $df = 2, 20$; (b) $F = 1.286/2 = .64$, $df = 2, 20$; (c) $F = 11.584/2 = 5.79$, $df = 2, 20$.

45.1 (a) $RSQ(2) = R_{rm-rc}^2 = .2446$; (b) $F = 0.15$, $df = 1, 16$; (c) $RSQ(3) = .1799$; (d) $RSQ(4) = .0647$; (e) $R_{rm-rows}^2 = .2518 - .2446 + .0647 = .0719$; (f) $F = 3.85$, $df = 1, 16$; (g) $R_{rm-cols}^2 = .2518 - .2446 + .1799 = .1871$; (h) $F = 1.38$, $df = 1, 16$.

45.2 (a) $RSQ(1) = R_{fm}^2 = .7784$; (b) $RSQ(2) = R_{rm-rc}^2 = .6486$; (c) $R_{rm-rows}^2 = .7784 - .6486 + .1287 = .2595$; (d) $R_{rm-cols}^2 = .7784 - .6486 + .5189 = .6487$; (e) $F = 5.268$, $df = 2, 18$; (f) $F = 42.146$, $df = 1, 18$; (g) $F = 5.268$, $df = 2, 18$.

46.1 (a) .8611; (b) .4444; (c) $F = 18.00$, $df = 1, 6$.

46.2 (a) .8162; (b) $F = 0.195$, $df = 2, 19$.

INDEX